Lecture Notes in Physics

Edited by J. Ehlers, München, K. Hepp, Zürich
R. Kippenhahn, München, H. A. Weidenmüller, Heidelberg
and J. Zittartz, Köln
Managing Editor: W. Beiglböck, Heidelberg

105

Dynamics and Instability of Fluid Interfaces

Proceedings of a meeting, held at the Technical University of Denmark, Lyngby, May 1978

Edited by T. S. Sørensen

Springer-Verlag
Berlin Heidelberg New York 1979

Editor

Torben Smith Sørensen
Fysisk-Kemisk Institut
DTH 206
DK–2800 Lyngby

ISBN 3-540-09524-1 Springer-Verlag Berlin Heidelberg New York
ISBN 0-387-09524-1 Springer-Verlag New York Heidelberg Berlin

Library of Congress Cataloging in Publication Data
Main entry under title: Dynamics and instability of fluid interfaces.
(Lecture notes in physics ; v. 105) Bibliography: p. Includes index. 1. Surfaces (Physics)--Congresses. 2. Surface chemistry--Congresses. 3. Fluids--Congresses. I. Sørensen, Torben Smith, 1945- II. Series. III. Title: Fluid interfaces.
QC173.4.S94D96 541'.3453 79-16836
ISBN 0-387-09524-1

Printing and binding: Beltz Offsetdruck, Hemsbach/Bergstr.
2153/3140-543210

EDITORS PREFACE

The present volume is the result of a meeting on the topic "Dynamics and Instability of Fluid Interfaces" held at the Technical University of Denmark, Lyngby (near Copenhagen), in May 1978. The participants were distinguished researchers in the field. They came from such diverse places as Belgium, USA, DDR, Brazil, Bulgaria, Spain, Sweeden and Denmark.

If one should describe the atmosphere at the meeting in just one word, this word could be: fascination. We were all deeply fascinated by the range and the scope of the physico-chemical problems covered by the lectures at the meeting. This diversity is very well represented in the content of the present volume. There are contributions on Marangoni effects during liquid-liquid mass transfer of significance for the chemical engineering of liquid-liquid extraction and for the illumination of surface chemical dissipative structures. There are other papers on the formation, thinning, instability and rupture of thin liquid films of utmost importance for foam stability, wetting phenomena and phenomena connected with the biological bilipid cell membrane. Still other papers deal with hydrodynamic, interfacial instabilities enforced by mechanical, chemical and electrical constraints. With the present industrial interest in electrically induced spontaneous emulsification, theoretical studies of this kind are strongly needed. Viscoelastic properties of surface layers of surfactants and macromolecules are also dealt with, and another paper tries to reconcile the theories on Marangoni instability with the theories of longitudinal interfacial waves. Thermally induced Marangoni instabilities and the influence of non-Boussinesquian effects and the Soret effect is the topic of another contribution. Finally, a paper deals with instability phenomena of spherical drops with surface chemical reactions and transfer of surfactants as a model for "kicking drops", spontaneous emulsification and certain cytokinetic phenomena, such as the division of the biological cell.

The common denominator for all the contributions is the simultaneous use of concepts from surface chemistry and physics and from hydrodynamics. Theoretic and experimental work is equally represented and even united in many of the papers. It is my hope that this volume will be of value for physicists, physico-chemists, chemical engineers, surface scientists and biologists.

I am greatful to the Technical University of Denmark for the supply of meeting facilities, to Undervisningsministeriets Internationale Kontor and Det Naturvidenskabelige Forskningsråd for economical support in the form of travel grants and to the participants for their most interesting contributions. Finally, the editor of "Lecture Notes in Physics" Dr. W. Beiglböck as well as the Springer-Verlag should be thanked for the assistance made in the publication of this congress report.

Torben Smith Sørensen

15-12 1978

Fysisk-Kemisk Institut
Technical University of Denmark.

TABLE OF CONTENT

INSTABILITIES INDUCED BY MASS TRANSFER, LOW SURFACE TENSION AND GRAVITY AT ISOTHERMAL AND DEFORMABLE FLUID INTERFACES.

by

Torben Smith Sørensen

Fysisk-Kemisk Institut, Technical University of Denmark, DK 2800, Lyngby, Denmark

1. Introduction

The beautiful hexagonal convection pattern observed by Bénard[1,2] in shallow pools of liquids heated from below inspired Rayleigh[3] to perform his classical stability analysis of buoyancy-driven convection flow. We now know that the driving force for the Bénard cells is not the buoyancy, but rather temperature induced gradients in surface tension[4,5]. For example, Block[6] observed also Bénard cells in shallow pools *cooled* from below, and Bénard cells were observed in experiments performed by astronauts in the Apollo 14 and Apollo 17 spacecrafts where the gravitational acceleration was 10^{-6} G.[7,8] Nevertheless, the identification between Bénard's observations and Rayleigh's analysis is very persistent even in standard reference texts on hydrodynamic and thermodynamic stability[9,10].

Lateral variations in surface tension can be generated by as well fluctuations in temperature as fluctuations in the adsorption of surfactive material to the interface. The first linear stability analysis taking surface tension gradients as the driving forces were performed independently by Pearson[11] (heat and mass transfer) and by Sternling and Scriven[12] (mass transfer of surfactant), although qualitative statements that surface tension is the driving force in many instability phenomena near surfaces were made already in the previous century by James Thomson[13], Marangoni[14] and Rayleigh[15].

In the analysis of Pearson as well as in the paper of Sternling and Scriven (S.&S.) the interface is assumed to be flat during convection. But it is well known that the interface deforms itself during convection. Bénard reported (ref.2 .p.120) that there was a depression of the surface where liquid was streaming towards the surface and an elevation of the surface above downflow regions. This is characteristic for surface tension-driven convection whereas the converse is true for buoyancy convection, see ref. 4 p.339 . The same type of interfacial depression above upflow regions was observed by Linde, Pfaff and Zirkel[16] in connection with transfer of ethanol from water to air, although it seemed to be a characteristic of the "secondary" roll cells rather than of the smaller "primary" roll cells.

Theoretically, the effects due to deformable interfaces were studied by Scriven and Sternling[4] in connection with heat transfer through a liquid-air boundary. They considered only stationary states of marginal stability, however, and excluded gravity effects (*e.g.* Rayleigh convection). The primary effect due to surface curvature is to introduce the absolute value of the surface tension through the law of Laplace for the overpresure behind a convex interface. K.A.Smith[17] studied the influence of gravity waves on the surface in heat transfer instabilities, but still did not consider the competition between surface tension gradients and buoyancy. As well S. & S. as Smith seem to agree that the effects due to deformation on the neutral stability of stationary disturbances only show up at very large wavelengths. The first authors maintain, however, that there does *not* exist a critical Marangoni-number below which all perturbations are stable, whereas Smith claims that the findings of S. & S. are due to their neglect of gravity in the equations of surface dynamics. Palmer and Berg[18] have studied the stabilising effects of surfactants on thermal instabilities in shallow pools of dilute binary solutions heated from below. They consider as well purely surface tension-driven flows as convection with competition between Marangoni-mechanisms and Rayleigh instability. They also discuss as well stationary as oscillatory instabilities, but they do not discuss the effect of simultaneous mass transfer through the interface.

In a recent series of papers[19-22], Sørensen, Hennenberg and their coworkers have carried out stability analysis on isothermal, deformable interfaces with transfer of surfactive substances as an outgrow of research in the role of chemical surface reactions in connection with motile events near cell membranes carried out by the Sanfeld-group[23-25]. They include gravity in the surface dynamics and analyse as well stationary as oscillatory instabilities with values of the parameters representative for the mass transfer of acetic acid from ethylene glycol to ethyl acetate, a system studied experimentally by Orell and Westwater[26]. Calculations were made for as well linear concentration profiles as for more realistic exponential profiles.

It is a callenge for the future to try to work together the theoretical results of S. & S. , Smith, Palmer and Berg and the Sørensen-Hennenberg-Sanfeld group with the experimental findings in the monumental work of Linde and coworkers on liquid motion in connection with mass- and heat transfer[16,27-40]. Also, P.Schwartz and H.Wilke from Linde's group in Berlin have constructed computer models being able to follow the instabilities into the non-linear realm, but without allowing for deformation of the interface[41-42]. In the present paper my task will be of a much more limited scope, however. The model describing the dynamics of an isothermal, deformable fluid interface will be introduced in § 2. The characteristic eqation for a deformable - originally plane - fluid interface with transfer of surfactant is shown in § 3. Some special solutions to the characteristic equation in the vicinity of critical wavenumbers for stationary instabilities are discussed in § 4, and possible interactions between the Marangoni-instability and the gravity-driven Rayleigh-Taylor instability are considered in § 5. In § 6 I give a summary of the assumptions behind and the results of S. & S.'s 1959-theory. In order to compare our results for exponential concentration profiles with S. & S.'s theory, a "modified S. & S. theory for exponential profiles" is constructed in § 7.

Computer studies and comparisons between the models are reported in § 8 where especially the ability of low surface tensions to destabilise certain oscillatory modes will be discussed. Such a destabilisation might have significance for spontaneous emulsification and perhaps even for the formation of microemulsions in the process of tertiary oil recovery by surfactant-cosurfactant-polymer slugs[43]. In § 9 a series of graphs for the determination of regions of stability and instability in terms of suitable dimensionless numbers is given for the realistic case of exponential concentration profiles (being an approximation to errorfunction profiles). An illustration of, how the graphs may be used, is given in § 10 in case of the ethylene glycol (EG) - ethyl acetate (EA) - acetic acid (HAc) system studied by Orell and Westwater[26]. Finally, a discussion of what to do in the future is included in § 11.

2. Dynamics of a Newtonian fluid interface

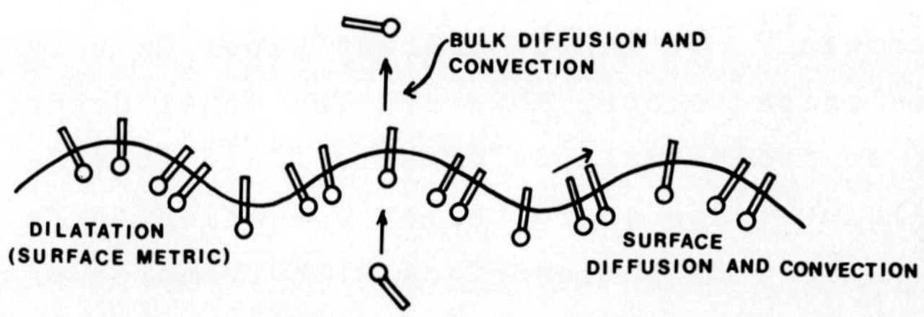

Fig. 1. Illustration of interfacial mass balance.

In fig. 1 we have illustrated the physics behind the balance of surfactant adsorbed to the interface. The interfacial concentration Γ varies due to convection at the interface, due to dilatation or compression of the interface (deformational change of the surface metric), due to surface diffusion and due to differences in the diffusion to or from the adjacent bulk solution. In symbols we have (see ref. 19 pp.50-52):

$$\frac{\partial \Gamma}{\partial t} + \frac{\dot{a}}{2a} \Gamma + \underset{\sim}{\nabla}_s \cdot (\Gamma \underset{\sim}{v}_s) - D_s \nabla_s^2 \Gamma = \Delta \{D \underset{\sim}{\nabla} c\} \cdot \underset{\sim}{n} \tag{1}$$

where a is the local value of the surface metric determinant and index s denotes "surface". The velocity in interfacial (in general Gaussian) coordinates is denoted v_s and the vector operators in the surface will in general be given by the tensorial description connected with the surface coordinates (see the paper by Scriven[44] and the book by Aris[45]. Δ stands for a difference between a property in fluid 2 and in fluid 1 when crossing the interface between 1 and 2. We assume constant diffusion coefficients, very small deformation of the interface δz and very small fluctuations in the interfacial concentration δΓ. Hereby we obtain the *linearised* interfacial mass balance (Γ_o is the unperturbed surface mass density).

$$\frac{\partial \delta\Gamma}{\partial t} = -\left(\frac{\dot{a}}{2a}\right)_o \cdot \Gamma_o - \Gamma_o \underset{\sim}{\nabla}_s \cdot \underset{\sim}{v}_s + D_s \nabla_s^2 \delta\Gamma + \Delta\left\{D \left.\frac{\partial \delta c}{\partial z}\right|_{z=o} + D \left.\frac{d^2 c_o}{dz^2}\right|_{z=o} \cdot \delta z\right\} \quad (2)$$

The last term on the r.h.s. containing the curvature of the *unperturbed* concentration profile was forgotten in some of our publications[19,20,22], but the mistake was corrected in ref. 21. This term was shown to be important for the destabilisation of oscillatory modes at low surface tensions. The term due to the change in the metric (number one at r.h.s. of (2)) will only contain second order terms in the surface deflection in case of small deformations of a plane interface, but in case of small deformations of a *spherical* interface, that term has to be taken into account (see ref. 25, p.198). For small deformations of a plane interface we have also the particular simple relations

$$\nabla_s^2 = \frac{\partial^2}{\partial x^2} + \frac{\partial^2}{\partial y^2} \quad (3)$$

$$\underset{\sim}{\nabla}_s \cdot \underset{\sim}{v}_s = \frac{\partial v_x}{\partial x} + \frac{\partial v_y}{\partial y} \quad (4)$$

where v_x and v_y are components of the bulk velocity adjacent to the interface.

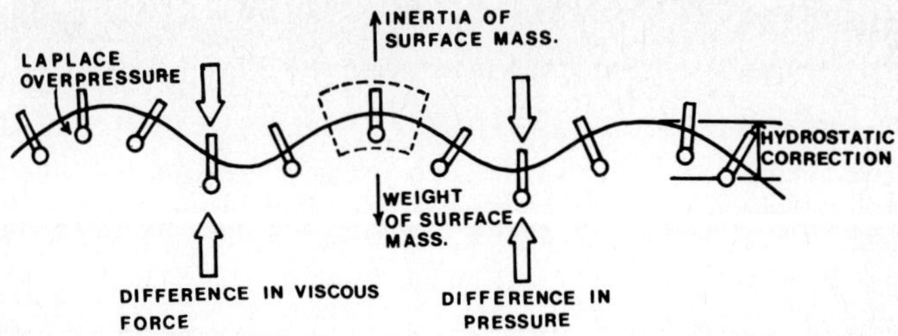

Fig. 2 Normal momentum balance for the interfacial layer.

Next we consider the *normal* momentum balance for the interfacial layer: the local interfacial mass times the normal acceleration must be equal to the net force on the surface mass in the vertical direction. Forces will belong to three classes. Some of the forces such as interfacial tension and forces from surface viscosity originate in the surface itself. Other forces stem from the adjacent bulk phases, such as pressure differences (including Laplace pressure) and differences in viscous stress. Finally, gravity (g) is an external force acting on the interfacial mass and producing hydrostatic pressure differences across deflected interfaces. The general tensorial forms of the momentum balances for arbitrary large deformations and for arbitrary geometries of the unperturbed surface are given by Scriven[44] and Aris[45] (for a brief account *vide* Hennenberg, Sørensen and Sanfeld[19] pp. 49-54). Here I shall limit myself to write down the linearised momentum balances for small deflections of a plane interface, small velocities and small fluctuations in surface mass density. In the normal direction (η = shear viscosity of bulk fluid, ρ density, σ_o unperturbed interfacial tension)

$$\Gamma_o \left(\frac{\partial v_z}{\partial t}\right) = \Delta\{(-\delta p) + g\rho\delta z + 2\eta\left(\frac{\partial v_z}{\partial z}\right)\}_{z=o} - g\delta\Gamma$$

$$+ \sigma_o \left(\frac{\partial^2}{\partial x^2} + \frac{\partial^2}{\partial y^2}\right) \delta z \qquad (5)$$

The traction due to interfacial tension gradients (Marangoni-effects) or to surface viscosity does *not* contribute to the normal balance for small deflections of the surface unit normal $\underset{\sim}{n}$ from the direction of the z-axis (perpendicular to undisturbed interface).

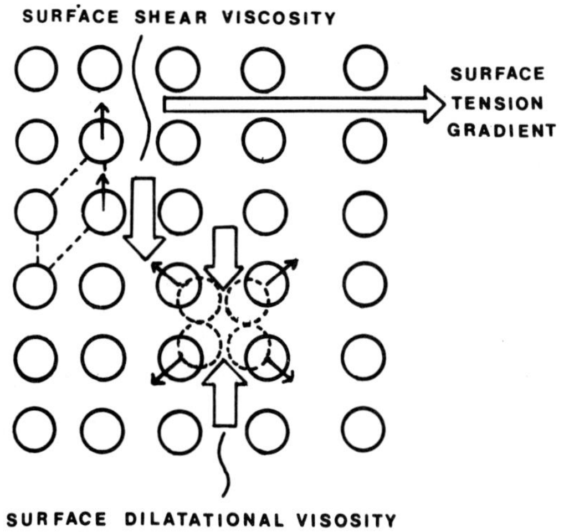

Fig. 3 Tangential momentum balance for the interfacial layer. (Surface in the plane of the paper).

The linearised *tangential* momentum balance for small deflections from a plane interface contains contributions from viscous stress tensors of the adjacent bulk fluids, surface tension gradients and net forces due to surface shear and dilatational viscosity. Boussinesq proposed already in 1913 a two dimensional analogue to the three dimensional Newtonian fluid[46]. The model surface was generalised by Oldroyd[47], Scriven[44] and

Slattery[48] to general surface coordinates and to surfaces undergoing continous change in shape and extent. Here we shall follow the analysis of Scriven (explained in detail in the book by Aris[45]) in which we distinguish between a dilatational interfacial viscosity (κ) and a shear interfacial viscosity (ε). It should be noticed that a "suface fluid" in case of a deformable surface can never be "incompressible", such as we have assumed for the bulk fluids. (Goodrich[49] has used a surface model with three elastic moduli and three viscosity coefficients, but the relation between the work of Goodrich and the work of other authors is not clear to me at present. Wasan and Mohan[43] describe a system with a dilatational and a shear elastic modulus in addition to κ and ε). The tangential momentum balance in our simplified case is given by

$$\Gamma_o \left(\frac{\partial v_x}{\partial t}\right) = \Delta\left\{\eta \left(\frac{\partial v_x}{\partial z} + \frac{\partial v_z}{\partial x}\right)\right\}_{z=o} + \frac{\partial \sigma}{\partial x} - \kappa \left(\frac{\partial}{\partial x}\frac{\partial}{\partial z} v_z\right)_{z=o} + \varepsilon \left(\frac{\partial^2}{\partial x^2} + \frac{\partial^2}{\partial y^2}\right)_{z=o} v_x \tag{6}$$

It should be mentioned that - strictly speaking - the Scriven-Aris surface momentum balance has been derived only for interfaces with no net mass exchange with the surroundings. Equations (5) and (6) can therefore only be exact at infinite dilution. I do not believe, however, that the errors committed hereby are more serious than the errors introduced by the neglect of the influence of the concentration gradients on the viscosities and densities of the bulk fluids and the neglect of the ability of the viscous dissipation in producing temperature fluctuations. In other words, I still believe that equations (5) and (6) may be used as first order approximations even in more concentrated mixtures with transfer of matter across the interface. There are also conceptual problems with the identification of the surface velocity in equation (1) with the center of mass velocity for the adjacent bulk phases in the case where we have transfer of solute appearing in high concentration. Generalised no slip conditions and a generalised surface mass balance should be constructed. My only excuse for applying (1) and (2) even in cases with mass transfer is the intractability of the equations resulting from the general procedure - rendering no simple stability considerations possible.

If we derive eqn.(6) with respect to x and add it to the analogous eqn. in v_y (derived with respect to y) we obtain by utilising the incompressibility condition for the bulk fluid

$$\frac{\partial v_x}{\partial x} + \frac{\partial v_y}{\partial y} + \frac{\partial v_z}{\partial z} = 0 \tag{7}$$

and eqn. expressed solely in terms of the normal velocity component:

$$\Gamma_o \frac{\partial}{\partial z}\left(\frac{\partial}{\partial t} v_z\right) = - \Delta \left\{\eta\left(\frac{\partial^2}{\partial x^2} + \frac{\partial^2}{\partial y^2}\right) v_z\right\}_{z=o} - \left[\frac{\partial^2}{\partial x^2} + \frac{\partial^2}{\partial y^2}\right] \delta\sigma$$
$$+ (\kappa + \varepsilon) \left[\frac{\partial^2}{\partial x^2} + \frac{\partial^2}{\partial y^2}\right] v_z \quad (z=o) \tag{8}$$

This equation will be used in constructing the characteristic equation for the stability problem together with eqns. (2) and (5).

3. Characteristic equation.

The procedure followed in deriving the characteristic equation is the procedure well known from the discipline of linear, hydrodynamic stability analysis[9]. Fluctuations from the stationary state of departure are characterised by ascribing an unperturbed value to the fluctuating variable and a small deviation due to fluctuation, *e.g.* for the surface concentration of surfactant

$$\Gamma(x,y) = \Gamma_o + \delta\Gamma(x,y) \tag{9}$$

All second order terms in the differential eqns. are discarded and solutions of the form

$$v_z = \tilde{v}_z(z) \cdot f(x,y) \cdot e^{\omega t} \tag{10}$$

are sought for. Due to the isotropy in the XY-planes the functions f(x,y) will have a wavelike character, *i.e.* the Helmholtz equation

$$\left[\frac{\partial^2}{\partial x^2} + \frac{\partial^2}{\partial y^2}\right] f = - k^2 f \tag{11}$$

will be satisfied. The separation constant k is called the *wavenumber*. The linearised Navier-Stokes equations for the velocities have the form (neglecting the differences in density ρ and kinematic viscosity ν produced by concentration differences)

$$\frac{\partial v_i}{\partial t} = - \frac{\partial \delta p}{\partial x_i} + \nu \nabla^2 v_i \qquad (i = 1,2,3) \tag{12}$$

Using the incompressibility condition (7) we obtain from (12)

$$\nabla^2 \delta p = 0 \tag{13}$$

with the solution

$$\delta p = \begin{cases} \tilde{p}_{20}\ e^{-kz} \cdot f(x,y) \cdot e^{\omega t} & ; \quad z \geq 0 \text{ phase 2} \\ \tilde{p}_{10}\ e^{+kz} \cdot f(x,y) \cdot e^{\omega t} & ; \quad z \leq 0 \text{ phase 1} \end{cases} \tag{14}$$

With the solutions δp, equation (12) for i=3 is an inhomogeneous differential equation in v_z which can be solved. Next, the *diffusion-convection equation* ($c_o^{(i)}(z)$ ∿ unperturbed conc.profile at time t)

$$\frac{\partial \delta c^{(i)}}{\partial t} = - v_z \cdot \left(\frac{\partial c_o^{(i)}}{\partial z}\right)_t + D_i\ \nabla^2\ \delta c^{(i)} \qquad (i=1,2) \tag{15}$$

for the dissolved surfactant can be solved with the knowledge of v_z. The solution c(x,y,z,t) to equation (15) has to be known for two reasons:

1. It is assumed that the fluctuation in local surface mass density of surfactant is determined through the concentration

fluctuations in the adjacent bulk solutions (local equilibrium, no diffusion barrier at interface)

$$\delta\Gamma = K_i \, [\delta c^{(i)} \, (z=o) + \frac{d}{dz} c_o^{(i)}\Big|_{z=o} \cdot \delta z] \qquad (16)$$

and $\delta\Gamma$ appears both in the interfacial mass balance (2) and in the tangential momentum balance, since

$$\delta\sigma = - \alpha \, \delta \, \Gamma \qquad \text{(central mechanochemical link)} \qquad (17)$$

where we assume again that α can be determined from equilibrium data. (Both K_i and α may depend on concentration of surfactant in adjacent bulk liquids).

2. The normal gradients $\partial\delta c/\partial z$ at the interface are required in the position of the interface in the interfacial mass balance (2), *cf.* last two terms on r.h.s. of (2).

I shall not go into details of the calculations here, since the reader may refer to references 19-22 for further information. The reader has to notice especially, however, the corrections made in ref.21, since the last term in eqn.(2) and the last term in eqn.(16) were omitted in the first papers. Those terms have certain consequences at low surface tensions and curved unperturbed concentration profiles, see later. The normal velocities are given by (10) with

$$\tilde{v}_z^{(2)} = \frac{k}{\omega\rho_2} \tilde{p}_{20} \, e^{-kz} + B_2 \, e^{-q_2 z}$$

$$\tilde{v}_z^{(1)} = - \frac{k}{\omega\rho_1} \tilde{p}_{10} \, e^{+kz} + B_1 \, e^{+q_1 z} \qquad (18)$$

with the parameters

$$q_i = \sqrt{k^2 + \omega/\nu_i} \quad ; \quad \mathrm{Re}\,(q_i) > 0 \quad (i = 1,2) \qquad (19)$$

The two arbitrary constants p_{10} and p_{20} can be eliminated by the requirement of no slip at the interface (in the normal and in the tangential directions). B_1 and B_2 remain arbitrary amplitudes, *i.e.* the energy of the perturbation is left unspecified.

As can be seen from the form of equation (15), the solution to the diffusion-convection equation depend on the assumptions with respect to the unperturbed concentration profiles $c_o(z,t)$. We shall here consider two cases:

Case A: Linear concentration profiles. We assume the following relations:

$$c_o^{(1)} = c_{oo}^{(1)} - \beta_1 z \quad ; \quad z < 0 \tag{20a}$$

$$c_o^{(2)} = c_{oo}^{(2)} - \beta_2 z \quad ; \quad z > 0 \tag{20b}$$

The concentration gradients are positive for transfer of matter from phase 1 to phase 2. Furthermore, we assume that there is no accumulation of surfactant on the interface in the unperturbed state

$$\beta_1 D_1 = \beta_2 D_2 \tag{21}$$

Notice, however, that we have *not* assumed that there is no accumulation of matter at the interface also in the *perturbed* state. *This condition was used by Sternling and Scriven* (ref.12 p.518, boundary condition xii) *at the same time as those authors consider the lateral concentration variations along the interface to be fundamental for the generation of Marangoni-effects.* This inconsistency - peculiar to the S. & S. theory - seems strangely enough not to have great influence on the results, see later.

The general solution to the diffusion-convection equation can be written in the form

$$\tilde{c}_i(z) = C_1^{(i)} e^{\pm r_i z} + \frac{\beta_i}{2k\omega}\left[(k - q_j)B_j - (k + q_i)B_i\right] e^{\pm kz} + \frac{\beta_i \nu_i B_i}{\omega(\nu_i - D_i)} e^{\pm q_i z} \qquad (i = 1,2;\ j = 2,1) \tag{22}$$

defining the parameter r as

$$r_i = \sqrt{k^2 + \frac{\omega}{D_i}} \qquad (i = 1,2) \tag{23}$$

The two arbitrary constants $C_1^{(i)}$ $(i = 1,2)$ can be expressed through the amplitude $\tilde{\Gamma}$ in the fluctuation of the surface mass density and through B_1 and B_2 using the local equilibrium assumption (16) and the connection

$$\frac{\partial}{\partial t}(\delta z) = v_z \quad (z = 0) \tag{24}$$

The result is

$$C_1^{(i)} = \frac{\tilde{\Gamma}}{K_i} - \frac{\beta_i}{\omega}\frac{D_i}{\nu_i - D_i} B_i \qquad (i = 1,2) \tag{25}$$

Using finally the interfacial mass and momentum balances (2), (5) and (8) we obtain the following system of linear, homogeneous equations

$$\underset{\approx}{M} \begin{bmatrix} \tilde{\Gamma} \\ B_1 \\ B_2 \end{bmatrix} = \underset{\sim}{0} \tag{26}$$

where the amplitude factors $\tilde{\Gamma}$, B_1 and B_2 are determined up to an arbitrary constant, if the characteristic equation

$$\det(\underset{\approx}{M}) = 0 \tag{27}$$

is satisfied. The first row in the matrix correspond to the interfacial mass balance, the second row to the normal momentum balances and the third row to the tangential momentum balances. The matrix elements are given by (after division of the 2.column by $(k-q_1)$, division of the 3. column by $(k-q_2)$ and after replacement of the 2. column by the sum of the 2. and 3. column and the 3. column by the difference between the 2. and the 3. column):

$$M_{11} = \omega + k^2 D_s + \Sigma \frac{r_i}{K_i} D_i \tag{28a}$$

$$M_{12} = \frac{\beta_1 D_1}{\omega} \Sigma \frac{q_i + k}{q_i + r_i} \tag{28b}$$

$$M_{13} = \Gamma^o - \frac{\beta_1 D_1}{\omega} \Delta \{\frac{q_i + k}{q_i + r_i}\} \tag{28c}$$

($\Delta \sim$ phase 2 - phase 1)

$$M_{21} = g \tag{28d}$$

$$M_{22} = \frac{1}{k} (\Gamma^o \omega - g \frac{\Delta\rho}{\omega} + \frac{k^2 \sigma_o}{\omega}) + \Sigma \eta_i (1 + \frac{q_i}{k}) + \frac{\omega}{k^2} \Sigma \rho_i \tag{28e}$$

$$M_{23} = \Delta \{\eta_i (1 - \frac{q_i}{k})\} \tag{28f}$$

$$M_{31} = \alpha k \tag{28g}$$

$$M_{32} = - M_{23} \tag{28h}$$

$$M_{33} = - \frac{1}{k} (\Gamma^o \omega + k^2 \eta_s) - \Sigma \eta_i (1 + \frac{q_i}{k}) \tag{28i}$$

In the last element we have introduced the symbol η_s instead of the sum of the two surface viscosities $\kappa + \varepsilon$.

Case B: Exponential concentration profiles. When a substance is present in phase 1 in a uniform concentration $c_\infty^{(1)}$ and in phase 2 in a uniform concentration $c_\infty^{(2)}$ at time $t = 0$ an errorfunction type of profile develops for $t > 0$. Assuming a Nernst distribution law with

$$K_d = \frac{K_2}{K_1} = \frac{c_o^{(1)} (z = 0)}{c_o^{(2)} (z = 0)} \tag{29}$$

we get by means of a slight generalisation of the formula for two-phase diffusion given in the book by Jost[50]:

$$c_o^{(1)} = \frac{1}{K_d + \sqrt{D_2/D_1}} \{K_d \left(c_\infty^{(2)}\sqrt{\frac{D_2}{D_1}} + c_\infty^{(1)}\right) + \sqrt{\frac{D_2}{D_1}} \left(K_d\, c_\infty^{(2)} - c_\infty^{(1)}\right) \operatorname{erf}\left(\frac{z}{2\sqrt{D_1 t}}\right)\} \quad (z<0) \tag{30a}$$

$$c_o^{(2)} = \frac{1}{K_d + \sqrt{D_2/D_1}} \{c_\infty^{(2)}\sqrt{D_2/D_1} + c_\infty^{(1)} + \left(K_d\, c_\infty^{(2)} - c_\infty^{(1)}\right) \operatorname{erf}\left(\frac{z}{2\sqrt{D_2 t}}\right)\} \quad (z> 0) \tag{30b}$$

It is easily controlled that eqns. (30a-b) give the proper boundary conditions for $z \to \pm\infty$ and satisfy eqn.(29) together with the no-accumulation boundary condition at $z = 0$. Since, however errorfunctions are quite intractable in the stability analysis we want to perform here, we approximate the erf-profiles by exponential profiles. We insert

$$1 - \operatorname{erf}(x) = 1 + \operatorname{erf}(-x) \cong e^{-1.444x} \tag{31}$$

in (30a-b) and obtain

$$c_o^{(1)} = c_\infty^{(1)} + c_{oo}^{(1)}\, e^{\beta_1 z} \quad (z<0) \tag{32a}$$

$$c_o^{(2)} = c_\infty^{(2)} + c_{oo}^{(2)}\, e^{-\beta_2 z} \quad (z>0) \tag{32b}$$

with

$$c_{oo}^{(1)} = \frac{K_d\, c_\infty^{(2)} - c_\infty^{(1)}}{K_d \sqrt{D_1/D_2} + 1} \tag{33a}$$

$$c_{oo}^{(2)} = -\sqrt{\frac{D_1}{D_2}}\, c_{oo}^{(1)} \tag{33b}$$

$$\beta_1 = \frac{1.444}{2\sqrt{D_1 t}} \tag{33c}$$

$$\beta_2 = \sqrt{\frac{D_1}{D_2}}\, \beta_1 \tag{33d}$$

It should be noticed that the approximation (31) coincides with the erf-profile at $x = 0$, $x = \infty$ and $x = 0.48$ (where $\mathrm{erf}(x) = 0.5$), and the deviation between $1 - \mathrm{erf}(x)$ and $\exp(-1.444x)$ is never more than 0.09.

For exponential profiles we have simple solutions to the diffusion-convection equation. Instead of (22) we obtain

$$\tilde{c}_i(z) = C_1^{(i)} e^{\pm r_i z} + C_2^{(i)} e^{\pm(k+\beta_i)z} + C_3^{(i)} e^{\pm(q_i+\beta_i)z} \tag{34}$$

with

$$C_1^{(i)} = \frac{\tilde{\Gamma}}{K_i} + B_i \frac{c_{oo}^{(i)}\beta_i(-1)^i}{D_i} \left[\frac{1}{(q_i+\beta_i)^2-r_i^2} - \frac{(k+q_i)/2k}{(k+\beta_i)^2-r_i^2} + \frac{k-q_i}{2k\omega} D_i\right]$$

$$+ B_j \frac{c_{oo}^{(i)}\beta_i(-1)^i}{D_i} \left[\frac{(k-q_j)/2k}{(k+\beta_i)^2-r_i^2} + \frac{k-q_j}{2k\omega} D_i\right] \tag{35a}$$

$$(i \neq j)$$

$$C_2^{(i)} = \frac{c_{oo}^{(i)}\beta_i(-1)^i}{D_i} \cdot \left\{\frac{1}{(k+\beta_i)^2-r_i^2}\right\} \cdot \left[\left(\frac{k+q_i}{2k}\right) B_i - \left(\frac{k-q_2}{2k}\right) B_j\right] \tag{35b}$$

$$C_3^{(i)} = \frac{c_{oo}^{(i)}\beta_i(-1)^{i+1}}{D_i} \cdot \frac{B_i}{(q_i+\beta_i)^2-r_i^2} \tag{35c}$$

The elements of the determinantal characteristic equations are the same as before, except for the following two elements:

$$M_{12} = \beta_1 c_{oo}^{(1)} \{\frac{D_1}{k\omega} \Sigma \frac{\beta_i(\beta_i+2k)}{k+\beta_i+r_i} - \frac{2D_1}{\omega} + \frac{1}{(k+\beta_1+r_1)(q_1+\beta_1+r_1)}$$

$$+ \frac{D_1/D_2}{(k+\beta_2+r_2)(q_2+\beta_2+r_2)}\} \quad (36a)$$

$$M_{13} = \Gamma^o + \beta_1 c_{oo}^{(1)} \{\frac{1}{(k+\beta_1+r_1)(q_1+\beta_1+r_1)}$$

$$- \frac{D_1/D_2}{(k+\beta_2+r_2)(q_2+\beta_2+r_2)}\} \quad (36b)$$

Note that the β's in the exponential case have the dimension $(\text{length})^{-1}$ whereas the β's in the linear case have the dimensions $(\text{mass})(\text{length})^{-4}$.

Strictly speaking we consider in case B perturbations from a time *dependent* initial state $c_o(z,t)$. However, as well the small signal diffusion-convection equation (15) as the linearised mass balance for the interface given in equation (2) are exact, since they can be obtained by subtracting from the equations for $\partial c(x,y,z,t)/\partial t$ and for $\partial\Gamma(x,y,\delta z,t)/\partial t$ - with $c = c_o(z,t)+\delta c(x,y,z,t)$ and $\Gamma = \Gamma_o(\delta z,t)+\delta\Gamma(x,y,\delta z,t)$ - the equations for $\partial c_o/\partial t$ and $\partial\Gamma_o/\partial t$. The derivative and the second derivative of the unperturbed concentration profile to be used in equations (2) and (16) have to be taken at a given time t, however. When that non-exponential time dependence is introduced in eqn.(2), the possibility of having solutions proportional to $e^{\omega t}$ - which is at the heart of the linear analysis of stability - is destroyed. We have therefore to assume that the convective perturbations are far more rapid than the slow changes due to diffusion. This will also most often be the case, but the solutions to the characteristic equation found should always be subjected to *a posteriori* verification with respect to this condition. This technique of "freezing" the concentration profiles were suggested already by S. & S. (ref.12 pp.522 - 523). It has previously been applied to the Rayleigh stability problem by Morton[51]. Case A in our analysis as well as the analysis of S. & S. do not suffer from the limitations mentioned here, but unfortunately a linear concentration profile is quite unrealistic in the case of mass transfer.

4. Solutions in the neighbourhood of stationary critical points.

In the neighbourhood of stationary, critical points we have

$$\omega_r \to 0 \qquad \omega_i = 0 \tag{37}$$

The characteristic equation (27) may therefore be expanded around the marginal state in a power series in ω retaining only terms proportional to ω^o and ω^1. After tedious calculations one obtains

$$P_1 \omega^* + P_o = 0 \tag{38}$$

with the dimensionless growth constant

$$\omega^* = \frac{\Sigma\eta_i}{k\sigma_o}\,\omega \tag{39}$$

and with the dimensionless coefficients given by (linear profile)

$$P_o = \Pi_1 + \delta_w \frac{g/\sigma_o}{1 + \frac{g}{\sigma_o}\left(\frac{-\Delta\rho}{k^2}\right)} \cdot \frac{2\beta_1 D_1 \Sigma\eta_i}{k^2\sigma_o}\left[2 + \frac{k\eta_s}{\Sigma\eta_i}\right] \tag{40}$$

$$\begin{aligned} P_1 = {} & \Pi_2 + \Pi_3 + \frac{2\Pi_1}{1-(g\Delta\rho/k^2\sigma_o)} - \frac{\beta_1 D_1 \alpha}{k^2\sigma_o}\,\frac{\Delta\rho}{\Sigma\eta_i} \\ & + \delta_w \cdot \frac{g/\sigma_o}{1+(g/\sigma_o)(-\Delta\rho/k^2)} \cdot \beta_1 D_1 \left[\left(2+\frac{k\eta_s}{\Sigma\eta_i}\right)\frac{\Sigma D_i^{-1}}{4k^3}\right. \\ & \left. - \frac{2}{k^2\Sigma\eta i}\left(\Gamma_o + \frac{\Sigma\rho_i}{2k}\right)\right] \end{aligned} \tag{41}$$

In the above formulae δ_w is a parameter which is equal to unity if the weight of the perturbation of the mass density of the adsorbed surfactant ($g\delta\Gamma$) is taken into account in the linearised normal momentum balance for the interface. All other authors have neglected this contribution which would not be expected to be important in view of the very small interfacial mass densities found in usual systems of surface chemistry ($\Gamma_o \approx 10^{-8} - 10^{-10}$ g/cm^2). In the famous book by Levich[52] on "Physicochemical Hydrodynamics", the dynamics of interfaces in connection with capillary waves in presence of surfactants is treated neglecting as well the interfacial weight term $g\delta\Gamma$ as the interfacial inertial terms $\Gamma_o\ \partial v/\partial t$. (More important, Levich ignored also interfacial viscosity in the tangential boundary conditions. On pp.420 - 423 he criticises even the concept of dilatational viscosity of Boussinesq[46] and ascribes all resistive forces due to stretching of the interface to the Gibbs-Marangoni effects due to disequilibrium between the stretched interface and the surrounding fluid. This confusion has been made also by several later authors, see ref.43, p.196).

Returning to eqations (40) and (41), the dimensionless numbers Π_1, Π_2, and Π_3 are given by

$$\Pi_1 = \frac{1}{\sigma_o}\{(2\Sigma\eta_i + k\eta_s)\cdot(kD_s + \Sigma\frac{D_i}{K_i}) + \alpha\Gamma_o - \frac{\alpha\beta_1}{4k^2}(1-\frac{D_1}{D_2})\} \tag{42a}$$

$$\Pi_2 = (1 + \frac{1}{2k}\Sigma K_i^{-1})(2 + \frac{\eta_s}{\Sigma\eta_i}k) + \frac{1}{\Sigma\eta_i}(kD_s + \Sigma\frac{D_i}{K_i})\cdot(\Gamma_o + \frac{1}{2k}\Sigma\rho_i) + \frac{\alpha\beta_1}{8k^3\Sigma\eta_i}(1-\frac{D_1}{D_2})\Sigma\frac{1}{D_i} \tag{42b}$$

$$\Pi_3 = \frac{3g\beta_1(1 + \frac{D_1}{D_2})(2 + \frac{\eta_s}{\Sigma\eta_i}k)}{4k^3\sigma_o(1 - \frac{g\Delta\rho}{k^2\sigma_o})} \tag{42c}$$

To be precise, the last term of Π_2 should be

$$-\frac{\alpha\beta_1}{8k^3\Sigma\eta_i}[\frac{D_1}{D_2}(\frac{1}{D_2} + \frac{1}{\nu_2}) - \frac{1}{D_1}(\frac{1}{D_1} + \frac{1}{\nu_1})]$$

(compare with the formulae (39a-b) and (40a-e) in ref.22 pp.67 - 68). In most cases, however, the kinematic viscosities are much greater than the diffusion coefficients, and the last term in equation (42b) is correct within that approximation.
In the case of exponential concentration profiles we have only calculated P_o in (38). This leads to the following equation for the determination of k_{cr} for the marginal states of stationary disturbances:

$$(2 + k\frac{\eta_s}{\Sigma\eta_i})(kD_s + \Sigma\frac{D_i}{K_i}) + \frac{\alpha\Gamma_o}{\Sigma\eta_i}[1-\frac{\beta_1^{exp}c_{oo}^{(1)}D_1}{\Gamma_o}\Delta\{\frac{1}{D_i(\beta_i^{exp}+2k)^2}\}]$$

$$= \delta_w \frac{g/\sigma_o}{1+(g/\sigma_o)(-\Delta\rho/k_{cr}^2)} \cdot \frac{(-\beta_1^{exp}c_{oo}^{(1)}D_1)}{k^2}[2 + k\frac{\eta_s}{\Sigma\eta_i}] \cdot [2 - \Sigma\frac{\beta_i^{exp}(k+\beta_i^{exp})}{k(2k+\beta_i^{exp})}] \quad (43)$$

5. Marangoni instability and Rayleigh-Taylor instability

In the present section we shall limit ourselves to discuss the case of a linear profile of surfactant leading to Marangoni instability. Simultaneously, our equations should permit a stability analysis of gravity-driven flows - arising when the denser fluid is placed on top of the less dense fluid. Such fluid motions are traditionally called Rayleigh-Taylor instability (ref.9, Chap.X), although it is important to note that the analysis of Rayleigh[53] and Taylor[54] did not include effects due to fluid viscosity or effects due to surface tensions. These effects were considered by Harrison[55], Bellman and Pennington[56] and Reid[57]. In principle, our analysis has lead to a characteristic equation where the Marangoni-instabilities and the Rayleigh-Taylor instabilities are intermingled. A pertinent question becomes therefore: Do the two kinds of instabilities interact with each other?

In equations (40) and (41) we put $\delta_w = 0$. We consider first the case

$$\Pi_1 \to 0 \quad (44)$$

In the neighbourhood of the states of marginal stability we have:

$$\omega^* \cong \frac{-\Pi_1}{\Pi_2 + \Pi_3 - \frac{\beta_1 D_1 \alpha}{k^2 \sigma_o} \frac{\Delta\rho}{\Sigma\eta_i}} \tag{45}$$

The equation determining the critical wavenumber for Marangoni-instabilities is $\Pi_1 = 0$ or

$$f(\bar{k}) = \bar{\eta}_s \bar{D}_s \bar{k}^4 + (2\bar{D}_s + \bar{\eta}_s)\bar{k}^3 + (2 + \bar{\Gamma})\bar{k}^2 = \frac{1}{4} \operatorname{sign}\left(\frac{\alpha\beta_1}{\Sigma \frac{D_i}{K_i}}\right) \cdot \left(1 - \frac{D_1}{D_2}\right) \tag{46}$$

where we have introduced "natural dimensionless numbers"

$$\bar{k} = k_{cr} \sqrt{\frac{(\Sigma\eta_i)|\Sigma D_i/K_i|}{|\alpha\beta_1|}} \tag{47a}$$

$$\bar{\eta}_s = \frac{\eta_s}{\Sigma\eta_i} \sqrt{\frac{|\alpha\beta_1|}{(\Sigma\eta_i)|\Sigma D_i/K_i|}} \tag{47b}$$

$$\bar{D}_s = \frac{D_s}{\Sigma D_i/K_i} \sqrt{\frac{|\alpha\beta_1|}{(\Sigma\eta_i)|\Sigma D_i/K_i|}} \tag{47c}$$

$$\bar{\Gamma} = \frac{\alpha\Gamma_o}{(\Sigma\eta_i)|\Sigma D_i/K_i|} \tag{47d}$$

Since all the parameters (47a-d) are positive (or zero) the fourth degree polynomium on the l.h.s. of (46) will assume any positive value for $\bar{k}$ running from 0 to infinity. The polynomium will be monotonously growing with $\bar{k}$, so if the r.h.s. is positive there will be one positive root $\bar{k}$ satisfying (46). If the r.h.s. is negative there will be no roots:

$$\mathrm{sign}\left(\frac{\alpha\beta_1}{\Sigma D_i/K_i}\right)\{1 - \frac{D_1}{D_2}\}\begin{cases} \geq 0 & 1 \text{ root } k_{cr} \\ < 0 & \text{no roots} \end{cases} \tag{48}$$

For a normal surfactant the surface tension will decrease with concentration, and α is positive, For diffusion from phase 1 to phase 2, β_1 is also positive. There will then be a possibility of having non-oscillatory Marangoni-instability only for $D_1<D_2$. Thus:

The necessary and sufficient condition for having marginal states of Marangoni-instability of the stationary kind during mass transfer of a surfactant between two immiscible fluids is that diffusion is out of the phase with the lower diffusion coefficient for the surfactant and into phase with the higher diffusion coefficient.

This result is in accordance with the theory of S. & S. and the experimental results of Linde and coworkers.

E.Schwarz[37] has even shown experimentally that for "inverse surfactants" with $\partial\sigma/\partial c > 0$, the direction of diffusion should be the reverse to have stationary instabilities. This is also in accordance with the S. & S. analysis and with the formula (46) above, since we have the following relations

$$\Gamma \cong - \frac{c^{(i)}M}{RT}\left(\frac{\partial\sigma}{\partial c^{(i)}}\right) \qquad (M = \text{molecular weight of surfactant}) \tag{49a}$$

$$K_i \equiv \left(\frac{\partial\Gamma}{\partial c^{(i)}}\right) \cong - \frac{M}{RT}\left(\frac{\partial\sigma}{\partial c^{(i)}} + c^{(i)}\frac{\partial^2\sigma}{\partial c^{(i)2}}\right) \tag{49b}$$

$$\alpha \equiv - \frac{\partial\sigma}{\partial\Gamma} \cong - \frac{1}{K_i}\left(\frac{\partial\sigma}{\partial c^{(i)}}\right) \tag{49c}$$

found by applying Gibbs' adsorption isotherm and assuming ideal behaviour of the dilute solutions of surfactant. Experimentally, it is found (see ref.37 figs.1 and 5) that for surfactants $\frac{\partial^2\sigma}{\partial c^2} < 0$ while for "reverse surfactants" $\frac{\partial^2\sigma}{\partial c^2} > 0$. For reverse surfactants (e.g. $CaCl_2$ in Butanol-water system or n-Hexanol in cyclohexanol/water system) the excess mass density in the surface Γ is negative and K_1 and K_2 are also negative, whereas α remains positive. From equation (48) follows now that diffusion has to be out of the phase with the *higher* diffusion coefficient to have stationary instabilities in the case of reverse surfactants.

S. & S. did not consider surface diffusion and they also neglected the influence of Γ_o (weight and acceleration terms connected with interfacial mass). Inserting in (46)

$$\bar{D}_s = \bar{\Gamma} = 0 \tag{50}$$

we obtain the simpler formula

$$f(\bar{k}) = \bar{\eta}_s \bar{k}^3 + 2\bar{k}^2 = \frac{1}{4} \operatorname{sign}\left(\frac{\alpha\beta_1}{\Sigma D_i/K_i}\right)\left(1 - \frac{D_1}{D_2}\right) \tag{51}$$

which may be translated to exactly the same formula as given by S. & S. (ref.12, formula (30n)) using the translation table between the present formalism and the S. & S. formalism given in ref.21, App.I. The fact that S. & S. have neglected the deformation of the interface and have used contradictory boundary conditions seems not to have any influence on the equation determining the marginal states of stationary disturbances. Neither have those intrinsic errors in the S. & S. model any influences upon the complete dispersion curves in case of linear profiles, as we shall se later on. We have also shown by computer sensitivity analysis that the surface diffusion coefficient does not play any significant role for realistic values of the parameters. This could hardly be known in advance, however. For example, Ivanov has shown both theoretically and experimentally that the influence of surface diffusion is paramount in the thinning and rupture of foam films (see ref.58, pp.42-43 and p.49 §3). From equation (51) it is easy to determine graphically the critical value of the wavenumber for a given system. Exponential concentration profiles are far more realistic than linear, however, and we shall give dimensionless graphs for that case-analogous to equation (51)-in §9.

We would expect another critical wavenumber at k = 0, since an infinite wavelength corresponds to uniform translation of the whole system. This situation is always neutrally stable and non-oscillatory. Therefore, equation (45) may be used taking the limit $k \to 0$. The result is

$$\omega^* \to \frac{2k\,\Sigma\eta_i}{\sigma_o\,\Sigma D_i^{-1}} \qquad k \to 0 \tag{52a}$$

or

$$\omega \to \left(\frac{2}{\Sigma D_i^{-1}}\right) k^2 \qquad k \to 0 \tag{52b}$$

From (52b) we see that we have stationary *instability* immediately after increasing the wavenumber from zero. We must therefore have a stationary instability region between k = 0 and $k = k_{cr}$ determined by (46) or (51). The slope of the dispersion curve $\omega^*(k)$ when passing through the critical point of Marangoni-instability k_{cr} is given by

$$\frac{d\omega^*}{dk} = \frac{-\left(\frac{d}{dk}\Pi_1\right)_{k_{cr}}}{(\Pi_2 + \Pi_3)_{k_{cr}} + \frac{\beta_1 D_1 \alpha}{k_{cr}^2\,\sigma_o}\,\frac{(-\Delta\rho)}{\Sigma\eta_i}} \tag{53a}$$

with

$$\left(\frac{d\Pi_1}{dk}\right)_{k_{cr}} = \frac{1}{\sigma_o}\left\{2\eta_s D_s k_{cr} + 2D_s\Sigma\eta_i + \eta_s\,\Sigma\frac{D_i}{K_i} + \frac{\alpha\beta_1}{2k_{cr}^3}\left(1 - \frac{D_1}{D_2}\right)\right\} \tag{53b}$$

If there are no oscillatory critical states inbetween k = 0 and $k = k_{cr}$ we would expect the slope given by (53a) to be negative. From (53b) we see that for normal surfactants we can be sure that the numerator in (53a) is negative if the diffusion direction required for instability is obeyed. The sum of Π_2 and Π_3 in the denominator of (53a) is positive in that case, if we further require Rayleigh-Taylor stability, *i.e.* $1-(g\Delta\rho/k^2\sigma_o) > 0$. To be sure that the last term in the denominator of (53a) is also positive we must further assume that diffusion takes place from the more dense to the less dense phase. This last term (which was omitted in ref.20 eqn.(3) due to the neglect of surface deformation) has no practical numerical significance for realistic values of the parameters, however.

The absolute value of the surface tension has obviously no influence on the determination of the critical Marangoni wavenumber according to (46). Neither has the value of the gravitational acceleration g. In principle, however, both parameters have an influence on the critical slope (53a), since we have

$$\left(\frac{d\omega}{dk}\right)_{k_{cr}} = \frac{k\sigma_o}{\Sigma\eta_i}\left(\frac{d\omega^*}{dk}\right)_{k_{cr}} = \frac{-A(k_{cr})}{\Pi_2(k_{cr}) + B(k_{cr})\phi(k_{cr}) - \dfrac{\beta_1 D_1 \alpha\Delta\rho}{\sigma_o k_{cr}^2 \Sigma\eta_i}} \qquad (54a)$$

with

$$A(k) = k\cdot\left\{2D_s\left[1+k\frac{\eta_s}{\Sigma\eta_i}\right] + \frac{\eta_s}{\Sigma\eta_i}\Sigma\frac{D_i}{K_i} + \frac{\alpha\beta_1}{2k^3\Sigma\eta_i}\left(1-\frac{D_1}{D_2}\right)\right\} \qquad (54b)$$

(**dimension** length/time)

$$B(k) = \frac{3\beta_1(1 + D_1/D_2)}{4k(-\Delta\rho)}\left(2 + \frac{\eta_s}{\Sigma\eta_i}k\right) \qquad (54c)$$

(dimensionless)

$$\phi(k) = \frac{(-\Delta\rho)}{k^2}\,\frac{g/\sigma_o}{1-\dfrac{g\Delta\rho}{k^2\sigma_o}} = \begin{cases} 0 & \text{for } g/\sigma_o = 0 \\ 1 & \text{for } g/\sigma_o = \infty \end{cases} \qquad (54d)$$

(dimensionless)

All the influence of gravity is concentrated in the $B\phi$-term in the denominator of (54a), and this term clearly represents a coupling between the Marangoni and the Rayleigh-Taylor instabilities. The influence is hyperbolic in g/σ_o through the function ϕ. This means that the effect of increasing gravity saturates at high values of the gravitational acceleration, but even with $\phi = 1$ it can be shown using realistic values of the parameters, that the $B\phi$-term is vanishingly small compared to $\Pi_2(k_{cr})$, so the influence of gravity is practically nil, see ref.20 pp.1591-1592 and fig.2. The last term in the denominator of (54a) stems from the Laplace-overpressure due to deformation and this term was not included in the analysis of ref.20. Even if this term is generally much smaller than $\Pi_2(k_{cr})$ there should be a possibility to make the term influential by making the surface tension converge towards zero.

After having discussed the Marangoni-instability and the possible interaction with the Rayleigh-Taylor instability we now turn towards the Rayleigh-Taylor instability and the influence on this instability from the Marangoni-instability. For that purpose we write equation (38) as ($\delta_w = 0$)

$$\omega^* = \frac{\Pi_{R.T.} - 1}{2 - \frac{B(k)\Pi_{R.T.}}{\Pi_1} + \frac{(1-\Pi_{R.T.})}{\Pi_1}\{\Pi_2 - \frac{\beta_1 D_1 \alpha}{k^2\sigma_o}\frac{\Delta\rho}{\Sigma\eta_i}\}} \tag{55}$$

with B(k) given by equation (54c) and the dimensionless "Rayleigh-Taylor number" defined by

$$\Pi_{R.T.} = \frac{g\Delta\rho}{k^2\sigma_o} \tag{56}$$

It is seen that the Rayleigh-Taylor number is unity for marginal stability:

$$\Pi_{R.T.}(k_{cr}) = 1 \tag{57}$$

Very close to the critical wavenumber for Rayleigh-Taylor instability equation (55) can be written

$$\omega^* \cong \frac{\Pi_{R.T.-1}}{2 - \frac{B(k_{cr})}{\Pi_1(k_{cr})}} \tag{58}$$

And from (58) we obtain for the critical slope

$$\left(\frac{d\omega}{dk}\right)_{k_{cr}} = \frac{k_{cr}\sigma_o}{\Sigma\eta}\left(\frac{d\omega^*}{dk}\right)_{k_{cr}} = \frac{-\frac{g\Delta\rho}{k_{cr}^2\Sigma\eta_i}}{1 - \frac{B(k_{cr})}{2\Pi_1(k_{cr})}} = \frac{-\frac{\sigma_o}{\Sigma\eta_i}}{1 - \frac{B(k_{cr})}{2\Pi_1(k_{cr})}} \tag{59}$$

The last term in the denominator of (59) is clearly an interaction term with the Marangoni-instability and it vanishes for vanishing mass transfer ($\beta_1 = 0$). In most realistic cases, however, we find that

$$\left|\frac{B(k_{cr})}{2\Pi_1(k_{cr})}\right| << 1 \tag{60}$$

and we have then the very simple relation for the Rayleigh-Taylor dispersion curve $\omega(k)$ in the neighbourhood of the critical wavenumber

$$\left(\frac{d\omega}{dk}\right)_{k_{cr}} \cong \frac{-\sigma_o}{\Sigma\eta_i} \tag{61}$$

Since the slope of the dispersion curve is negative, we have instability for $k < k_{cr}$ and since (52b) is also valid in this case we have another critical point at $k = 0$ and an instability region in the interval $k(0/k_{cr})$ with a state of maximum instability - $\omega(k_{max}) = \omega_{max}$ - inbetween, just as we saw for the Marangoni-instability. It is remarkable that neither the critical wavenumber nor the critical slope depend on the *surface viscosity*. The Marangoni-instability, on the other hand, is found to be profoundly affected by surface viscosity both with regard to the critical wavenumber and with regard to the whole dispersion curve, see equations (46) and (51) and calculations given in §§8-9. However, the Marangoni-instability is not influenced in practice by the value of the gravitational acceleration or the absolute value of the surface tension, whereas both parameters play a fundamental role in the Rayleigh-Taylor instability. So even if the dispersion relations $\omega(k)$ look quite similar in teh two cases, it must be stated that the hydrodynamic modes seem to be quite different and complementary to each other. This is in harmony with the well-known fact that the Rayleigh-Taylor instability never leads to hexagonal or polygonal tesselated structures such as do the Marangoni-instabilities.

There is a possibility of interaction between the two kinds of instabilities, however, which should not be left unmentioned. If the critical wavenumber for the R.T. instability is situated very closely to the critical wavenumber of the Marangoni-instability, $\Pi_1 \to 0$ in (59) and the B-term in the denominator may become important and thereby an influence of the surface viscosity is also brought in (*vide* equation (54c)). It is hardly surprising that the two instabilities interact when their critical wavenumbers are very close to each other. A similar effect is seen in the expression (54a) for the critical slope of the Marangoni-dispersion curve, since the ϕ-function has a singularity at the critical wavenumber for the R.T. instability, so that we have mutual interaction between the Marangoni- and the R.T. instability in that case.

6. The assumptions and the results of the Sternling-Scriven model.

Since the model of Marangoni-instabilities proposed by Sternling and Scriven[12] are simpler than the model described here and since it gives in many cases the same or almost the same results, I shall give here a short account of the assumptions behind the model and of the basic results delivered by the model. The assumptions are the following:

1. The "Boussinesqian" point of view is taken and all fluid properties are taken as constants (no dependence of concentration). Also, heat evolution due to viscous dissipation is neglected.

2. The concentration gradients in each phase are assumed linear. There is no accumulation at the interface in the undisturbed state.
3. Only two-dimensional infinitesimal disturbances are considered. Due to the linearity only a single Fourier-component has to be considered at the time (roll cells) and the time dependence of all disturbances can be factored out as $e^{\omega t}$.
4. It is assumed that the dominant wavelength (corresponding to k_{max} and ω_{max}) will be closely related to the wavelength found in the non-linear, mature, tesselated convection structure.
5. The interfacial tension is assumed great enough for the interface to remain planar (hereby σ drops out of the problem).
6. Gravity effects are neglected. Normal stress balance at the interface is not considered at all.
7. Local equilibrium is assumed, *i.e.* the two phases and the interface are assumed in thermodynamic equilibrium at all points of contact.
8. Interfacial mass balance is not used. Rather, no accumulation at the interface is assumed in the *perturbed* state as a boundary condition joining the two solutions to the diffusion convection equations in the bulk. This is highly contradictory to the assumption of the importance of the Marangoni-forces produced by lateral variations of the concentration in the adjacent bulk fluids and to the assumption of local equilibrium.
9. S. & S. do not separate the effects of surface diffusion disequilibrium and other non-equilibrium surface concentration perturbations of σ (generally called Gibbs-Marangoni effects) from the intrinsic dilatational viscosity at the surface. This seems to be somewhat connected with the assumption no.8.
10. The temperature fluctuations generated by uneven release of heat of solution passes the interface are neglected. S. & S. present some evidence that for a benzene-water system with acetone as surfactant, the interfacial forces generated by temperature variations are roughly a thousandfold less than those generated by concentration variations. Nevertheless,

interfacial turbulance in systems with exothermic neutralisations at the interface is well known[59-60].

11. The diffusing surfactant is assumed non-ionic, since electric phenomena at the interface are not accounted for.
12. Transfer at the interface is diffusion-limited and not determined by an adsorption activation energy barrier.

Assumptions 5, 6, 8, and 9, are the really peculiar ones for the S. & S. theory, while the other assumptions have also been used in our model. In case of a single salt diffusing through an interface, the S. & S. theory may probably be used without regard to assumption 11, since electroneutrality requires that the two ions diffuse in a concerted manner. (E.Schwartz[37] has as mentioned before confirmed the S. & S. theory in the case of the "inverse" surfactants $CaCl_2$, NaCl, $CuCl_2$, and $NaNO_3$ diffusing between butanol and water). However, when more than one electrolyte is present - such as in the $C_{18}Cl$/KCl/nitrobenzene/water system studied by Dupeyrat and Nakache[61] - difficulties are met in solving the highly non-linear electrodiffusion equations which may give rise to electric triple layers[62] and to time varying diffusion potentials[63]. Also, the fascinating and industrially useful phenomena of electrocapillary emulsification[64] will only be fully understood with the help of hybrid hydrodynamic and electrodynamic stability theories[65,66,67].

The result of the analysis of S. & S. is summarized in the following dispersion relation (ref.12 eqn.(30) translated to our formalism by means of the translation table in ref.21, Appendix I)

$$\frac{D_2 \eta_2}{\alpha K_2 \beta_2} k^2 = \frac{L_1}{L_2} \tag{62}$$

with the dimensionless numbers

$$L_1 = \frac{D_2}{D_1} \frac{k^2}{(r_1+k)(r_1+q_1)} - \frac{k^2}{(r_2+k)(r_2+q_2)} \tag{63a}$$

$$L_2 = (K_d \frac{D_1}{D_2} \cdot \frac{r_1}{k} + \frac{r_2}{k})[\frac{\eta_1}{\eta_2} (\frac{k+q_1}{k}) + (\frac{k+q_2}{k}) + \frac{k\eta_s}{\eta_2}] \tag{63b}$$

Notice that the growth constant ω is implicit present in equation (62) in the parameters r_i and q_i. S. & S. have made a qualitative survey of the loci of $\omega_r = 0$ and found results which may be formulated as the following rules of thumb:

1. With transfer *from* the phase with low D *to* the phase with high D we have always a stationary instability ($\omega_i = 0$) as discussed in the previous section and shown in fig.4. With the opposite direction of diffusion we can *never* have a *stationary* critical wavenumber in the S. & S. and in our model, c.f. the discussion in the previous paragraph.

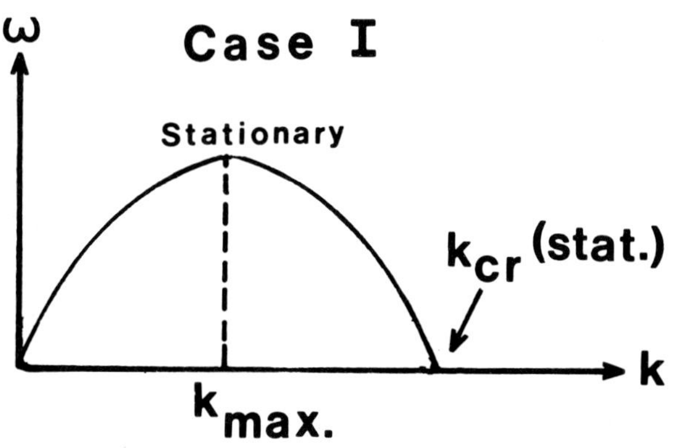

Fig.4 S. & S. case I. Diffusion from the phase with lower D to the phase with higher D.

2. If we have the case I *and* diffusion is out of the phase with *higher* kinematic viscosity (and lower D), diffusion in the reverse direction is completely *stable*. (case II).
3. If we have the case I *and* diffusion is out of the phase with *lower* ν (and lower D) the reverse direction of diffusion is *unstable*. For

$$\frac{D_{to}}{D_{from}} \geq \frac{\nu_{to}}{\nu_{from}} > 1 \qquad (64)$$

the *reverse* direction of diffusion leads to a purely oscillatory instability (indicated as a dashed line in fig.5, case III), whereas for

$$\frac{\nu_{to}}{\nu_{from}} > \frac{D_{to}}{D_{from}} > 1 \qquad (65)$$

the *reverse* direction of diffusion gives rise to a double branch of stationary instabilities joining the oscillatory branch in a saddle point as shown in fig.5, case IV.

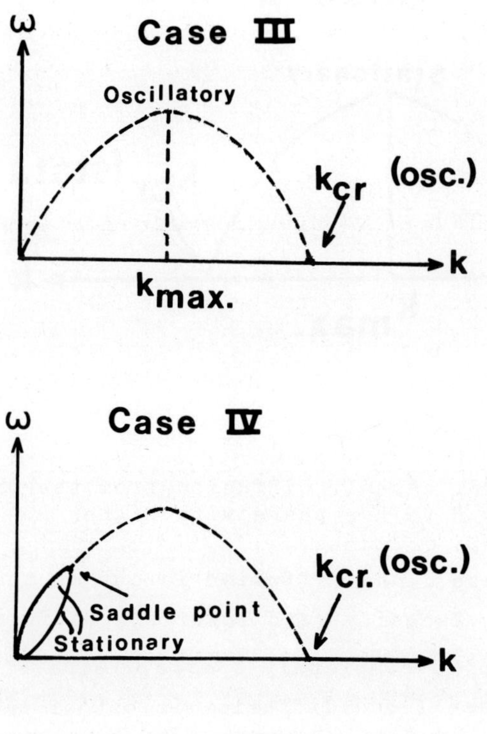

Fig.5 S. & S. case III and case IV.

4. For reverse surfactants (giving rise to elevated surface tensions instead of lowered surface tensions) all the conclusions given in 1-3 hold for *reverse* directions of diffusion.

The four rules given here cover all the investigated cases of S. & S. except for a few quite special cases of no practical interest. The combination case I + case II is the most frequent situation in practical examples of mass transfer, cf. Linde and Schwartz[28]. Due to the approximate Stokes' law relation between viscosity and diffusion coefficient, it is not so easy to have both the diffusion coefficient and the kinematic viscosity lower in the same phase. However, one might imagine some cases where a polymer changes conformation from one phase to the other. Also, in cases of diffusion from a gas phase into a liquid phase with not too high viscosity, the inequality $D_{to}/D_{from} < \nu_{to}/\nu_{from} < 1$ is valid for not too high viscosities of the liquid. According to fig.5, case III we should then have oscillating instabilities, and Linde and Kunkel[36] found indeed instabilities in the form of travelling waves in the cacase of transfer of ether across an air- 1,1,2,2-tetrachlor ethane interface. They also found oscillatory instabilities in cases where the solute is dimer in an organic phase and monomer in a water phase. During passage from the organic phase to the water phase the dimer splits into two monomers causing oscillatory instabilities. In such cases one has to extend the S. & S. theory to encompass surface reactions[68,23,24,25] and this is outside the scope of the present investigation.

7. The Sternling-Scriven model for exponential profiles.

One of the unrealistic features of the S. & S. model when dealing with mass transfer is that it treats only linear concentration profiles. This draw-back can easily be remedied, however, by allowing for exponential concentration profiles, but maintaining all the other characteristic features of the S. & S. theory. Performing that analysis we end with a characteristic equation which is an extension of equations (62, 63a-b), but still simpler than the full determinantal eqn.(27).

In this paragraph I shall follow S. & S. quite closely, and therefore only the outlines will be given here. Considering two-dimensional convection cells, the velocity components may be derived from a streamfunction

$$v_z = -\frac{\partial\psi}{\partial x} \qquad v_x = +\frac{\partial\psi}{\partial z} \tag{66}$$

Solutions are sought in the form

$$\psi = \phi(z)\, e^{ikx} \cdot e^{\omega t} \tag{67}$$

where the amplitude $\phi(z)$ satisfies the Orr-Sommerfeld equation

$$\phi'''' - 2\phi'' + \phi = \frac{\omega}{k^2\nu}(\phi'' - \phi) \tag{68}$$

The general solution is given by

$$\phi = A_1 e^{kz} + A_2 e^{-kz} + A_3 e^{qz} + A_4 e^{-qz} \tag{69}$$

with the proper boundary conditions in phase 1 and phase 2 and at the interface we obtain

$$\phi_1(z) = -A_2 \frac{k-q_2}{k-q_1}(e^{kz} - e^{q_1 z}) \tag{70a}$$

$$\phi_2(z) = +A_2 (e^{-kz} - e^{-q_2 z}) \tag{70b}$$

The tangential stress balance at the interface is given by

$$\tau_{zx}^{(1)} - \tau_{zx}^{(2)} = \frac{\partial\sigma_{xx}}{\partial x} \quad \text{at } z = 0 \tag{71}$$

where the shear stress components from the adjacent bulk liquids are given by

$$\tau_{zx}^{(i)} = \eta_i \left(\frac{\partial v_z^{(i)}}{\partial x} + \frac{\partial v_x^{(i)}}{\partial x}\right) \quad (i = 1,2) \tag{72}$$

The components of the "dynamic surface tension" are given by the Boussinesq expression

$$\sigma_{xx} = \sigma_o + \kappa \left(\frac{\partial v_x}{\partial x} + \frac{\partial v_y}{\partial y}\right) + \quad \left(\frac{\partial v_x}{\partial x} - \frac{\partial v_y}{\partial y}\right) \tag{73}$$

Thus we have

$$\frac{\partial \sigma_{xx}}{\partial x} = \left(\frac{\partial \sigma_o}{\partial_c(2)}\right) \frac{\partial_c(2)}{\partial x} + \eta_s \left(\frac{\partial^2 v_x}{\partial x^2}\right) \qquad (z = 0) \tag{74}$$

with

$$\eta_s = \kappa + \varepsilon \tag{75}$$

and

$$\frac{\partial \sigma_o}{\partial c^{(2)}} = - K_2 \alpha \tag{76}$$

Using (70a) and (70b) we have from (71) and (74)

$$- K_2 \alpha \left(\frac{\partial c^{(2)}}{\partial x}\right)_{z=0} = A_2 \eta_2 k^2 e^{ikx} \cdot e^{\omega t} \frac{q_2 - k}{k} \cdot \left[\left(\frac{k+q_2}{k}\right) + \frac{\eta_1}{\eta_2}\left(\frac{k+q_1}{k}\right) + \frac{k\eta_s}{\eta_2} \right] \tag{77}$$

The linearised diffusion-convection equations are given by

$$\frac{\partial \delta c}{\partial t} - D\left(\frac{\partial^2 \delta c}{\partial z^2} + \frac{\partial^2 \delta c}{\partial x^2}\right) = - v_z \frac{dc_o}{dz} \tag{78}$$

where we assume

$$c_o^{(1)} = c_\infty^{(1)} + c_{oo}^{(1)} e^{\beta_1 z} \tag{79a}$$

$$c_o^{(2)} = c_\infty^{(2)} + c_{oo}^{(2)} e^{-\beta_2 z} \tag{79b}$$

Seeking solutions in the form

$$\delta c = \tilde{c}(z)\, e^{ikx} \cdot e^{\omega t} \tag{80}$$

we obtain

$$[\frac{d^2}{dz^2} - r_\ell^2]\ \tilde{c}^{(\ell)} = ikA_2 \cdot \begin{cases} \frac{1}{D_1}(\frac{k-q_2}{k-q_1})(e^{kz}-e^{q_1 z})\beta_1 c_{oo}^{(1)} e^{\beta_1 z} & (\ell=1) \\ \frac{1}{D_2}(e^{-kz}-e^{-q_2 z})\beta_2 c_{oo}^{(2)} e^{-\beta_2 z} & (\ell=2) \end{cases} \quad (81)$$

Solutions are of the form of equation (34) in §3 with

$$C_2^{(1)} = + \frac{\beta_1 c_{oo}^{(1)}}{D_1} \cdot \frac{k-q_2}{k-q_1} \cdot \frac{ikA_2}{(k+\beta_1)^2 - r_1^2} \quad (82a)$$

$$C_3^{(1)} = - \frac{\beta_1 c_{oo}^{(1)}}{D_1} \cdot \frac{k-q_2}{k-q_1} \cdot \frac{ikA_2}{(q_1+\beta_1)^2 - r_1^2} \quad (82b)$$

$$C_2^{(2)} = + \frac{\beta_2 c_{oo}^{(2)}}{D_2} \cdot \frac{ikA_2}{(k+\beta_2)^2 - r_2^2} \quad (82c)$$

$$C_3^{(2)} = - \frac{\beta_2 c_{oo}^{(2)}}{D_2} \cdot \frac{ikA_2}{(q_2+\beta_2)^2 - r_2^2} \quad (82d)$$

The constants $C_1^{(i)}$ are still arbitrary. They are fixed by the requirements

$$K_d = \frac{K_2}{K_1} = \frac{\tilde{c}^{(1)}\ (z=0)}{\tilde{c}^{(2)}\ (z=0)} \quad (83)$$

$$D_1 \left.\frac{d\tilde{c}^{(1)}}{dz}\right|_{z=0} = D_2 \left.\frac{d\tilde{c}^{(2)}}{dz}\right|_{z=0} \quad (84)$$

The boundary condition (84) is the quite strange condition of the S. & S. theory discussed under point 8 in the beginning of §6. We obtain for $\tilde{c}^{(2)}(z=0) = C_1^{(2)} + C_2^{(2)} + C_2^{(2)} + C_3^{(2)}$:

$$\frac{\tilde{c}^{(2)}(z=0)\ \{r_2 D_2 + K_d r_1 D_1\}}{ikA_2} = \quad (85)$$

$$\beta_1 c_{oo}^{(1)} \{\frac{k-q_2}{k-q_1}\}\{\frac{1}{q_1+\beta_1+r_1} - \frac{1}{k+\beta_1+r_1}\} + \beta_2 c_{oo}^{(2)} \{\frac{1}{q_2+\beta_2+r_2} - \frac{1}{k+\beta_2+r_2}\}$$

Introducing this result into equation (77) we get the dispersion relation between ω and k (whereby the arbitrary constant A_2 cancels).
We obtain

$$L_2 \cdot \frac{D_2}{D_1} \cdot \frac{q_2-k}{r_1} =$$

$$\frac{K_2\alpha}{D_1 r_1 \eta_2}\{\beta_1 c_{oo}^{(1)}\left(\frac{k-q_2}{k-q_1}\right)\left(\frac{1}{q_1+\beta_1+r_1} - \frac{1}{k+\beta_1+r_1}\right) \tag{86}$$

$$+ \beta_2 c_{oo}^{(2)}\left(\frac{1}{q_2+\beta_2+r_2} - \frac{1}{k+\beta_2+r_2}\right)\}$$

where the dimensionless number L_2 is the one given by equation (63b) in §6. Since we have no accumulation at the interface in the unperturbed state we have

$$\beta_1 c_{oo}^{(1)} D_1 = - \beta_2 c_{oo}^{(2)} D_2 \tag{87}$$

and the r.h.s. of (86) may be simplified. We obtain finally

$$\frac{\eta_2 D_2 k^2}{\beta_2^{exp} c_{oo}^{(2)} K_2 \alpha} = \frac{L_1^{exp}}{L_2} \tag{88}$$

with

$$L_1^{exp} = \frac{D_2}{D_1} \cdot \frac{k^2}{(q_1+\beta_1^{exp}+r_1)(k+\beta_1^{exp}+r_1)} - \frac{k^2}{(q_2+\beta_2^{exp}+r_2)(k+\beta_2^{exp}+r_2)} \tag{89}$$

If we take the limit $\beta_1/k \to 0$, $\beta_2/k \to 0$ and $\beta_2^{exp} c_{oo}^{(2)} = \beta_2^{linear}$ the dispersion relation (88) reduces to the one given by S. & S. for a linear profile, see equations (62) and (63a-b) in §6. In case of marginal, non-oscillatory instability we have $\omega \to 0$ and q and r $\to$ k. Then we have

$$\frac{\eta_2 D_2 k_{cr}^2}{\beta_2^{exp} c_{oo}^{(2)} K_2 \alpha} = \frac{L_1^{exp}(cr)}{L_2\ (cr)} \tag{90a}$$

$$L_1^{exp}(cr) = \frac{D_2}{D_1} \cdot \frac{k_{cr}^2}{(2k_{cr}+\beta_1^{exp})^2} - \frac{k_{cr}^2}{(2k_{cr}+\beta_2^{exp})^2} \qquad (90b)$$

$$L_2(cr) = (1 + K_d \frac{D_1}{D_2}) \frac{1}{\eta_2} (2\Sigma\eta_i + k_{cr}\eta_s) \qquad (90c)$$

In ref.21 equation(27a) L_1^{exp} was erratically given in the form (90b) instead of the general expression (89). However, in the calculations where the modified S. & S. theory was used in that paper, the correct dispersion relation has been applied.

Equation ((90a-b) for the marginal states of non-oscillatory instability may also be written

$$(2\Sigma\eta_i + k\eta_s)(\Sigma \frac{D_i}{K_i}) = \alpha\beta_1^{exp} c_{oo}^{(1)} D_1 \Delta \left[\frac{1}{(D_i(2k+\beta_i)^2}\right] \qquad (91)$$

We now see that equation (91) is identical to equation (43) in §4 if we put

$$D_s = \Gamma_o = \delta_w = 0 \qquad (92)$$

We see that we obtain largely the same result for the non-oscillatory critical states from the modified S. & S. theory and from the theory of Sørensen *et al.*

8. Comparison between different models of Marangoni-instability and reality. The influence of curvature of concentration profile and of low interfacial tension.

Only the marginal, non-oscillatory states can be found by simple methods. The general form of the dispersion relation is

$$F(k, \omega_r, \omega_i, \text{parameters}) = 0 \qquad (93)$$

where F may be in the form of a 3x3 determinant or a simpler expression of the S. & S. type. If we fix the parameters (diffusion coefficients, viscosities and so on) and choose a given wavenumber k, we still have for arbitrary ω_r and ω_i that

$$F^* \cdot F \geq 0 \tag{94}$$

where F^* is the complex conjugate of F. It can be shown (see ref.20, Appendix) that if F is an analytic function of ω, there can be no local minima or maxima in the absolute square $F^2 = F^* \cdot F$ with nonzero values of F^2. The roots of (93) may therefore be found by a minimization procedure.

In some of the numerical computations given below the following values of the parameters have been used:

$$\begin{aligned}
&\alpha = 4.13\cdot 10^8 cm^2 s^{-2}; \quad K_1 = 1.27\cdot 10^{-7} cm; \quad K_2 = 1.93\cdot 10^{-7} cm \\
&\eta_1 = 0.103 \text{ poise} \quad ; \quad \eta_2 = 6.36\cdot 10^{-3} \text{ poise} \\
&D_1 = 0.21\cdot 10^{-5} cm^2 s^{-1}; \quad D_2 = 3.4\cdot 10^{-5} cm^2 s^{-1}; \quad D_s = 1.8\cdot 10^{-5} cm^2 s^{-1} \\
&\rho_1 = 1.064 gcm^{-3} \quad ; \quad \rho_2 = 0.9142 gcm^{-3} \\
&\sigma_o = 1.44 dyn\ cm^{-1} \quad ; \quad \Gamma_o = 2.67\cdot 10^{-10} gcm^{-2}
\end{aligned} \tag{95}$$

Those values correspond approximately to the system

phase 1: ethylene glycol   phase 2: ethyl acetate

with acetic acid as diffusing surfactant. The values are for very low concentrations of acetic acid in both phases, see ref. 19 pp.69-70 for an evaluation. The ethylene glycol-ethyl acetate-acetic acid system was studied by Orell and Westwater[26] with respect to the types of convection structure and dimensions of the convection cells, so a comparison should be possible.

We consider first the non-oscillatory case (case I, fig.4) where the diffusion direction is from the phase with the lower diffusion coefficient (phase 1 = ethylene glycol) to the phase with the higher. Dispersion curves calculated by means of (94) are given in fig.6 for linear concentration profiles with a

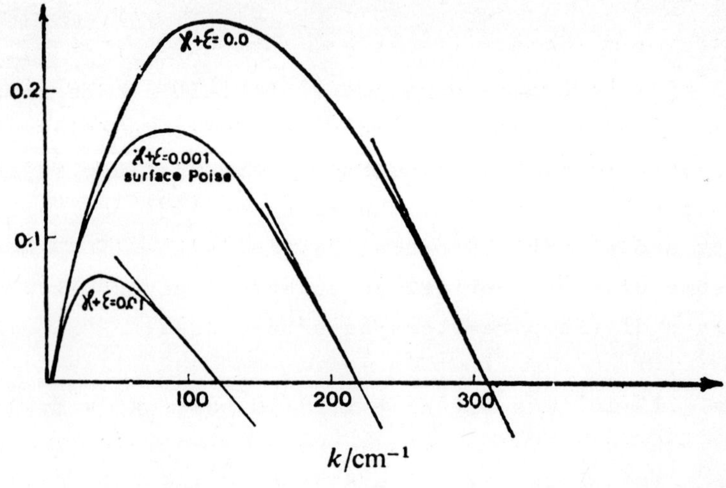

Fig.6 Dispersion curves for non-oscillatory instabilities with linear concentration profiles (case I) and $\beta_1 = 0.042$ gcm^{-4}. Other data as (95)

slope $\beta_1 = 0.042$ gcm^{-4} (corresponding to a drop from a subphase concentration of 0.5 w% acetic acid to zero in the superior phase over a distance of 1 mm). The dispersion curves as well as the critical wavenumbers are deeply dependent on the assumed (unmeasured) surface viscosity $\eta_s = \kappa + \varepsilon$. As well k_{cr} as k_{max} and ω_{max} decrease with increasing surface viscosity. Surface viscosity naturally damps the rate of growth of the instabilities and the damping is relatively greater for small wavelengths $\lambda = 2\pi/k$ than for great.

The critical wavenumbers found can also be checked by calculation by equations (38) + (40), (46) or (51). The three equations give the same results as the computer solutions which means that surface diffusion, weight and inertia of interface, gravitational acceleration and absolute value of surface tension play no significant role, and the simpler S. & S. theory with an undeformable interface may be used as well. The critical slopes $d\omega/dk$ indicated by straight lines on fig.6 are also consistent with equation (54a-b) and (42b) neglecting the two last terms in the denominator of (54a). Therefore g and σ_o have neither any influence on the critical slope. Actually, the whole dispersion curves in fig.6 are completely identical to the curves calculated by the S. & S. dispersion relation (62) where g and σ are lacking. In ref.20 we found no deviations from the S. & S. curves with the determinantal equation and $g = \cdot 10^{-6}$, 981, 10^6, 10^9 and 10^{18} cm/s^2 and with $\sigma = 10^{-6}$, 1.44, 10 and 100 dyn/cm. However, we cannot be completely sure on those results since the last term in equation (16) was erratically omitted in the evaluation of the characteristic determinant in refs.19, 20, and 22. In ref.21 the calculations were repeated with the correct determinant giving the same results, but g and σ were kept as 981 cm/s^2 and 1.44 dyn/cm, respectively. It is probably correct that g has no influence, since it appears only in the saturating function ϕ in the denominator of (54a) given by (54d). However, it is seen that the last term in the denominator of (54a) - which is a term precisely linked to the correction introduced in ref.21 - should be influential at least for very low surface tensions. For the system considered we obtain for $\eta_s = 0$ surface poise (k_{cr}=310cm^{-1})

$$\left(\frac{d\omega}{dk}\right)_{k_{cr}} = \frac{-777}{3.59\cdot 10^5 + 1.44\cdot 10^{-3}\phi + (5.21\cdot 10^{-4}/\sigma_o)} \qquad (96)$$

The slope $-777/3.59\cdot 10^5 = -2.16\cdot 10^{-3}$ cm/s is the slope indicated in fig.6.

Since ϕ runs from 0 to 1, when g/σ_o runs from 0 to ∞ (when $\Delta\rho < 0$), the influence of gravity is seen to be completely negligible. For the last term in the denominator to become important we should have

$$\sigma \sim 10^{-9} \text{ dyn/cm}$$

a quite low number, indeed! It is somewhat surprising that the effect of lowering the surface tension below that value seems to be stabilizing, since it lowers the numerical value of

$d\omega/dk$. Low surface tension usually promotes instability and spontaneous emulsification. It might be due to a sign error in the derivation of equation (54a) and the region around 10^{-9} dyn/cm should be checked by direct computer calculations. However, I do not believe those effects to be really significant, since the possible surface tensions are so small that the rate of any Marangoni-effects would be exceedingly slow.

With respect to the numeric values in fig.6, the order of magnitude fits very well with the data reported by Orell and Westwater. The wavelengths corresponding to the wavenumbers of maximum instability are found from fig.6 to be 0.185, 0.0766, and 0.0561 cm for surface viscosities 0.01, 0.001, and 0 surface poise, respectively. Orell and Westwater find cell dimensions ranging between 0.02 and 0.13 cm for concentrations between 0.1 w% to 10 w% acetic acid in ethylene glycol and with elapsed time of contact between phases ranging from few minutes to 50 h. (see ref.26, figs. 7 and 8). The initial growth constant was not checked experimentally, but the fact that polygonal cells appeared almost immediately after establishment of the phase contact seems to make plausible the growth times of 5 to 10 s shown in fig.1.

For diffusion in the opposite direction Orell and Westwater found no instability in accordance with as well our calculations and the theory of S. & S. (case II). Except for the highest concentrations of acetic acid, Orell and Westwater observed propagating polygonal cells and not stationary cells. According to theory we should *not* have possibility of instabilities with $\omega_i \neq 0$ in the present case, and we should therefore not expect propagating cells. Comparing the description of Orell and Westwater with the more detailed and structured explanations given by Linde and coworkers (see *e.g.* ref.28) we may conclude, however, that the "propagating" cells are the same as the "substructure cells" ("Unterstruktur") described by Linde *et al.* Those cells are one or two orders of magnitude smaller than the "secondary cells" ("grossräumige Rollzellen") and the substructure cells are either "created" in the middle of the secondary cells and move radially to the periphery ("Z-Zellen") or they are created at the periphery and move towards the center ("P-

Zellen"). At small concentrations of surfactants Linde and Schwartz observed only the substructure (ref.28, figs. 2-4). Thus, there is little doubt that the polygonal cells observed by Orell and Westwater are the "Sternling-Scriven cells" and that the propagating character is a secondary feature stemming from non-linear interactions between the primary Marangoni-instabilities. The substructure is intrinsically stationary, but may be carried along by secondary flow.

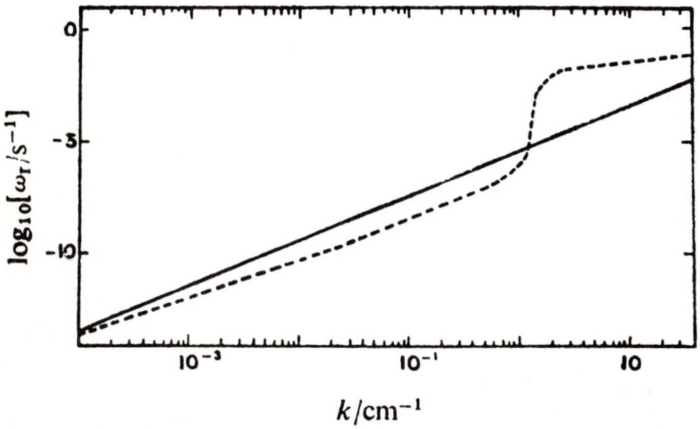

Fig.7 The asymptotic behaviour of ω_r for small wavenumbers. η_s = 0.001 surface Poise and β_1 = 0.042 gcm^{-4}.
Dashed line computed values. Full line: asymptotic parabola.

In connection with our previous analysis there is still one point to settle about the slope $d\omega/dk$ at k = 0. According to equation (52b) the dispersion curve $\omega(k)$ should be parabolic in k for small k's and the slope $(d\omega/dk)_{k=0}$ should be zero. It does not appear so in fig.6, but the more detailled analysis in fig. 7 shows that this is just a question of scale.

Fig.8 shows calculations of the Rayleigh-Taylor instability with the values of ρ_1 and ρ_2 interchanged. The dispersion curve

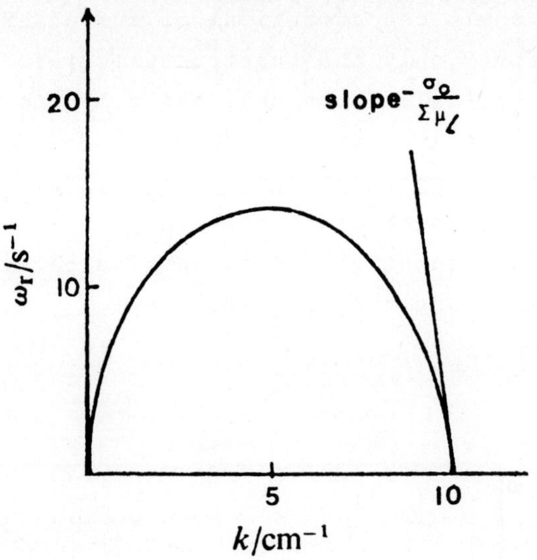

Fig.8 Rayleigh-Taylor instability with mass transfer of a surfactant. Data as (95) with the only change that the values of ρ_1 and ρ_2 have been interchanged. The dispersion curves are identical for η_s = 0, 0.001 and 0.01 surface Poise. β_1 = 0.042 gcm^{-4}

resembles that of the Marangoni-instability, but the resemblance is deceptive, since the curve is *completely unaffected by mass transfer of surfactant and the value of the surface viscosity has no influence, either*. Thus, the mode of liquid motion must be completely different in the two cases. The critical wavenumber for the R. & T. instability can be exactly calculated by equation (57) and the slope $(d\omega/dk)_{k_{cr}}$ by equation (61) as indicated on fig.8.

Exponential concentration profiles are more apt to describe realistic situations in the case of mass transfer than linear profiles. Fig. 9 shows a calculation pertaining to the data given in (95). It is characteristic for the exponential profiles that there are two critical wavenumbers with an instability region inbetween. We shall return to the question of their determination in the next prargraph. Here it should be pointed out

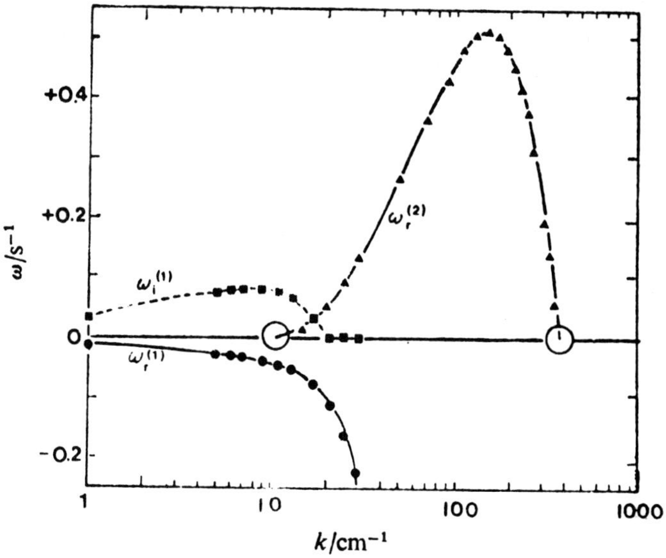

Fig.9 Real and imaginary components of ω for an exponential concentration profile. Solution (2) is purely real. Circles indicate analytically calculated valves of k_{cr}. Solution (1) is a damped capillary wave.

η_s = 0.001 surface Poise. $\beta_1^{exp} = 100\ cm^{-1}$

$c_{oo}^{(1)} = -\ 2.1\cdot10^{-3}\ gcm^{-3}$, $\beta_2^{exp} = 9.40\ cm^{-1}$,

$c_{oo}^{(2)} = 1.38\cdot10^{-3}\ gcm^{-3}$.

that the instability region may be qualitatively explained by means of an "exergy release-excess dissipation" principle which might be of value in the discussion also of other dissipative structures (see ref.22 p.74. For discussion about exergy *vs.* energy see for example refs.69-72). When $k > k_{cr}$(high) the chemical exergy released by smoothing out of concentration is not sufficient to overcompensate the dissipation due to surface viscosity, which is dominant at small wavelengths. On the other hand, when k decreases the "exergy harvest" due to increased scale of convection (wavelength increases) is subject to what economists call "diminishing returns to scale", since the perturbations reach out into the regions of constant concentrations. Below k_{cr}(low) it is therefore no more profitable for the system to set up convection currents. This lower limit will not be present in the unrealistic linear case, since in principle an infinite amount of exergy may be harvested when $\lambda \to \infty$.

By chance a stable oscillatory solution was also found in the analysis behind fig.9. This solution may be characterised as a capillary wave which becomes critically damped at a certain wavelength due to surface viscosity. It should be mentioned, however, that while the purely real solution on fig.9 was unaffected by the corrections of the determinantal elements due to deformation (discussed in ref.21 and in §3 of this paper), the capillary wave indicated on fig.9 could not be found with the corrected determinant. This appears to be somewhat mystical and should be investigated more closely in the future.

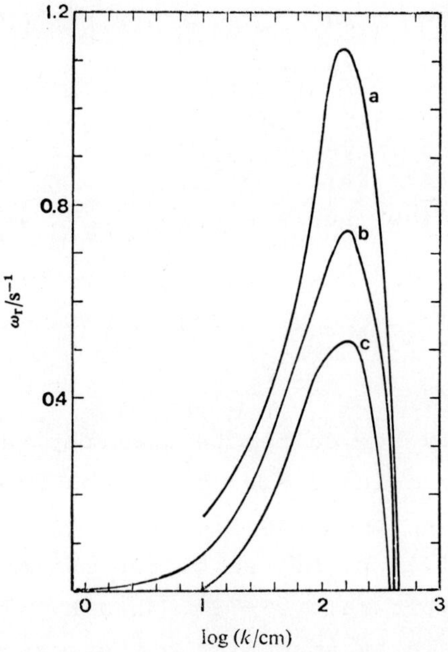

Fig. 10 Influence of thickness of diffusion zone and other factors on non-oscillating instabilities. Data as (95).
$\eta_s = 10^{-3}$ sP. $c_\infty^{(1)} = 0.0042$ gcm^{-3}; $c_\infty^{(2)} = 0$.
a) Linear profile $\beta_1^{lin} = 0.3$ gcm^{-4} and exponential profile with $\beta_1^{exp} = 1$ cm^{-1}, $c_{oo}^{(1)} = -0.3$gcm^{-3}
b) Exp.profile $\beta_1^{exp} = 100$ cm^{-1}, $c_{00}^{(1)} = -3\cdot10^{-3}$ gcm^{-3}
c) The same as solution (2) in fig.9.

In fig.10 the effect of decreasing the thickness of the diffusion zone is illustrated. Data as (95) are used, and it is assumed that all the acetic acid is initially in the phase 1 with predominance of ethylene glycol. The concentration 0.042 g acetic acid/cm^3 corresponds to a solution of 0.5 w% acetic acid in ethylene glycol. In curve a and b it is assumed that the concentration profiles have evolved from a stepfunction by simple diffusion at the moment where the instability sets in. Therefore, the connections given by (33a-d) for the erf-profiles have been used. The same mass transfer rate at the interface is assumed in case a and case b on fig.10, but the thickness of the diffusion zone is decreased from a to b, since β_1^{exp} is increased from 1 cm^{-1} to 100 cm^{-1}. With the same mass transfer rate at the interface curve a is identical to the curve calculated by the S. & S. theory and by the more complete theory with linear profiles. This is because the extension of the diffusion zone (1 cm) is so large in comparison to the wavelengths of instability (10^{-2} cm) so that we have linearity of the concentration profiles in the perturbation region. Going from a to b a certain stabilisation is observed. This is not due to the increased curvature of the concentration profile *per se*. The reason is rather that the concentration threshold has to be diminished if one wants to maintain the mass transfer rate at the interface fixed and to continue to satisfy the relations (33a-d) see fig.11.

Returning to fig.10, it is seen that curve c is identical to solution (2) on fig.9. The mass transfer rate and the thickness of the concentration profile in phase 1 are the same in case b and c, but the relations (33a-d) have not been used, whereas - of course - the mass conservation equation

$$\beta_1^{exp} c_{oo}^{(1)} D_1 = - \beta_2^{exp} c_{oo}^{(2)} D_2 \qquad (97)$$

is obeyed in case c as well as in case a and b. The upper critical wavenumber is seen to be quite unaffected by these changes, but the lower critical wavenumber is displaced towards quite low values in the cases a and b. A careful analysis shows

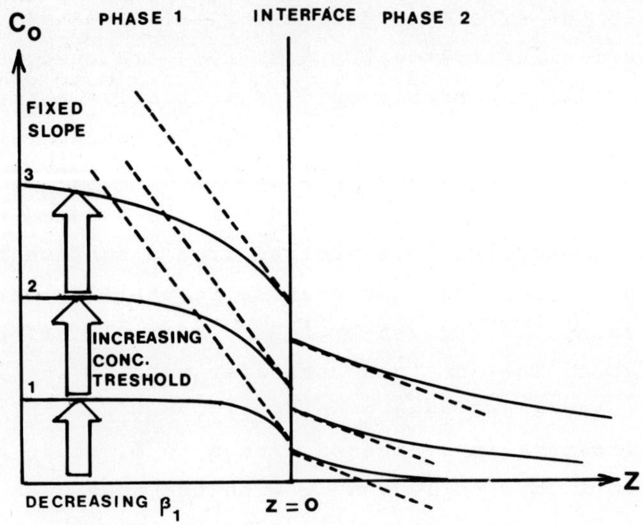

Fig.11 When the thickness of the diffusion zone decreases, concentration threshold also decreases with fixed mass transfer rate.

that the dispersion curve (b) actually crosses the k-axis around 0.045 cm^{-1}. All dispersion curves are *identical* when calculated by means of the complete determinantal equation or the modified S. & S. theory described in §7.

In §6 we discussed the possibility according to the S. & S. theory of having *oscillating* instabilities with the reverse direction of diffusion, *i.e. from* the phase with the *higher* diffusion coefficient *to* the phase with the *lower* diffusion coefficient (case III and IV). The condition is only that the diffusion is *also* in the direction *from high to low* kinematical viscosity.

On fig.12 results are given from an analysis of a situation corresponding to case IV on fig.5. The region of the double, stationary branch was not investigated, since it did not appear to have any great importance. Dispersion curve (a) is identical to the dispersion curve given in the S. & S. paper fig.7.

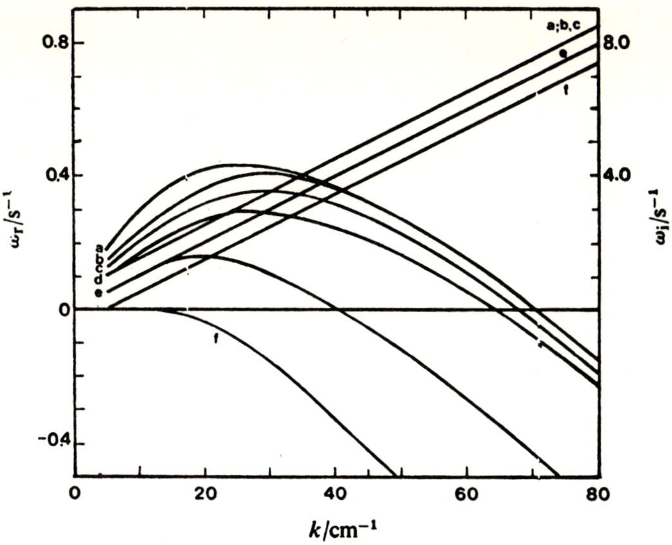

Fig.12 Influence of thickness of diffusion zone on Oscillatory instabilities.
Data: $D_1/D_2/D_s = 1.333/2/1.8\cdot10^{-5}\ cm^2s^{-1}$.
$\nu_1/\nu_2 = 0.5/1\cdot10^{-2}\ cm^2/s$. $K_1 = K_2 = 10^{-5}$ cm.
$\Gamma_o = 2.67\cdot10^{-9}\ g/cm^2$. $\rho_1 = \rho_2 = 1 g/cm^3$. $\alpha = 10^7$ erg/g.
$\eta_s = 0$ sP. $g = 981\ cm/s^2$.

(a) S. & S. model ($\sigma_o=\infty$) and complete model with linear concentration profiles and σ_o=1.44 and 10^{-6} dyn/cm. $\beta_1^{lin} = -1.5\ g\ cm^{-4}$.
(b) Complete model and modified S. & S. model with exponential profiles. $\beta_1^{exp} = 1\ cm^{-1}$; $c_{oo}^{(1)} = 1.5\ g\ cm^{-3}$; $\sigma_o = \infty$, 1.44 and 10^{-6} dyn/cm.
(c) $\beta_1^{exp} = 5\ cm^{-1}$; $c_{oo}^{(1)} = 0.3\ gcm^{-3}$; $\sigma_o = \infty$, 1.44 dyn/cm.
(d) $\beta_1^{exp} = 10\ cm^{-1}$; $c_{oo}^{(1)} = 0.15\ gcm^{-3}$; $\sigma_o = \infty$, 1.44 dyn/cm.
(e) $\beta_1^{exp} = 50\ cm^{-1}$; $c_{oo}^{(1)} = 0.03\ gcm^{-3}$; $\sigma_o = \infty$, 1.44 dyn/cm.
(f) $\beta_1^{exp} = 100\ cm^{-1}$; $c_{oo}^{(1)} = 0.015\ gcm^{-3}$; $\sigma_o = \infty$, 1.44 dyn/cm.

No difference is found between the S. & S. model with a flat interface ($\sigma_o = \infty$) and the complete model with linear profiles and $\sigma_o = 1.44$ dyn/cm or 10^{-6} dyn/cm. The family of curves (b) through (f) arise by using exponential profiles and decreasing the thickness of the concentration zone while maintaining the interfacial mass transfer flux. A stabilising effect similar to the one observed for the case I instability is seen on fig.12. The explanation is again the diminishing concentration threshold (fig.11). In the case of the dispersion curve (f) the extension of the diffusion zone is around 0.1 mm and the system is completely stable. However, we have to do with a "masked instability" which will reveal itself again at low interfacial tension, *vide* fig.13. For all the cases (b) through (f) we found found no difference between the results when using the complete model with exponential profiles or the modified S. & S. model ($\sigma_o = \infty$).

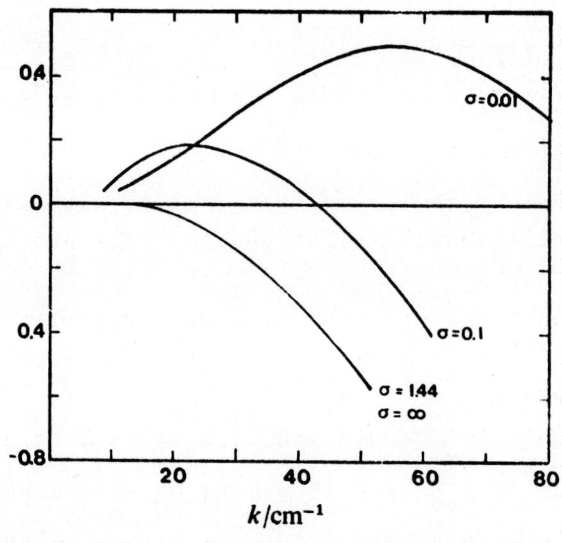

Fig.13 Destabilisation of masked oscillatory instability (curve (f) in fig.12) by low surface tensions using the complete model with deformable interface. Values of σ_o in dyn/cm. The frequency ω_i stays inside the "linear belt" (straight lines a through f) shown on fig.12.

On fig.13 the destabilisation of a masked oscillatory instability by low surface tensions is shown. This is an effect which cannot be explained by the modified S. & S. theory. Neither 0.1 nor 0.01 dyn/cm are unrealistically low interfacial tensions and it is seen that a marked instability is produced already at 0.1 dyn/cm. The wavenumber of maximum instability is seen to be shifted towards higher values as σ_o goes down. The wavelength of the perturbations are accordingly shifted towards smaller values. Typical wavelengths on fig.12 are 0.2 - 0.3 cm whereas the wavelength has decreased to 0.1 cm for σ_o = 0.01 dyn/cm on fig.13. For extremely small surface tensions it seems not impossible that the dominating wavelength could decrease even further. Instabilities of that kind might be the reason for certain kinds of *spontaneous emulsification*. (For a description of various mechanisms of spontaneous emulsification, see Davies and Rideal[73]).

It is generally known that low - or even negative - interfacial tensions promote spontaneous emulsification. As a classic example, Ilkovic[74] showed that the interfacial tension between mercury and an aqueous solution of a quaternary ammonium salt could be greatly reduced by applying a negative potential to the mercury drop. At around - 2 V the electrocapillary curve reaches zero surface tension and at more negative potentials a brown cloud of colloidal mercury in water is formed. It is also known that the most intense interfacial agitation in connection with exothermic reactions between two solutes at the interface is found in systems with low surface tensions[75]. Local surface tension lowerings were suggested as an explanation for spontaneous emulsification during soap formation (lauric acid in oil + NaOH in water) as early as 1888 by Quincke[76].

It should be remarked, however, that the masked case IV instability discussed here is a quite special case, and it is surprising that there is not a similar effect on the non-oscillatory instability or on the oscillatory instability with linear or almost linear profiles (case a and b on fig.12). The solution space of the full determinantal equation should be somewhat more carefully investigated in the future for possible solutions completely different from the S. & S. cases.

9. Dimensionless graphs for determination of regions of stationary Marangoni-instabiliteis with exponential concentration profiles.

In §5 the minimum number of "natural" dimensionless numbers was introduced in order to find the region of stationary Marangoni-instability with linear concentration profiles. For normal surfactants with diffusion *from* the phase with the lower *to* the phase with the higher diffusion coefficient we find always an instability region between $k = 0$ and $k = k_{cr}$ with k_{cr} determined by equation (46). For exponential profiles the situation is somewhat more complicated. The equation for stationary, marginal stability was written down in §4 equation (43). Neglecting the weight of the interfacial mass perturbation ($\delta_w = 0$) this equation may be rewritten in natural dimensionless numbers:

$$g(\hat{k}) = f(\hat{k}) \tag{98a}$$

$$g(\hat{k}) = \hat{\Gamma} + [2\hat{\eta} + \hat{\eta}_s \hat{k}][\hat{D} + \hat{D}_s \hat{k}] \tag{98b}$$

$$f(\hat{k}) = \frac{1}{[2\hat{k} + 1]^2} - \frac{D_1/D_2}{[2\hat{k} + (\beta_2/\beta_1)]^2} \tag{98c}$$

with

$$\hat{\Gamma} = \frac{\Gamma_o \ \beta_1}{(-\ c_{oo}^{(1)})} \tag{99a}$$

$$\hat{\eta} = \frac{(\Sigma\eta_i)\ \beta_1}{(-\ c_{oo}^{(1)})\sqrt{\alpha}} \tag{99b}$$

$$\hat{\eta}_s = \frac{\eta_s \ \beta_1^2}{(-\ c_{oo}^{(1)})\sqrt{\alpha}} \tag{99c}$$

$$\hat{k} = k/\beta_1 \tag{99d}$$

$$\hat{D} = \frac{1}{\sqrt{\alpha}} \ \Sigma \ \frac{D_i}{K_i} \tag{99e}$$

$$\hat{D}_s = \frac{D_s \ \beta_1}{\sqrt{\alpha}} \tag{99f}$$

For the data given in (95) we see that $\hat{\Gamma}/(2\hat{\eta}\hat{D}) \sim 10^{-2}$ and $\hat{D}_s/\hat{D} \sim 10^{-5}$. At least it seems to be a good approximation to neglect the influence of the surface diffusion, but also the influence of the surface mass density may probably often be neglected. However, $\hat{\Gamma}$ might play a role for *monolayers of macromolecules*. Neglecting $\hat{D}_s$ and introducing the relation (33d) between β_1 and β_2 we obtain

$$g(\hat{k}) = \hat{A} + \hat{B} \cdot \hat{k} \tag{100a}$$

$$f(\hat{k}, \frac{D_1}{D_2}) = \frac{1}{[2\hat{k} + 1]^2} - \frac{D_1/D_2}{[2\hat{k} + \sqrt{D_1/D_2}]^2} \tag{100b}$$

with

$$\hat{A} = \hat{\Gamma} + 2\hat{\eta}\,\hat{D} \cong 2\hat{\eta}\,\hat{D} \tag{101a}$$

$$\hat{B} = \hat{D}\,\hat{\eta}_s \tag{101b}$$

A general analysis is now feasible, since there are only three parameters left: D_1/D_2, $\hat{A}$ and $\hat{B}$. For normal surfactants and for diffusion from phase 1 to phase 2 ($c_{oo}^{(1)} < 0$), $g(\hat{k})$ will always be a positive, monotonously increasing function of the positive dimensionless wavenumber $\hat{k}$. It is seen that for $D_1/D_2 < 1$, $f(\hat{k})$ will be positive whereas for $D_1/D_2 > 1$, $f(\hat{k})$ will be negative. To have any roots we therefore need to have $D_1/D_2 < 1$. Thus, we find qualitatively the same result as in the simple S. & S. theory: *A necessary condition for having states of marginal, stationary Marangoni-instability is that diffusion is in direction from the phase with the lower to the phase with the higher diffusion coefficient.*

It is seen that f tends towards zero for $\hat{k} \to 0$ and $\hat{k} \to \infty$. Furthermore for $D_1/D_2 < 1$, f has a maximum for

$$\hat{k}_o = \frac{\sqrt[3]{D_1/D_2} - \sqrt[2]{D_1/D_2}}{2[1 - \sqrt[3]{D_1/D_2}]} \quad (> 0) \tag{102}$$

This maximum is peculiar for the case of exponential profiles compared to the linear case. The *necessary* condition for stationary instability given above is thus *not sufficient,* since

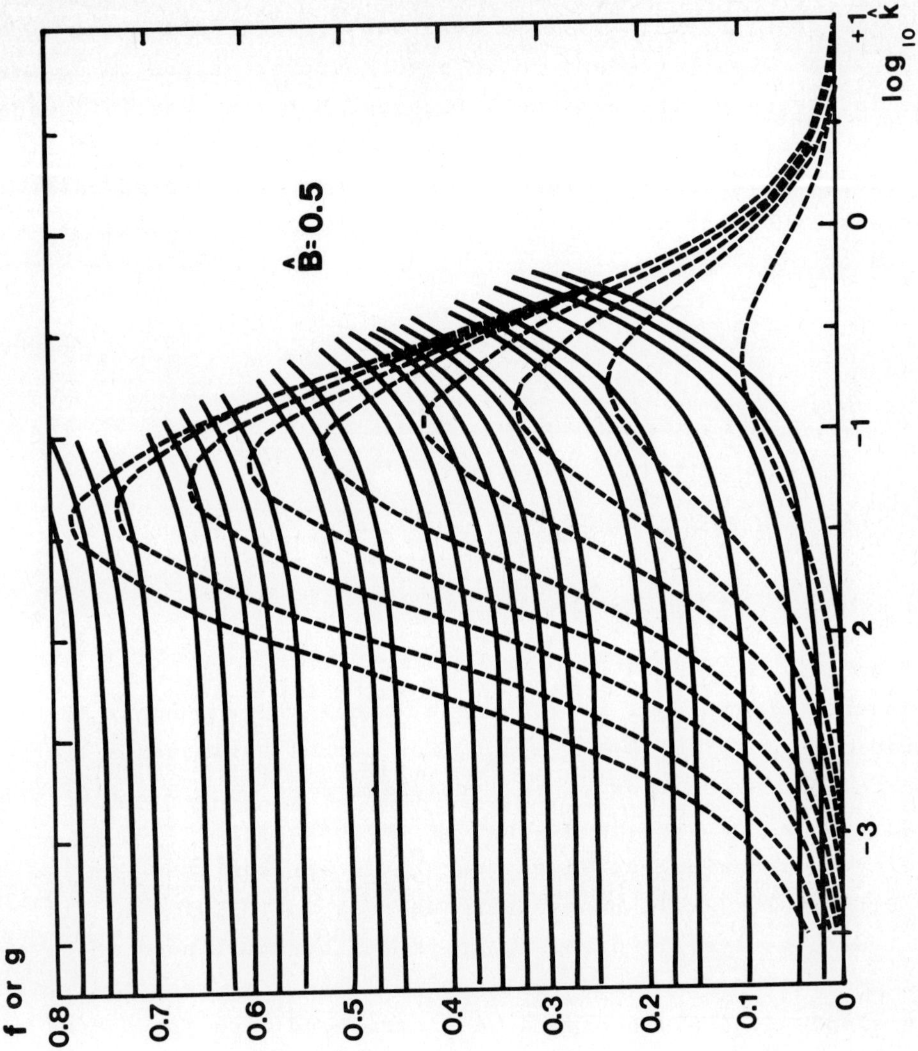

Fig.14 Dimensionless graph for determination of stationary Marangoni-instability regions for diffusion zones of finite thickness. Dashed curves correspond to $f(\hat{k})$ for (from bottom) D_1/D_2 = 0.5; 0.2; 0.1; 0.05; 0.02; 0.01; 0.005; 0.002; 0.001. Diffusion from phase 1 to phase 2. Solid curves correspond to $g(\hat{k})$. Value of $\hat{A}$ can be read on the ordinate axis following the curves to the left. $\hat{B}$ = 0.5 for all curves.

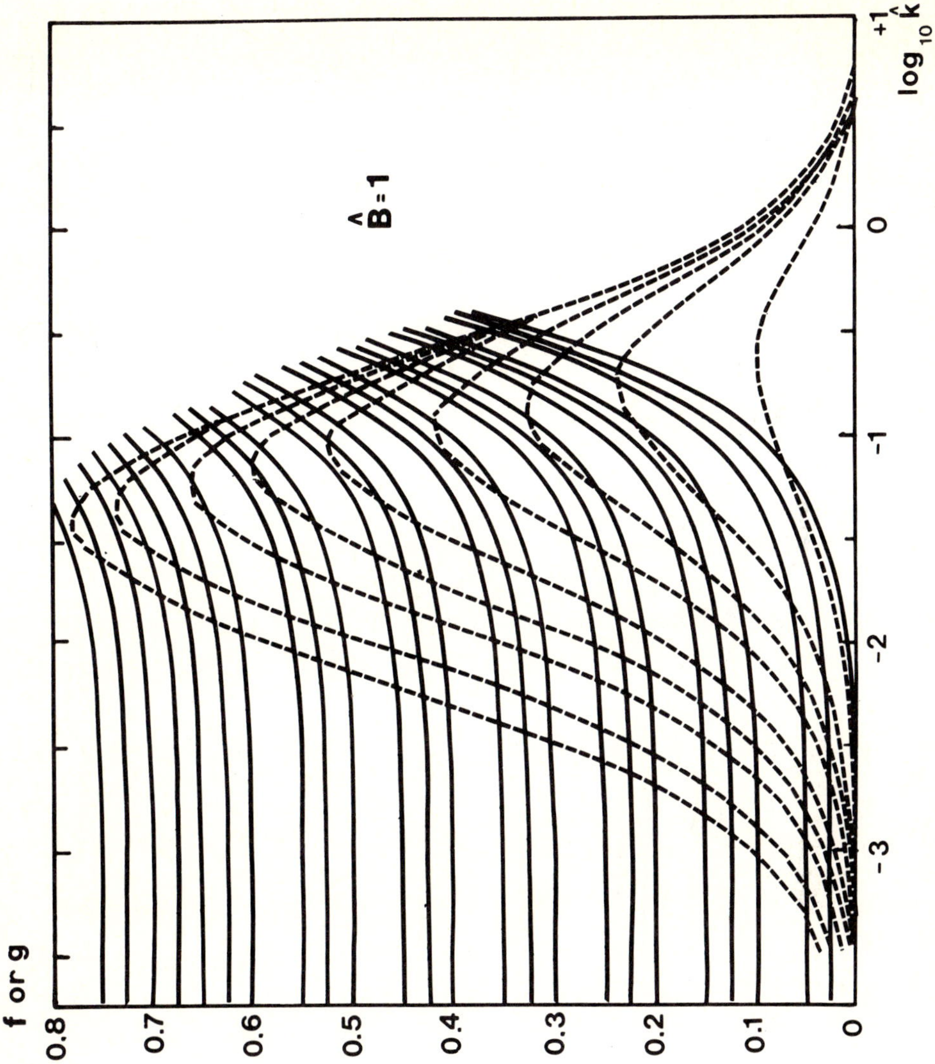

Fig.15 Dimensionless graph for determination of stationary Marangoni-instability regions. $\hat{B}$ = 1. See text below figure 14.

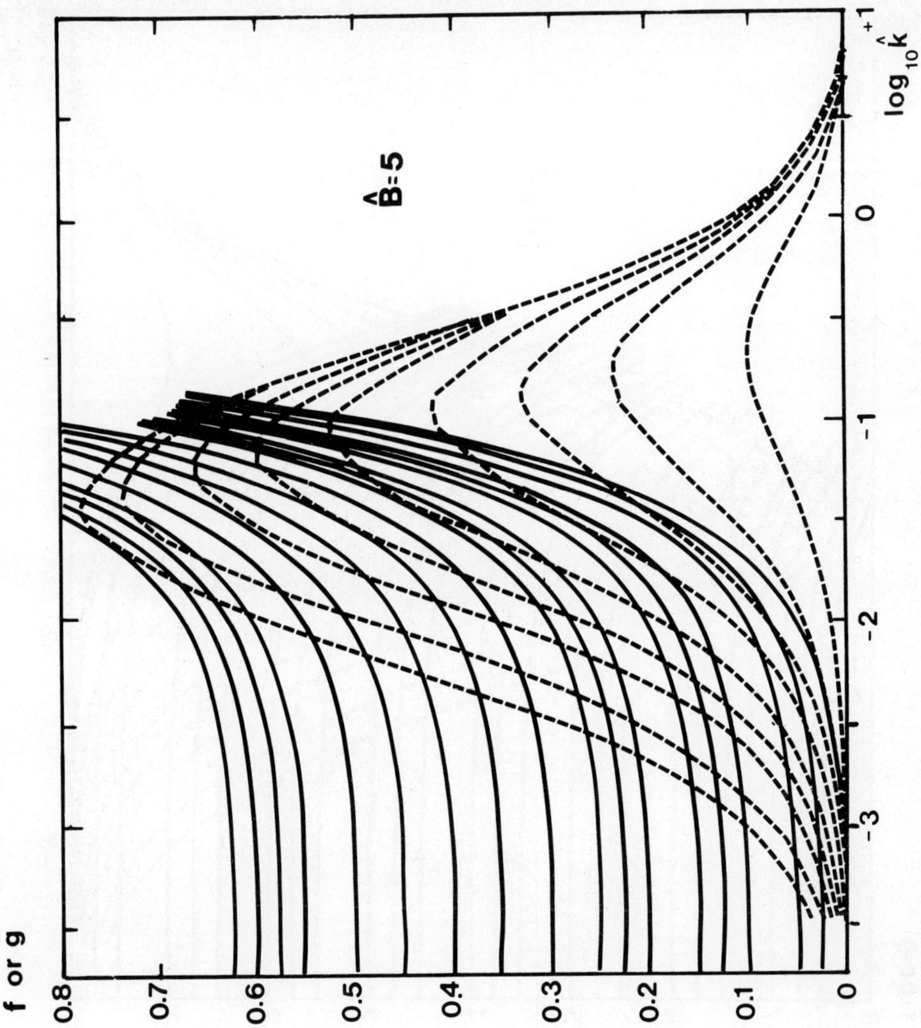

Fig.16 Dimensionless graph for determination of stationary Marangoni-instability regions. $\hat{B}$ = 5. See text below figure 14.

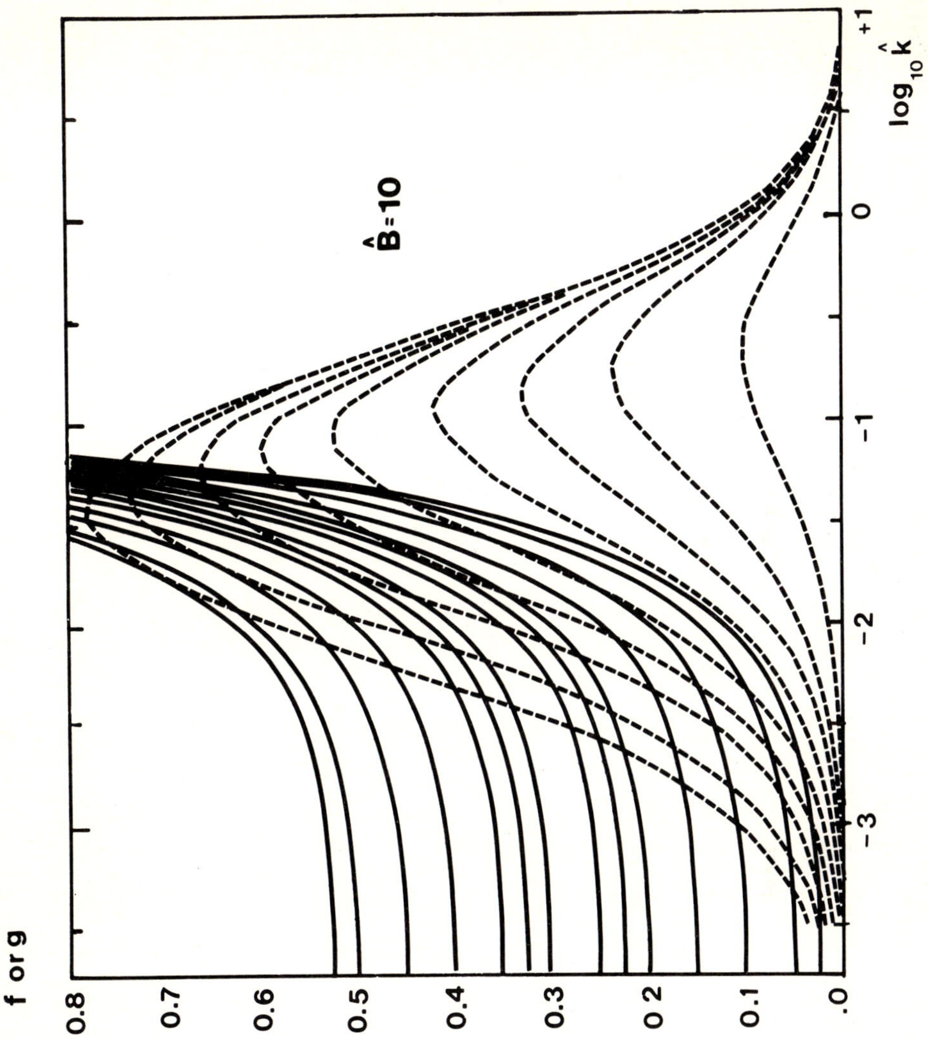

Fig.17 Dimensionless graph for determination of stationary Marangoni-instability regions. $\hat{B}$ = 10. See text below figure 14.

$g(\hat{k})$ may lie completely above the $f(\hat{k})$ curve. Thus, there may be no intersection, touching or two intersections between f and g dependent on the parameters A and B. Consulting equation (33a) we see that $|K_d c_\infty^{(2)} - c_\infty^{(1)}|$ measures the "distance from equilibrium" for fixed β_1 and t and the greater this distance is made the more negative $c_{\infty}^{(1)}$ will be. This means that $\hat{\eta}$ and $\hat{\eta}_s$ and hereby $\hat{A}$ and $\hat{B}$ will decrease, and if the distance from equilibrium is great enough, the $g(\hat{k})$ curve will move into the region of the $f(\hat{k})$ curve. Therefore, we will have instability between the two intersection points. Hereby, we have cast the phenomenon of Marangoni-instability into the usual framework discussed for other types of "dissipative structures" in the monography of Glansdorff and Prigogine[10].

Dimensionless graphs for the determination of instability regions for $\hat{B}$ = 0.5, 1,5 and 10 are given as figs. 14, 15, 16, and 17. The dashed curves correspond to $f(\hat{k}, D_1/D_2)$ for different values (less than unity) of the parameter D_1/D_2. The full curves correspond to $g(\hat{k})$ for different values of $\hat{A}$.

Of especial importance are the points where the g-curves are touching the f-curves. Imagine a given system with a given $D_1/D_2 < 1$ and diffusion from phase 1 to phase 2 with a given "distance from equilibrium" $|K_d c_\infty^{(2)} - c_\infty^{(1)}|$. For small values of the time of contact t, the profiles are close to the step-function and β_1 is large, *vide* equation (33c). From equations (99b-c) one observes that $\hat{\eta}$ and $\hat{\eta}_s$ will then also be large and hence $\hat{A}$ and $\hat{B}$ from (101a-b). The g-curves will therefore initially be situated above the given f-curve. As time passes the g-curve will move downwards, and the first instability generated will have a wavelength corresponding to $2\pi/k_{cr}$ where k_{cr} is calculated from the value of $\hat{k}$ in the touching point. Soon afger, convection has smeared out the concentration profile even further, and the β's drop considerably, so we move into the realm with two critical wavenumbers and an instability region. The result is a more or less Gaussian distribution of convection cell dimensions around the wavelength of maximum instability such as actually observed by Orell and Westwater (ref.26, fig.6). The distribution will be broader when time passes.

Strictly speaking, we have left the domain of application of our model, however, as soon as the first convection has set in. The experimental findings of Linde[77] of the splitting into *two* concentration profiles of the erf-type of the original concentration profile when the convection sets in and the *relaxation-oscillations* of the convection pattern linked to this fact remain challenging problems for future theoreticians.

In figs.18 and 19 the critical values of $\hat{k}$ and $\hat{A}$ for the touching points are given for various values of $\hat{B}$ as functions of the ratio D_1/D_2. The points of the curves have been found directly from the graphs in figs.14 - 17 (and a similar one for $\hat{B} = 7$ not shown here). It is observed from fig.18 that $\hat{k}_{cr}$ tends to increase with increasing ratio D_1/D_2, *i.e.* the relative size of the wavelength to the extension of the diffusion zone *decreases* when D_1/D_2 increases towards 1 from below. When surface viscosity plays an increasing role ($\hat{B}$ increases), the relative wavelength will only decrease with increasing D_1/D_2 up to a certain point, then it will *increase* sharply with increasing D_1/D_2. The values of $\hat{A}_{cr}$ decrease with increasing D_1/D_2, however, for all values of $\hat{B}$. It decreases also with increasing values of $\hat{B}$, so that surface viscosity and a D_1/D_2 closer to unity naturally enough stabilise the system, since the "distance from equilibrium" or the time of contact has to be greater for the instability to appear.

10. Concentration dependence of stationary Marangoni-instability.

The dimensionless quantities $\hat{A}$ and $\hat{B}$ in the preceeding paragraph will be dependent on the concentration of surfactant adjacent to the interface, since the parameters $c_{oo}^{(1)}$, α, K_i, Γ, and η_s will be functions of surfactant concentration.

In some cases η_i, D_i and ρ_i will also be quite strong functions of concentration. This will be the case in the system of Orell and Westwater at the higher concentrations of acetic acid, since there is partial miscibility between the ethylene glycol phase and the ethyl acetate phase. When the concentration of acetic acid rises, the composition of the two phases merge to-

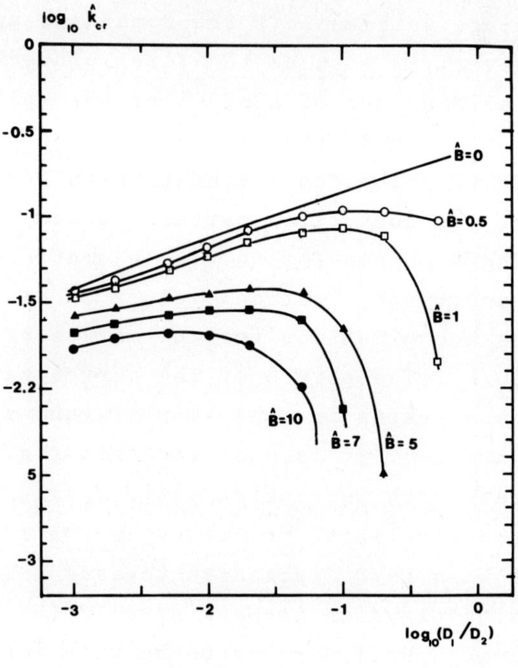

Fig. 18 Wavenumber of the touching points as a function of D_1/D_2 for various values of $\widehat{B}$.

gether and at the point where $\sigma = 0$, the two phases are identical. The viscosities, diffusion coefficients and densities will therefore also merge together. To be precise, the Orell and Westwater system cannot be described simply as two phases with interdiffusion of acetic acid, since ethyl acetate and ethylene glycol also have to diffuse. The Orell and Westwater system is therefore not the best one suited for detailed comparisons between theory and experiments. Unfortunately, this system is probably the one studied in most quantitative detail in literature. One can hope, however, that at least the behaviour of that system at the lower concentrations of acetic acid may be related to simple formulae proposed in the preceding section.

In order to predict the variation of α and K_i with the concentration one has to consider a specific model of adsorption to the interface. I shall consider here only the simplest model

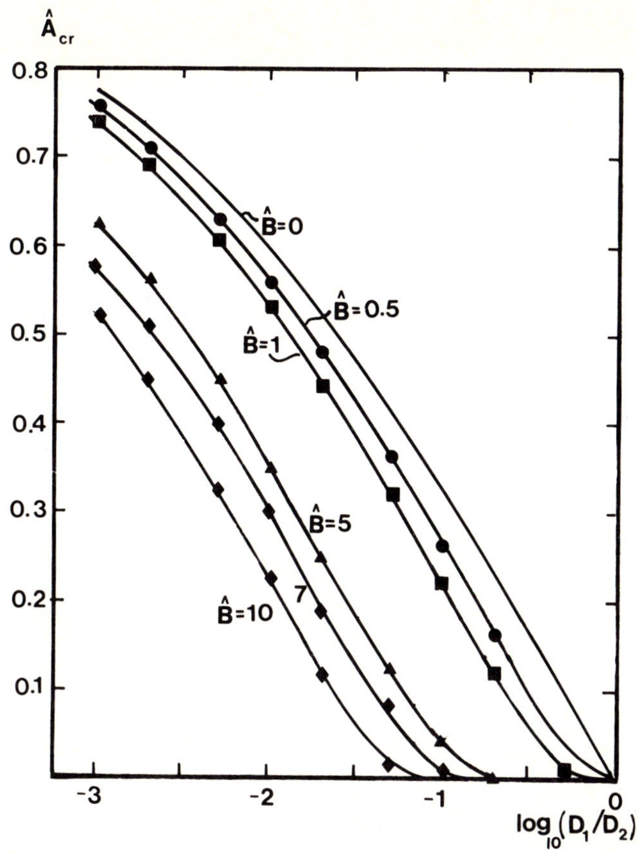

Fig. 19 The "inverse distance from equilibrium number" $\hat{A}_{cr}$ of the touching points as a function of D_1/D_2 for various values of $\hat{B}$.

possible: the Langmuir adsorption isotherm. In that case the matter adsorbed to the interface is given by:

$$\Gamma = \Gamma_\infty \,(1 + c^*) \;; \quad c^* \equiv c^{(1)}_{adj}/a_1 \tag{103}$$

with $c^{(1)}_{adj}$ = adjacent concentration in phase 1 and a_1 = concentration of half saturation = inverse adsorption coefficient. Using (103) together with the Gibbs' adsorption isotherm one derives easily the Szyszkówski equation for the surface pressure (see ref.78 pp.114-117):

$$\Pi = \frac{RT\Gamma_\infty}{M} \ln [1 + c^*] = \sigma_o - \sigma \qquad (104)$$

Γ_∞ in g/cm^2, R in erg/mol · K)

This equation was used in 1908 by Szyszkowski[79] to fit Π(c) curves for fatty acids of not too big chain lengths at air water interfaces. Using (49b-c) together with (104) one gets immediately:

$$K_1 = \frac{\partial\Gamma}{\partial c^{(1)}_{adj}} = \frac{\Gamma_\infty}{a_1} \cdot \frac{1}{(1+c^*)^2} \qquad (105)$$

$$\alpha = \frac{RT}{M} (1+c^*) \qquad (106)$$

If a Langmuir isotherm can similarly be applied to the adsorption from phase 2, we have:

$$K_2 = \frac{\Gamma_\infty}{a_2} \cdot \frac{1}{\left(1+ \frac{c^{(2)}_{adj}}{a_2}\right)^2} = \frac{\Gamma_\infty}{a_2} \cdot \frac{1}{\left(1+ \frac{a_1}{a_2 K_d} c^*\right)^2} \qquad (107)$$

where K_d is the ratio between adjacent concentration in phase 1 divided by adjacent concentration in phase 2 according to (29). By the Szyszkowski equation we have:

$$\Pi = \frac{RT}{M} \Gamma_\infty \ln[1 + c^*] = \frac{RT}{M} \Gamma_\infty \ln \left[1 + \frac{a_1}{a_2 K_d} c^*\right]$$

from which equation we have automatically the requirement:

$$K_d = \frac{a_1}{a_2} = \frac{K_2}{K_1} \qquad (108)$$

Thus the Nernst distribution law is satisfied, but K_1 and K_2 vary with concentration.

For simplicity, I consider here only the case with negligible surface viscosity, i.e. $\hat{B} = 0$. (The dimensionless number $\hat{\Gamma}$ is also neglected in comparison to $2\hat{\eta}\hat{D}$ in A, see eqn.(101a). In that case the critical value of $\hat{A}$ is just given by the maximum value of the f-function, when $D_1/D_2 < 1$ and we have transfer from phase 1 to phase 2:

$$\hat{A}_{cr} = f_{max} \tag{109}$$

where f_{max} is obtained by insertion of (102) into (100b), i.e.

$$f_{max} = \frac{\{1 - \sqrt[3]{D_1/D_2}\}^3}{\{1 - \sqrt[2]{D_1/D_2}\}^2} \tag{110}$$

f_{max} decreases with increasing D_1/D_2 as shown on the curve $\hat{B}=0$ on fig.19. For $\hat{A}_{cr}$ we have:

$$\hat{A}_{cr} \cong 2 \frac{(\Sigma\eta)\ \beta_1^{cr}\ \Sigma \frac{D_i}{K_i}}{\alpha\ (-\ c_{oo}^{(1)})} \tag{111}$$

Assuming the bulk concentration in phase 2 to be zero we have from (33a):

$$c_{oo}^{(1)} = -\frac{c_\infty^{(1)}}{1 + K_d \sqrt{D_1/D_2}} \tag{112}$$

The adjacent concentration $c_{adj}^{(1)} = c_\infty + c_{oo}^{(1)}$, so we have for the dimensionless adjacent concentration:

$$c^* = \{\frac{K_d \sqrt{D_1/D_2}}{1 + K_d \sqrt{D_1/D_2}}\}(\frac{c_\infty^{(1)}}{a_1}) \tag{113}$$

Inserting (105), (106), (108), (112), and (113) in (111) we obtain:

$$\hat{A}_{cr} = \frac{2(\Sigma\eta)\ \beta_1^{cr}\ D_1 \cdot K_d \sqrt{\frac{D_1}{D_2}}\ (1 + \frac{D_2}{D_1})}{(RT\Gamma_\infty/M)} (\frac{1 + c^*}{c^*}) \tag{114}$$

Assuming that neither η_i nor D_i depend on c^* we obtain the following concentration dependence for β_1^{cr}:

$$\beta_1^*(cr) = \beta_1^*(cr,max) \cdot \left(\frac{c^*}{1 + c^*}\right) \qquad (115)$$

The dimensionless reciprocal penetration depth β_1^* is defined by:

$$\beta_1^* \equiv \frac{(\Sigma\eta)\ D_1}{(RT\Gamma_\infty/M)} \cdot \beta_1 \qquad (116)$$

and the maximum value of the critical, reciprocal penetration depth in (115) is given by:

$$\beta_1^*(cr,max) = \frac{1}{2K_d} \cdot \frac{\sqrt[2]{D_2/D_1}}{\left(1 + \frac{D_2}{D_1}\right)} \cdot \frac{\{1 - \sqrt[3]{D_1/D_2}\}^3}{\{1 - \sqrt[2]{D_1/D_2}\}^2} \qquad (117)$$

With increasing $c_\infty^{(1)}$, c^* increases proportionally and β_1^{cr} increases hyperbolically towards the maximum value. Since the critical, dimensionless wavenumber $\hat{k}_o = k^{cr}/\beta_1^{cr}$ is fixed by the ratio D_1/D_2 (vide eqn.(102)), the critical wavenumber will also increase towards a maximum value k_{max}^{cr} corresponding to a minimum wavelength $\lambda_{max}^{cr} = 2\pi/k_{max}^{cr}$.

According to fig.8 in the paper of Orell and Westwater there is indeed a sharp initial decrease in the average center-to-center distance of propagating cells as the acetic acid concentration in the ethylene glycol phase is increased. According to (115) β_1^{cr} should go to zero when the concentration tends towards zero, and the same should be the case with k^{cr}. The cell dimensions should therefore tend towards infinity for low concentrations. Fig.8 in Orell and Westwater's paper seems indeed to support that conclusion. It should be stressed once more that I consider the "propagating cells" in Orell and Westwater's paper to be in reality stationary, primary cells moved by secondary flow, induced by interaction between primary convection cells.

On fig.20 the hyperbolic dependence of k^{cr} on $c_\infty^{(1)}$ for the Orell and Westwater system is clearly brought out as a linear dependence of average wavelength measured for the convection cells on $1/c_\infty^{(1)}$. The regression line is given by $\lambda_{cr} = \lambda_{cr}(\min) + \lambda_{cr}(\min) \frac{a_1}{c_\infty^{(1)}} = 0.0118 \left(\frac{1}{c_\infty^{(1)}}\right) + 0.0164$ from which one obtains $a_1 = 0.0072$ g/cm^3 and $\lambda_{min}^{cr} = 0.0164$ cm.

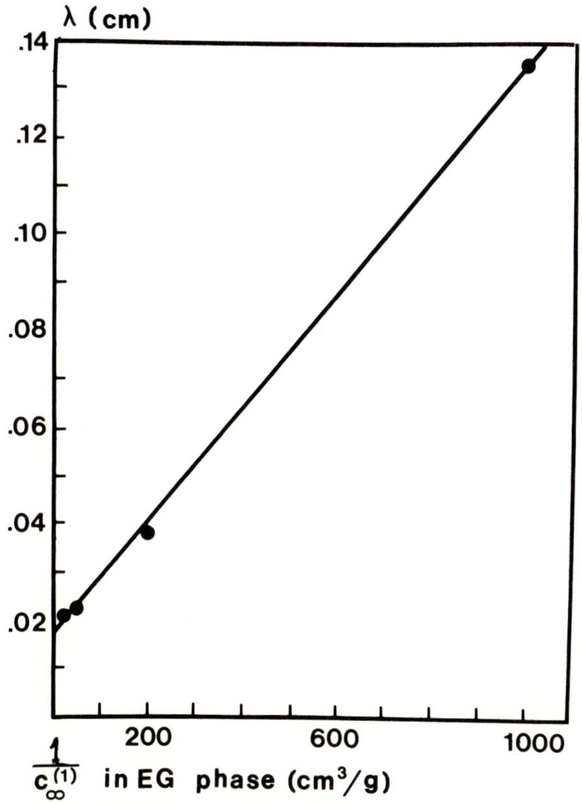

Fig.20. Plot of the average center-to-center distance of convection cells after 1 hr of contact vs. inverse concentration of acetic acid in the ethylene glycol phase for the Orell and Westwater system.

In fig.21 it is shown that the $\sigma(c)$ data for the ethylene glycol phase are well fitted by the Szyszkowski isotherm, except for the concentrations near to the closing point of the solubility gap. We find $a_1 = 0.012$ g/cm^3 which value is at least of the same order of magnitude as the value (0.0072) found independently above. We find also $(RT\Gamma_\infty/M) = 0.74$ dyn/cm. From the estimated data given in (95) we calculate from (117)

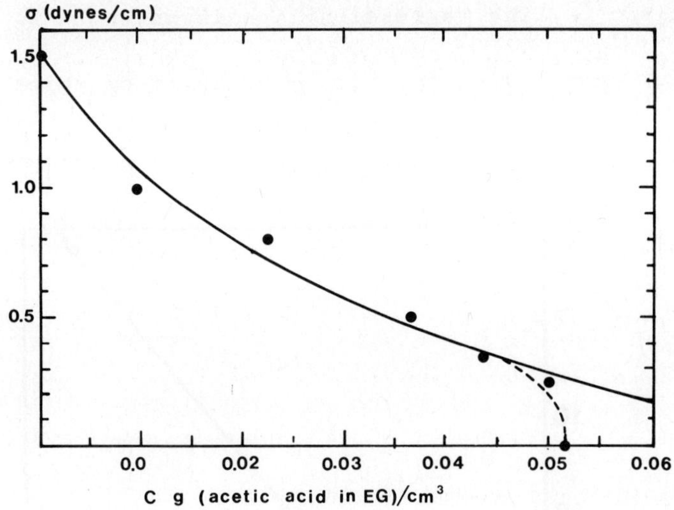

Fig.21. Interfacial tension in Orell and Westwater's system. Values taken from ref.26 fig.12. Full line:

Szyszkowski isotherm $1.5 - \sigma = 0.74 \ln \{1 + \frac{c^{(1)}_{adj}}{0.012}\}$

$\beta_1^*(cr,max) = 0.0302$ and from (116) $\beta_1(cr,max) = 3.23 \cdot 10^6 \cdot 0.0302 = 9.75 \cdot 10^4\ cm^{-1}$. From (102) we obtain $\hat{k}_o = 0.121$ and therefore $k_{cr}^{max} = 0.121 \cdot 9.75 \cdot 10^4\ cm^{-1} = 1.19 \cdot 10^4\ cm^{-1}$. The corresponding wavelength is $\lambda_{cr}^{min} = 2\pi/1.19 \cdot 10^4 = 0.00053$ cm. This value is very much below the value 0.0164 cm obtained by the plot given on fig.20. We have seen in the preceding paragraphs, however, that surface viscosity has a pronounced influence on Marangoni-instabilities. The assumption that $\hat{B} = 0$ may therefore be wrong in this case.

It is possible to utilise the graphs 18 - 19 in connection with the Szyszkowski isotherm found in fig.21 and the formulae for α and K_i to fit the experimentally found $\lambda(c)$ curve. Hereby the surface viscosity η_s is used as the parameter of adjustment and is thus found indirectly. This analysis will not be performed here, however, for the following reasons: a) The Orell and Westwater system will exhibit varying D's and varying η's with increasing concentration of acetic acid, because of closure of the miscibility gap. b) The average cell dimensions measured by Orell and Westwater do not really correspond to the critical wavelength, since the measurements were performed 1 hr after the formation of phase contact whereas the first roll cells appear already a few seconds after the formation of the contact. (Nevertheless I think that the values are not too far from critical values, since the secondary roll cells will perform a continuous surface renewal, so that the instabilities start afresh at any time). c) The estimation of the parameters (95) for the Orell and Westwater system is quite uncertain, <u>vide</u> ref.22. This is especially the case for the values of the diffusion coefficients.

It is to be hoped that careful measurements will be made in the future of the cell-dimensions of the first cells appearing after contact and of the dimensions of the concentration profile at the outbreak of instability. The systems used should be any with very little partial miscibility and careful measurements of surface tension as function of concentration and of diffusion - and distribution coefficients should be made together with measurements of the viscosities as function of concentration in the two phases. Then there should be a hope of calculating correct values of the surface viscosity at different concentrations. Here I shall delimit myself to a determination of the order of magnitude of the surface viscosity at the lowest acetic acid concentration for the Orell and Westwater system just to show how the graphs 18 - 19 may be employed.

For the concentration 0.001 g acetic acid/cm^3 EG-phase we have from the Szyszkowski isotherm $\alpha = 4.46 \cdot 10^8\ cm^2/s^2$ and $K_1 = 1.27 \cdot 10^{-7}$ cm. The K_1-value is identical to the value estimated in (95) whereas α is slightly different. We shall therefore use the data (95), but with the new value for α. Using the definitions (101a) and (101b) for $\hat{A}$ and $\hat{B}$ we have:

$$\beta_1^{cr} = 7.68 \cdot 10^3\, \hat{A}_{cr} \tag{118}$$

$$\eta_s = 1.68 \cdot 10^3\, \hat{B}/(\beta_1^{cr})^2 \tag{119}$$

Using the graphs on figs.18 and 19 one obtains:

$\hat{B}$	$\hat{A}_{cr}$	$\beta_1^{cr}(cm^{-1})$	$\hat{k}_{cr}$	$k_{cr}(cm^{-1})$	$\lambda_{cr}(cm)$
0	0.37	$2.84 \cdot 10^3$	0.126	358	0.0176
1	0.27	$2.07 \cdot 10^3$	0.0851	176	0.0357
5	0.07	$0.54 \cdot 10^3$	0.0316	17.1	0.367
7	0.03	$0.23 \cdot 10^3$	0.0200	4.6	1.37

The value of λ measured by Orell and Westwater for freshly generated roll cells after 1 hr of contact is 0.135 cm, so $\hat{B}$ should be between 1 and 5. According to (119) η_s will therefore be between $3.9 \cdot 10^{-4}$ and $2.9 \cdot 10^{-2}$ surface Poise. The range of surface viscosities (shear) reported in many oil-water systems with surfactants in the paper of Wasan and Mohan[43] is also 10^{-4} - 10^{-1} surface Poise, so the above figures do not appear unrealistic.

11. Discussion.

From the preceding considerations the following conclusions can be drawn:

a) The effect of gravity and Laplace-overpressure has not a great effect on the Marangoni-instability problem for an initially plane, but deformable interface.

b) The simple model of Sternling and Scriven works quite well, especially when extended to diffusion zones of finite thickness. A difference has only been found in case of oscillatory instabilities with low interfacial tensions, where a destabilisation has been observed in a special case.

c) For zones of finite thickness, the stationary instabilities appear in the following way: When the concentration profile is very steep the situation is stable, but when the concentration gradient reaches a critical value, instability sets in with a definite wavelength λ_{cr}. Later, we have a distribution of wavelengths around this value and the "stationary cells" are moved by secondary flow. In the linear theory we obtain now not only a lower λ_{cr} as in the Sternling and Scriven theory, but also an upper λ_{cr}. In between we have a wavelength λ_{max} with maximum instability.

d) The wavelength λ_{cr} tends to infinity for the surfactant concentration tending towards zero. (When λ_{cr} exceeds the dimensions of the vessel no convection cells will be seen). For higher concentration λ_{cr} <u>decreases</u> with <u>increasing</u> concentration towards a minimum value, if the viscosities and the diffusion coefficients are not changed by addition of surfactant.

e) Surface viscosity has a pronounced effect on Marangoni-instabilities whereas the gravity-driven Rayleigh-Taylor instability seems unaffected by surface viscosity. Although there is in principle a coupling between Rayleigh-Taylor effects and Marangoni effects, this interference seems to be negligible except when the wavelengths of the two instabilities are very closely situated.

f) Surface diffusion seems not to be important for Marangoni-effects of the discussed type. (For film thinning and rupture this parameter is fundamental[58]).

In spite of a certain progress, there remain many unclear points before the gap between theory and experiments can be said to be properly closed. Sawistowski[80] has given an excellent account of the state of the art from the point of view of liquid-liquid extraction which can be very much enhanced by Marangoni-effects. Many of the statements and conclusions in Sawistowski's paper deserve further elucidation, however. For example, Sawistowski puts forward some observations of Jebsen-Marwedel[81] and of Jettmar and Roesler[82]. Jebsen-Marwedel found that movements produced at the surface of non-homogeneous glass melts might penetrate into the bulk fluid. He then tried to find a pair of liquids which might give rise to the same effects at room temperature. He found that when acetic acid was spread on water, spontaneous mixing occurs accompanied by strong eruptive movements. Jettmar and Roesler continued the studies with other completely miscible pairs of liquids. They found that those effects only occur when $(\Delta\rho) \cdot (\Delta\sigma) < 0$ where the differences between densities and surface tensions (against air) have to be taken in the same direction. Here we have apparently a strong coupling between Marangoni- and gravity effects which is completely un-understandable in terms of the present theory. But the effects seem to be confined to miscible and partially miscible[83] liquids.

Another interesting thing discussed in Sawistowski's paper is a model of Marsh, Sleicher and Heideger[84] which should be an extension of the Sternling-Scriven model to incorporate and erf-profile. The conclusions reached by those authors are not in accordance with the analysis given here. For example, they have found that their model gives rise to oscillatory instabilities in much broader ranges of the parameters than predicted by the Sternling and Scriven theory and that the dominant wavelengths of the oscillatory instabilities differ between the two models. In the present paper we have seen that the effects of finite thickness are not very profound - except for providing an upper λ_{cr} in addition to the lower. Marked differences are only found for deformable interfaces and low surface tensions, but this situation is not covered in the model of Marsh, Sleicher and Heideger. Sawistowski concludes also that the model of Marsh, Sleicher and Heideger seems to give results less consistent with experiments than the Sternling and Scriven model. In contrast, the model for finite thickness presented in this paper seems to work quite well and to yield physically reasonable results, so the theoretical foundation of the Marsh, Sleicher and Heideger model should be scrutinised in future.

The work of Linde <u>et al</u>., Sawistowski <u>et al</u>. (see ref.80 pp. 344-358) as well as the work of Bakker <u>et al</u>.[85] contain a lot of empirical information which is difficult to decipher theoretically in a clear cut fashion. I would like here to express the hope that experimentalists would concentrate somewhat more on the problem of measuring dimensions and growth times very shortly after the formation of liquid-liquid contact, to test the predictions of the linear theories. From a practical point of view those very first convection cells are not very interesting, but for a theoretical progress to take place, the simplest theories have to be checked the first. It was made evident in § 10 that surface viscosities - which are largely unknown today - play an important rôle. Independent measurements of $\eta_s = \kappa + \varepsilon$ should therefore be made on the systems applied, e. g. by following the damping of capillary waves on the interface and using a special case of the characteristic eqn. discussed in § 3.

From the point of view of giving some new explanations of the important phenomena of spontaneous emulsification the present study has been somewhat deceiving, since it is shown that the absolute value of the surface tension plays only a role under very specialized circumstances, even if it is well known experimentally that low surface tension promotes spontaneous emulsification. At present I have strong reasons to believe, however, that those findings are artefacts linked to the linearisation of the deformations around an initially strictly plane interface. For spherical interfaces (see the paper of Sørensen and Hennenberg in the present book) the low values of the interfacial tension can be shown to have a pronounced effect even on stationary Marangoni-instabilities. During vigorous interfacial turbulence,deformations of the interface can easily become so large, that a local spherical geometry is more appropriate than a plane geometry. We shall therefore expect strong interactions from low interfacial tensions in the non-linear realm of Marangoni instability even for initially plane interfaces.

REFERENCES.

1. H. Bénard, Rev.Gen.Sci.pures et appl., 1900, 11, 1261, 1309
2. H. Bénard, Ann.Chim.Phys., 1901, 23. 62
3. Lord Rayleigh, Phil.Mag., 1916, 32, 529
4. L. E. Scriven and C. V. Sternling, J.Fluid Mech.,1964, 19, 321
5. C. Normand, Y. Pormeau, M. G. Velarde, Rev.Mod.Phys., 1977, 49, 581
6. M. J. Block, Nature, Lond., 1956, 178, 650
7. P. G. Grodzka and T. C. Bannister, Science, 1972, 176, 506
8. P. G. Grodzka and T. C. Bannister, Science, 1975, 187, 165
9. S. Chandrasekhar, Hydrodynamic and Hydromagnetic Stability (Oxford University Press, 1961).
10. P. Glansdorff and I. Prigogine, Thermodynamic Theory of Structure, Stability and Fluctuations (Wiley-Interscience, 1971).
11. J. R. A. Pearson, J.Fluid Mech., 1958, 4, 489

12. C. V. Sternling and L. E. Scriven, A.I.Ch.E.Journal, 1959, 5, 514
13. J. Thomson, Phil.Mag., 1855, [4], 10, 330
14. C. Marangoni, Nuovo cimento, 1871, [2], 5 - 6, 239; 1878, [3], 3 97, 193
15. Lord Rayleigh, Proc.Roy.Soc.(London), 1890, 47, 281; 48, 127, 363
16. H. Linde, S. Pfaff and C. Zirkel, Zeitschr.f.physik.Chem., 1964, 225, 72
17. K. A. Smith, J.Fluid Mech., 1966, 24, 401
18. H. J. Palmer and J. C. Berg, J.Fluid Mech., 1972, 51, 385
19. M. Hennenberg, T. S. Sørensen and A. Sanfeld, J.C.S.Faraday II, 1977, 73, 48
20. T. S. Sørensen, F. Y. Hansen, J. Nielsen and M. Hennenberg, J.C.S. Faraday II, 1977, 73, 1589
21. T. S. Sørensen, M. Hennenberg and F. Y. Hansen, J.C.S. Faraday II, 1978, 74, 1005
22. T. S. Sørensen, M. Hennenberg and A. Sanfeld, J.Colloid Interface Sci., 1977, 61, 62
23. M. Hennenberg, T. S. Sørensen, A. Steinchen and A. Sanfeld, J.Chim.Phys., 1975, 72, 1202
24. T. S. Sørensen, M. Hennenberg, A. Steinchen and A. Sanfeld, Progr.Colloid Polymer Sci., 1976, 61, 64
25. T. S. Sørensen, M. Hennenberg, A. Steinchen and A. Sanfeld, J.Colloid Interface Sci., 1976, 56, 191
26. A. Orell and J. W. Westwater, A.I.Ch.E.Journal, 1962, 8, 350
27. H. Linde, Fette, Seifen, Anstrichmittel, 1958, 60, 826, 1053
28. H. Linde und E. Schwartz, Zeitschr.f.physik.Chem., 1963, 224 331
29. H. Linde und D. Thiessen, Zeitschr.f.physik.Chem., 1964, 227 223
30. H. Linde und E. Schwartz, Monatsberichte d.Deutsch.Akad. Wissenschaft, Berlin, 1964, 6, 330
31. H. Linde und B. Sehrt, Monatsberichte d.Deutsch.Akad.Wissenschaft, Berlin, 1965, 7, 341
32. H. Linde und K. Winkler, Zeitschr.f.physik.Chem., 1965, 230, 207
33. H. Linde und K. Loeschcke, Chemie-Ingenieur-Technik, 1966, 39, 65
34. D. Thiessen, Zeitschr.f.physik.Chem., 1966, 232, 27
35. H. Linde, E. Schwartz und K. Gröger, Chem.Eng.Sci., 1967, 22, 823
36. H. Linde und M. Kunkel, Wärme- und Stoffübertragung, 1969, 2, 60
37. E. Schwartz, Wärme- und Stoffübertragung, 1970, 3, 131
38. H. Linde und N. Shulewa, Monatsberichte d. Deutsch.Akad. Wissenschaft, Berlin, 1970, 12, 883
39. H. Linde und P. Friese, Z.phys.Chem., Leipzig, 1971, 247, 225
40. E. Schwartz, Monatsberichte d. Deutsch.Akad.Wissenschaft, Berlin, 1971, 13, 507
41. H. Wilke, Ein numerisches Verfahren zur Behandlung grenzflächendynamischer Probleme beim gekoppelten Impuls- und Wärmetransport über fluide Phasengrenzen, Thesis, Akad.d. Wissenschaft, DDR, Berlin, 1973

42. H. Linde und P. Schwartz, Chem.Techn., 1974, 26, 455
43. D. T. Wasan and V. Mohan, in Improved Oil Recovery by Surfactant and Polymer Flooding, pp 161 - 203 Academic Press, New York, 1977
44. L.E.Scriven, Chem.Eng.Sci., 1960, 12, 98
45. R. Aris, Vectors, Tensors, and the Basic Equations of Fluid Mechanics, (Prentice-Hall, London, 1962)
46. J. Boussinesq, Ann.Chem.Phys., 1913, 29, 349, 357, 364
47. J. G. Oldroyd, Proc.Roy.Soc.(London), 1955, A232, 567
48. J. C. Slattery, Chem.Eng.Sci., 1964, 19, 379, 453
49. F. C. Goodrich, Proc.Roy.Soc.(London), 1961, A260, 481, 490, 503
50. W. Jost, Diffusion in Solids, Liquids, Gases, (Academic press, New York and London, 3rd ed., 1960), pp. 68 - 69
51. B. R. Morton, Quart.J.Mech.Appl.Math., 1957, 10, 433
52. V. G. Levich, Physiochemical Hydrodynamics, (Prentice-Hall, Englewood Cliffs, N.J., 2.ed., 1962) Chap.XI
53. Lord Rayleigh, Scientific Papers, Vol.2, pp. 200 - 207, Cambridge, England, 1900
54. G. Taylor, Proc.Roy.Soc.(London), 1950, A201, 192
55. W. J. Harrison, Proc.London Math.Soc., 1908, 6, 396
56. R. Bellman and R. H. Pennington, Quart.Appl.Math., 1954, 12, 151
57. W. H. Reid, Proc.Cambridge Phil.Soc., 1961, 57, 415
58. I. B. Ivanov, Physico-Chemical Hydrodynamics of Thin Liquid Films, Synopsis of dissertation, University of Sofia, 1977
59. T. K. Sherwood and J. C. Wei, Ind.Eng.Chem., 1957, 49, 1030
60. J. C. Wei, Thesis, (M.I.T., Cambridge, 1955)
61. M. Dupeyrat and E. Nakache, in Electrical Phenomena at the Biological Membrane Level, (Elsevier, Amsterdam, 1977) pp. 377 - 388
62. T. S. Sørensen and K. F. Jensen, J.C.S.Faraday II, 1975, 71, 1805
63. N. O. Østerberg, J. B. Jensen and T. S. Sørensen, Acta Chemica Scandinavica A 32 - 33 (in press)
64. A. Watanabe, K. Higashitsuji and K.Nishizawa, J.Coll.Int. Sci., 1978, 64, 278
65. C. A. Miller and L. E. Scriven, J.Coll.Int.Sci., 1970, 33, 360, 371
66. P. M. Bisch and A. Sanfeld, in Les instabilités hydrodynamiques en convection libre, forcée et mixte, J-C Legros & J. K. Platten eds. (Lecture Notes in Physics Vol.72, Springer Verlag, Berlin, Heidelberg, New York, 1978) pp.112-119
67. A. Sanfeld, Introduction to the Thermodynamics of Charged and Polarized Layers, (Wiley-Interscience, London, 1968)
68. E. Ruchenstein and C. Berbente, Chem.Eng.Sci., 1964, 19,329
69. T. S. Sørensen, Acta Chem.Scand., 1976 A30, 555
70. T. S. Sørensen, ibid, 1977 A31, 347
71. T. S. Sørensen, ibid, 1977 A31, 437
72. T. S. Sørensen, ibid, 1977 A31, 892
73. J. T. Davies & E. K. Rideal, Interfacial Phenomena, 1963, pp. 360 - 366
74. Ilkovic, Coll.Trav.Chim.Tchécosl. 1932, 4, 480
75. T. K. Sherwood and J. C. Wei, Ind.Eng.Chem., 1957, 49, 1030
76. Quincke, Wiedemans Ann., 1888, 35, 593

77. H. Linde, Ber.d.deut.Akad.d.Wiss,m Berlin, DDR, 1960, Untergruppe B/I/1, Nr.3, pp. 23 - 35
78. A. Sheludko, "Colloid Chemistry", Elsevier, Amsterdam-London-New York, 1966
79. Szyszkowski, Z.physik.Chem., 1908, 64, 385
80. H. Sawistowski, "Interfacial Phenomena", Chap.9 in "Recent Advances in Liquid-Liquid Extraction", (C. Hanson, ed.) Pergamon Press, Oxford, New York, Toronto, Sydney, Braunschweig, 1971
81. H. Jebsen-Marwedel, Kolloidzeitschrift, 1948, 111, 46
82. W. Jettmar and F. C. Roesler, Kolloidzeitschrift, 1951, 123 2
83. R. S. Burdon, "Surface Tension and the Spreading of Liquids" Cambridge University Press, Cambridge, 1949
84. B. D. Marsh, C. A. Sleicher and W. J. Heideger, Paper presented at 57th Annual Meeting of the Am.Inst.of Chem.Engineers, Philadelphia, 1965
85. C. A. P. Bakker, F. H. Fentener van Vlissingen and W. J. Beek, Chem.Eng.Sci., 1967, 22, 1349

DISSIPATIVE STRUCTURES AND NONLINEAR KINETICS OF THE MARANGONI-INSTABILITY

by

H. Linde and P. Schwartz

Zentralinstitut für physikalische Chemie der Akademie der Wissenschaften der DDR, 1199 Berlin, DDR

and

H. Wilke

Zentralinstitut für Mathematik und Mechanik der Akademie der Wissenschaften der DDR, 108 Berlin, DDR

Summary

The Marangoni-instability with the driving force of heat- or mass transfer across fluid interphases causes a self-amplification and self-organisation of movements at a fluid interphase which develop a spectrum of multiform dissipative structures. There exist manifold substructured spatial periodic systems of a hydrodynamic kind with a time depending behaviour which can degenerate to a typical relaxation oscillation.

The latter differs completely from the spatial and temporal period structure of a two-parameter-oscillation with a wave-like behaviour.

These instabilities lead to a nonlinear acceleration in the mass- and heat transfer. Even the effect of Marangoni-stability is provable as a damping of the eddies of the forced surface renewal.

Under the condition of the balance between the shear stress and the surface tension gradient at a surface, which is covered with adsorbed surface active agents, there exists at low viscous surfaces a more turbulence-analogic hydrodynamic instability with stationary spatial periodic and instationary spatial and temporal periodic dissipative structures.

1. Introduction

To introduce graphically the fundamental problem, a very simple qualitative proof is possible in the case of the cooling of a liquid surface by evaporation. In pure liquids, the surface tension σ decreases with increasing temperature T. Fig.1 shows at the left the temperature distribution versus the distance x from the interphase and at the right two geometrical dimensions x and y of a vessel with the interphase (or surface) between liquid and gas.

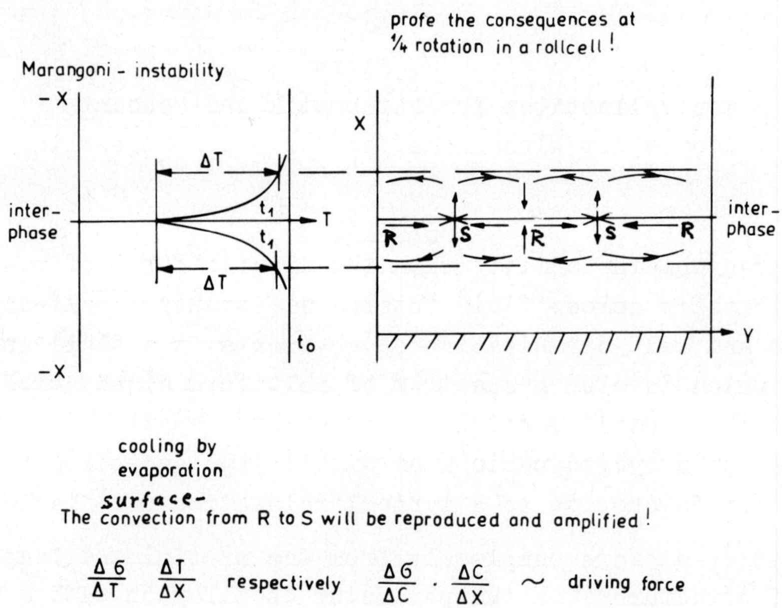

Fig. 1. Graphical analysis of stability by the method of small disturbances.

If we open the closed system of a vessel, which is filled with an easy vaporable liquid, the temperature at the surface will decrease to a depression because of the exchanged latent heat by the evaporation. We realize graphically a small disturbance in the form of roll cells. At a very short interval after the start of this disturbance, the spots R will be heated directly by the convective transport of heat from the warmer boundary layers at both sides of the surface, whereas the spots S will be influenced insignificantly of the initial temperature at the spots R. (R, S are spots at the interphase!)

From an equality of the temperature at the surface at the beginning at t_0, we result at t_2 a distribution like $T_R > T_S < T_R > T_S$ and because of $\frac{\Delta\sigma}{\Delta T} < 0$ a σ-distribution $\sigma_R < \sigma_S > \sigma_R < \sigma_S$ of the reversed sign. The Marangoni-effect leads to a surface convection from the smaller σ_R to the greater σ_S, which is the proof of the reproduction and amplification of the initial disturbance in the form of the roll cells of the Marangoni-instability. This roll cell system with a structure of convection, of temperature and of surface tension is a typical dissipative structure. The driving force of this instability is $\frac{\Delta\sigma}{\Delta T} \cdot \frac{\Delta T}{\Delta X}$.

With K. Gröger and E. Schwarz we made a serious mathematic analysis /1/, which is in agreement with the experimental results and this estimation: roll cell instability by a depression of heat, stability by a source of heat. This work followed the method of the linear analysis of the Marangoni-instability at the conditions of heat or mass transfer by Sternling and Scriven /2/, which also lets expect oscillation in addition to roll cell instability and stability.

2. Dissipative structures of the Marangoni-instability

The scheme shows graphically the mainline of the regime of the Marangoni-instability, which we confirmed experimentally /3/.

It is valid for mass transfer as well as for heat transfer. If we change the sign of $\frac{\Delta\sigma}{\Delta T}$ or of $\frac{\Delta\sigma}{\Delta c}$, we have to change the direction of mass or heat transfer too in order to get the same regime.

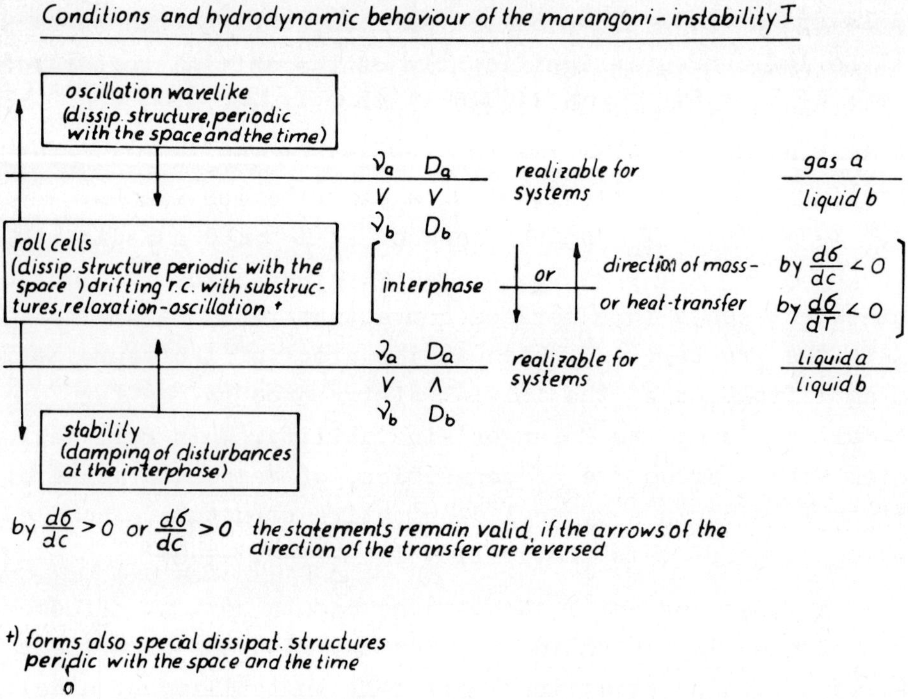

Fig. 2. Scheme for the conditions of the Marangoni-instability

In the idealized form, the roll cell system is a dissipative structure periodic with the space. Only if we proof very thin layers of liquids and gases in the form of a wet solid or a capillary split, the roll cells can stay at the spot.

The shadow-schlieren-picture 3 shows roll cells in a capillary split, which are growing up during the concentration gradient decreases because of the instationary mass transfer. The change of the cell size is according to the theory.

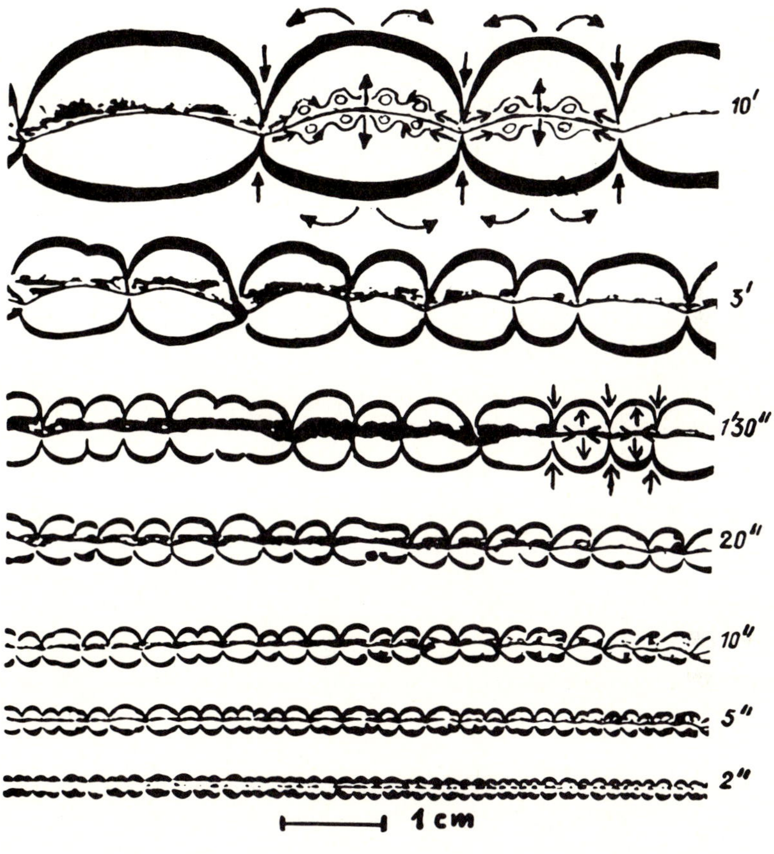

Fig. 3. Shadow-schlieren-picture of roll cells in a capillary split.

$$\frac{1\ \%\ C_{16}H_{33}OSO_3Na + C_5H_{11}OH}{H_2O}$$

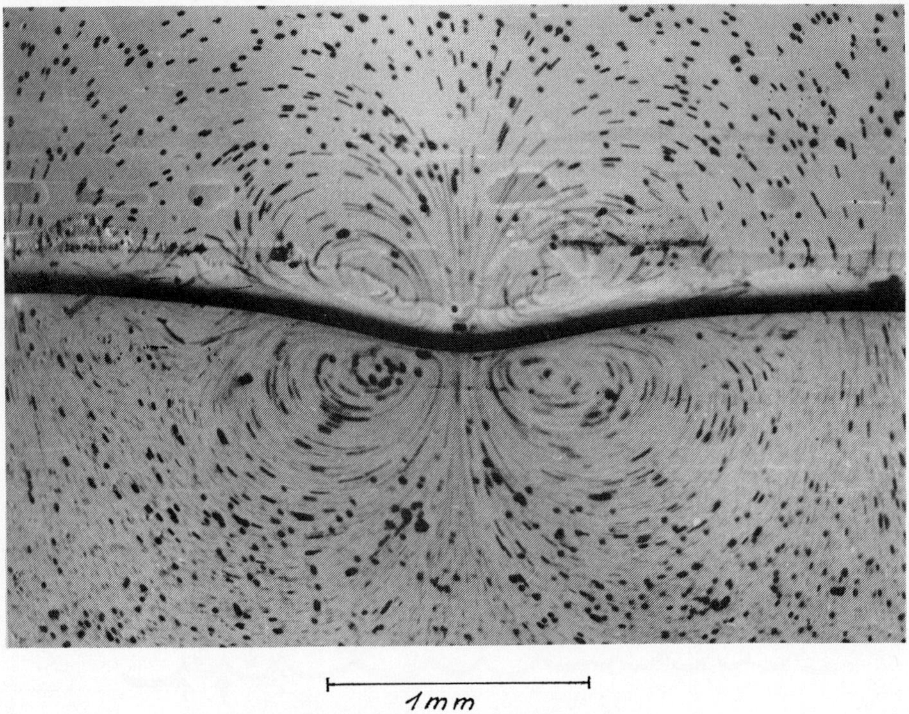

Fig. 4. Picture of the traces of small particles in the roll cell.

$$\frac{1\ \%\ C_{16}H_{33}OSO_3Na + C_5H_{11}OH}{H_2O}$$

15 s after the start
1/5 s time of exposition

The picture 4 of the traces of small particles in the capillary split shows the streamlines. From the curvature of the meniscus we calculated the development of the surface tension along the surface with a modified method of the rising meniscus (Fig. 5).

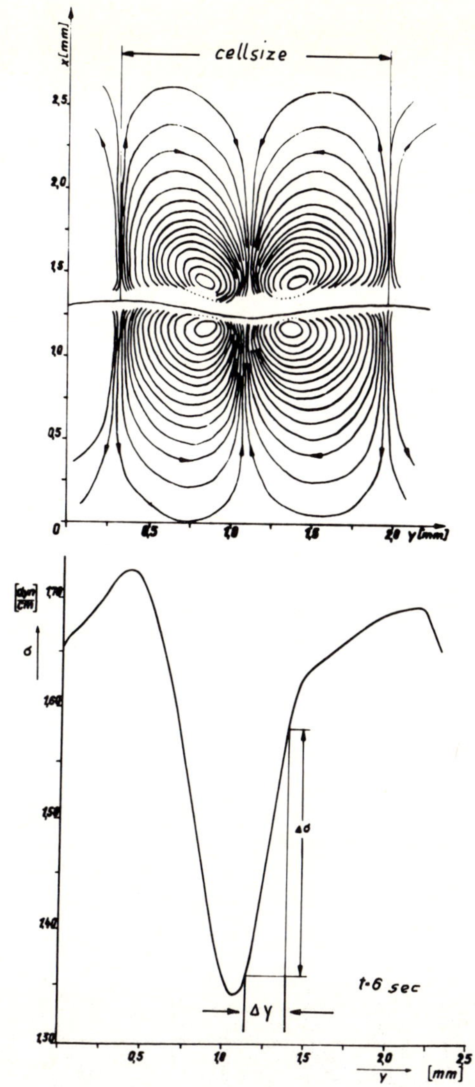

Fig. 5. Streamlines and σ-distribution along the surface for a system like Fig. 4, experimental interpretation.

With the GOEM++ /4/ and the MIMEG++ /5/ we have an analytical and a numerical approach for the roll cell instability (GOEM) and for roll cells and oscillation (MIMEG).

++ See the appendix.

The experimental result of the velocity distribution in a mass transfer case is shown in comparison on principle with the case of heat transfer (calculation with GOEM and MIMEG respectively) (Fig. 6, 7).

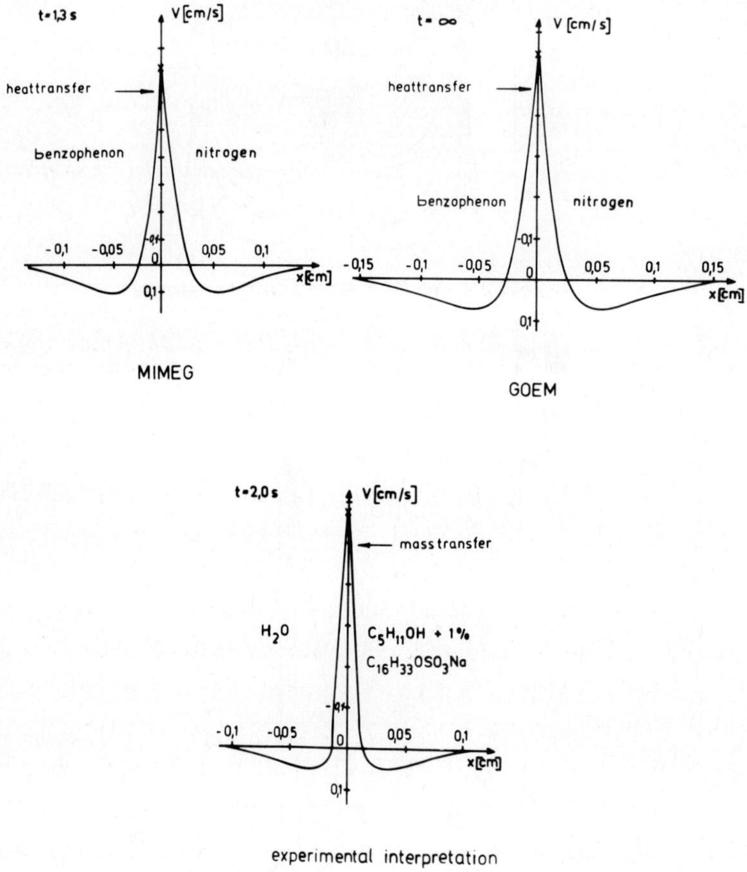

Fig. 6. Velocity V parallel to the interphase versus the distance x from the interphase, y at the center of the eddy.

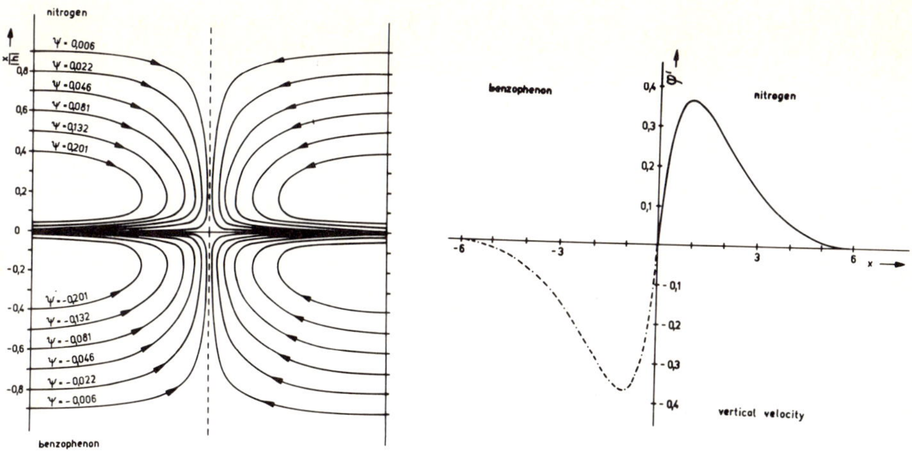

Fig. 7. Streamlines for a roll cell, test system III with GOEM /4/

Vertical velocity at the vertex of the roll cell versus the distance from the surface from the experiment.

The theoretic values of the velocity parallel and vertical to the interphase correspond to the experimental values.

The Fig. 8 shows with the MIMEG- respectively the GOEM-calculation the evolution of the initial small temperature distribution at the surface until the roll cell is completely amplified. If we take into consideration the dependence of the surface tension from the temperature, we recognize the analogy to the development of the surface tension along the surface in the experimental interpretation of Fig. 5. Until t = 0,128 s, the roll cell is amplified in the initially assumed size z. But until t = 0,19 s, the size of the roll cell decreases to the half in this size because of the solving of another eigenvalue for z. We remember, that the roll cell is characterized by a surface convection from the spot with higher temperature (smaller σ) to the spot with smaller temperature.

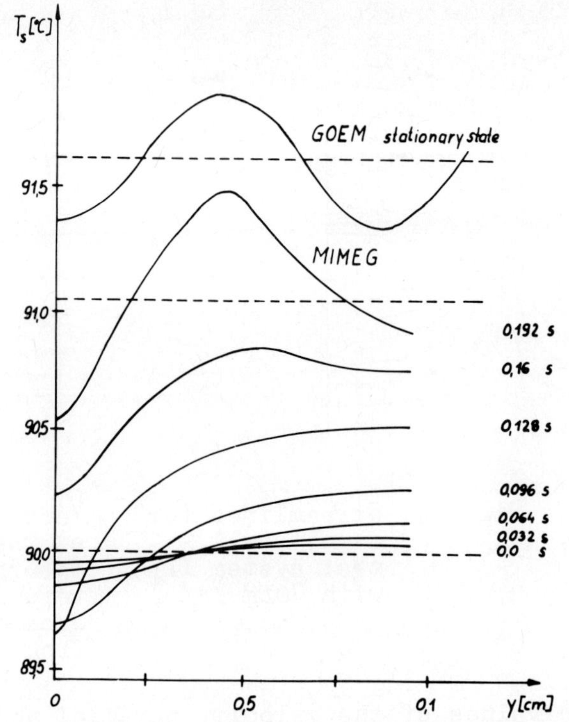

Fig. 8. Distribution of the temperature along the surface T_s (Y) during the amplification of a roll cell by the heat transfer from benzophenone to nitrogen, MIMEG-calculation.

Further we see, that the average level of the temperature at the surface increases during the amplification. This is also seen in the comparison of the average temperature at the initial state and at the state of the amplified roll cell in the analogous heat transfer system n-octane-nitrogen (Fig. 9). The higher temperature at the surface leads to a higher driving force ΔT in the phase with the smaller heat conductivity (nitrogen), whereas the convective transport of the roll cell leads to a smaller temperature gradient near to the sur-

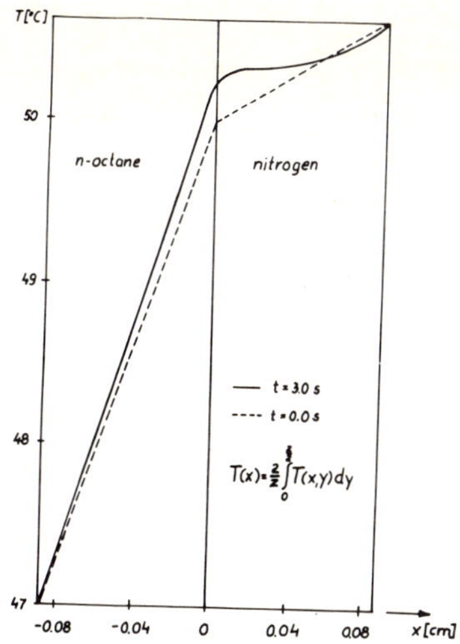

Fig. 9. Average temperature distribution $\bar{T}(x)$ at the initial state ($t = 0,0$ s) and at the completely amplified roll cell ($t = 3,0$ s). Heat transfer from n-octane to nitrogen in the stationary state.

face and to a higher temperature gradient far from the surface in the phase with the smaller temperature conductivity.

In the mass transfer case the analogous concentration distribution is very sharply developed in both liquid phases because of the small diffusion coefficients. We can confirm it indirectly from the shadow picture (Fig. 3) and directly with the Philpot-Svensson-method (Fig. 10). At the interphase, the concentration gradient is very smooth. Only in the distance, which is given by the size of the roll cell, there are sharp gradients of the concentration.

At extended surfaces or interphases, the roll cells have two different time depending behaviours during an interval, in which the concentrations at the external boundaries of the relevant part of the system are practically constant.

At first the roll cells (of first order) are not stationary with the space. They are moving along the surface together with both boundary layers and form roll cells of a

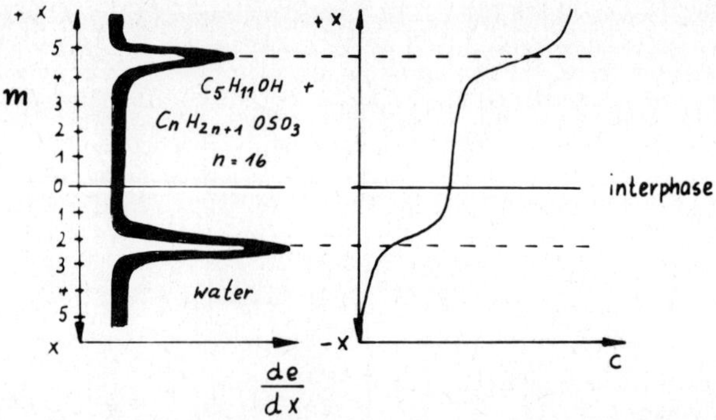

Fig. 10. The Philpot-Svensson method represents the concentration gradient versus the distance from the interphase, graph of a photo.

Corresponding concentration distribution.

greater dimension (of second order), which themselves also are moving in the same manner forming still greater large scaled roll cells (of third order). Therefore we observe a hierarchy of structures and substructures. The shadow-schlieren-pictures 11, 12, 13, which are taken vertically to the interphase, show a lot of details of the structural hierarchy.

With the GOEM it is possible to calculate the maximal tangential velocity V_0 as a function of the dimensionless "spatial periodicity number" k.

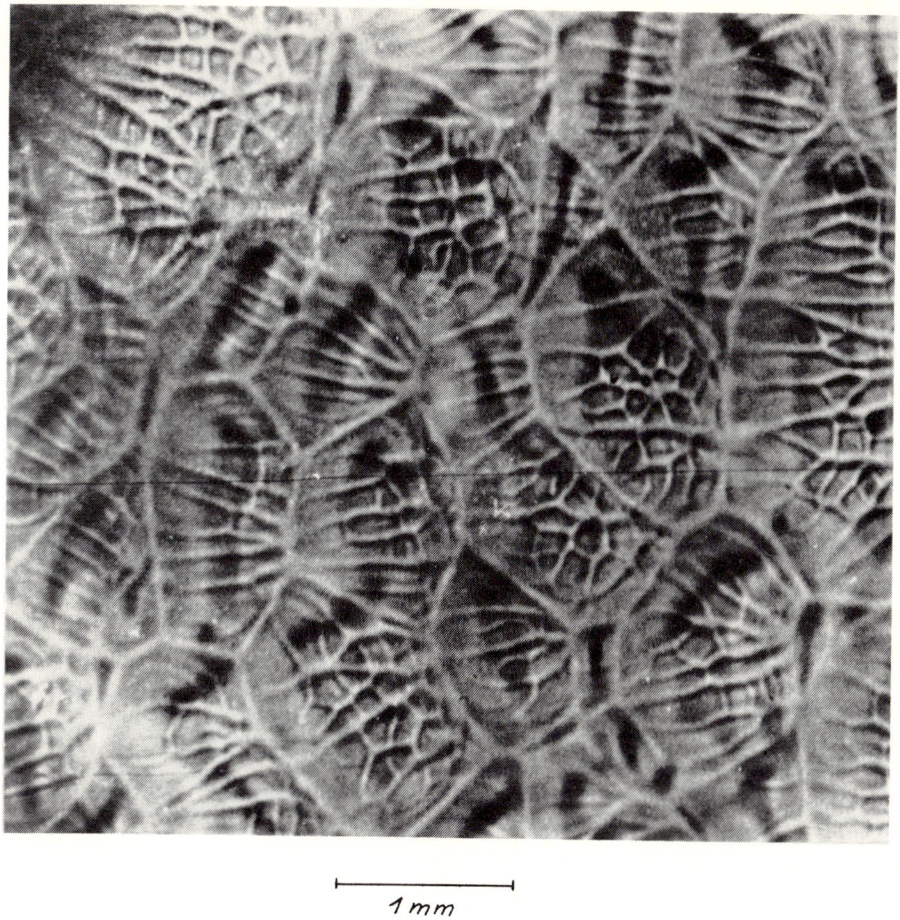

Fig. 11. One- and twomodal roll cells of first order as a substructure of a roll cell of second order Z-type = surface drift from the center to the periphery.

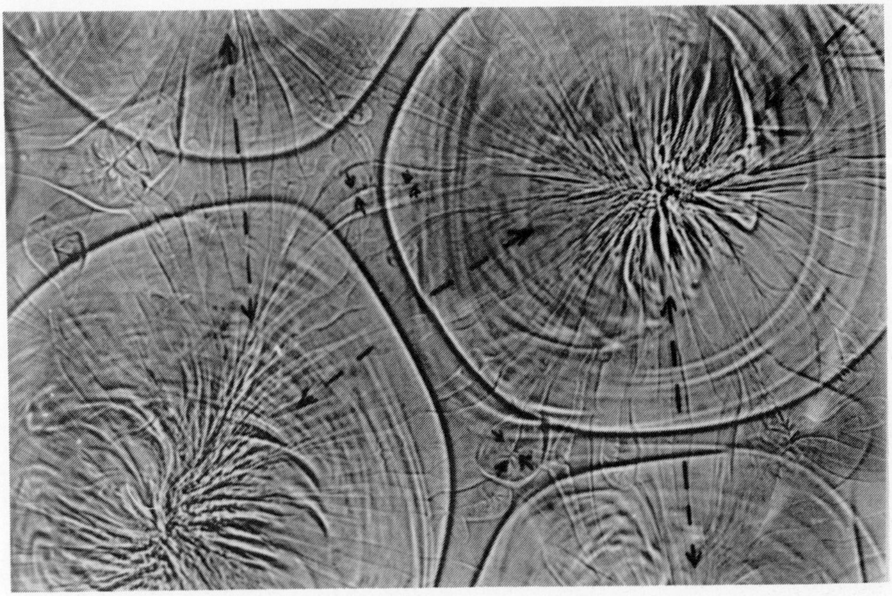

5mm

Fig. 12. Roll cells of third order,
P-type = surface drift from the periphery to the center .

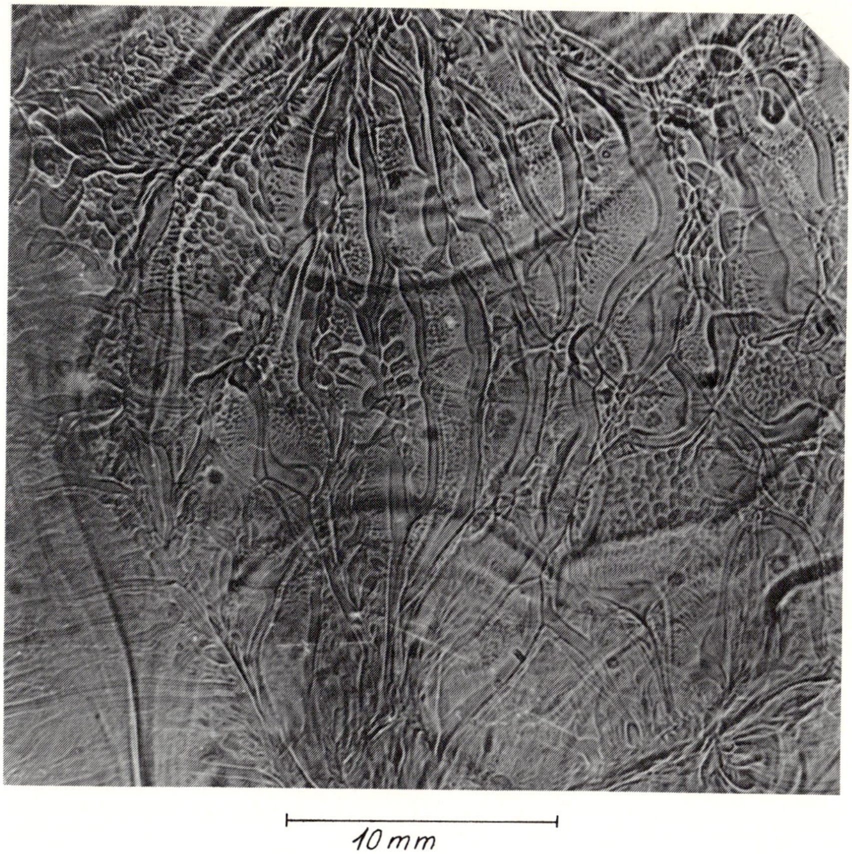

Fig. 13. Substructure of a P-roll cell of third order.

At Fig. 14 we recognize two different extreme values of V_o at two areas of the roll cell sizes Z, which differ about one order of magnitude from each other. These maxima at $|h| \sim 6$ and $0,2 \leq |h| \leq 0,8$ are the result of the specific amplification of roll cells of first and of second order respectively.

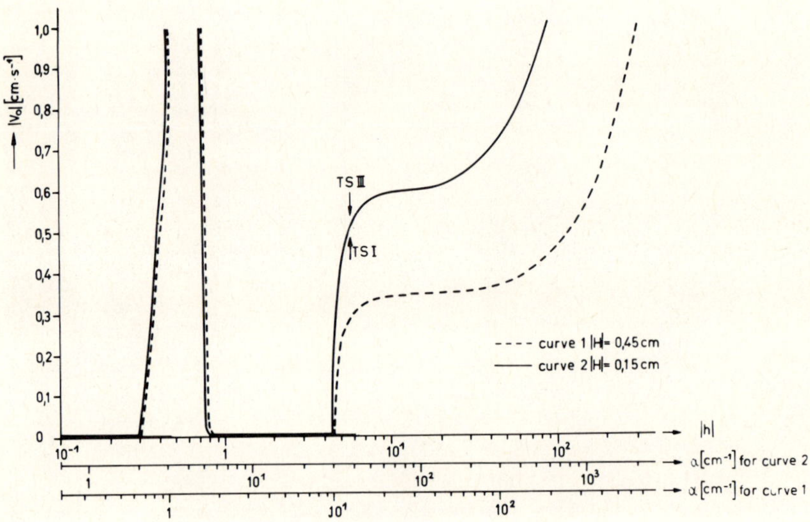

Fig. 14. Maximal tangential velocity V_o at the surface versus the dimensionless "spatial periodicity number" $h = \alpha \cdot H$

$$\alpha = \frac{2\pi}{Z}$$

Z = size of roll cell or spatial periodicity of forced convection

H = thickness of the layer

The second kind of time depending behaviour of the roll cells often occurs at a higher level of instability in liquid-liquid mass transfer systems: the roll cells are not able to become stationary. Before the circle of convection is closed, which is necessary for a stationary self-reproducing roll cell, the sufficient supply of the surface active agent to fulfil the condition of feedback is brokendown. Then the roll cell is stopped suddenly and at other places in the neighbourhood the same procedure repeats. This is the mechanism of relaxation oscillation, which occurs in a wide range of frequencies and areas of the spreading circles.

The relaxation oscillation with a very low frequency transits to a periodic break down and new amplification of mostly large scaled roll cells with a substructure. The mechanism is analogous to the well-known "sawtooth oscillation", an one-parameter oscillation of an electric system. It can exist in a not synchronized structure of many processes (Fig. 15) as well as in a single process at plane surfaces or drops.

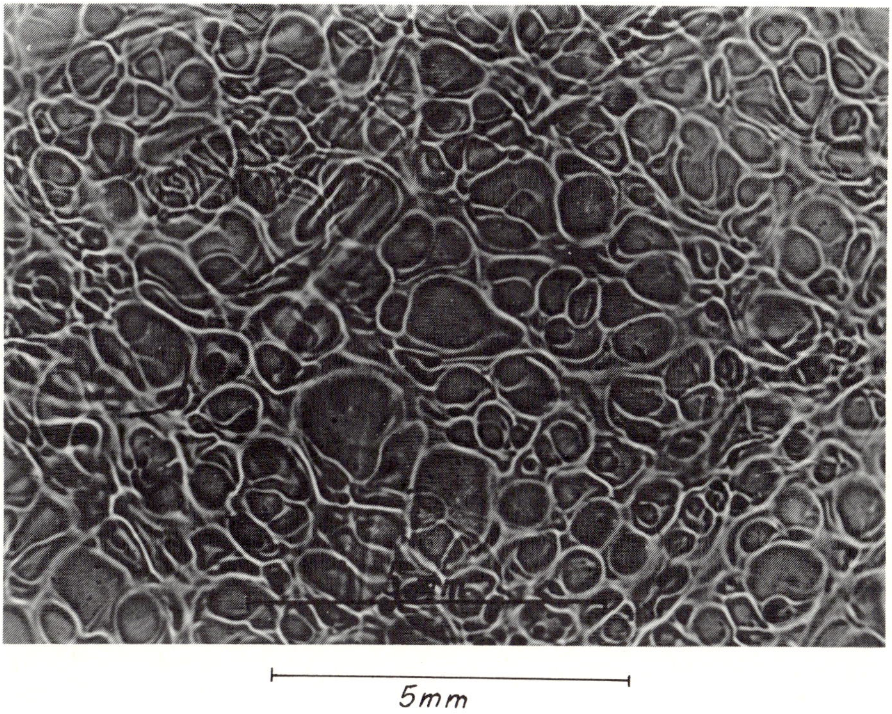

Fig. 15. Relaxation oscillation, shadow picture vertical to the surface.

It is commodious to give a graphic explanation of a single process with a special mass transfer example at a Hg-electrode (Fig. 16) /6/.

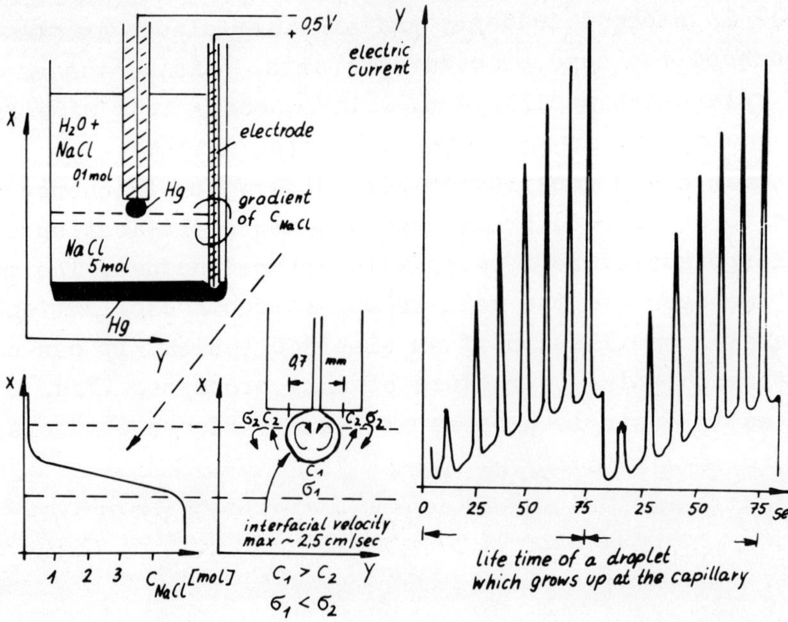

Fig. 16. Relaxation oscillation of a single roll cell at a Hg-droplet.

A Hg-drop of 0,7 mm diameter on a capillary of 0,3 mm diameter is put near to the boundary between 0,1 mol NaCl and 5 mol NaCl in pure water. NaCl is an example for a surface active agent, which is only soluble in water and adsorbable at the Hg-surface. NaCl is adsorbed at the high concentrated spot of the Hg-surface (lower surface tension) and desorbed at the low concentrated spot, whereby a surface convection takes place from the spot of lower surface tension to the spot of higher tension.

If we approach the droplet to the boundary of concentration, we observe a short periodic amplification andbreak down of a roll cell at the Hg-surface, which occurs with as well as without an additional electric potential of about -0,5 V at the Hg-electrode. The latter condition - and this is of relevance for polarographic problems - is very favourable to show the additional transport by the maxima in the electric current. During the quick amplification the supply of NaCl from the under layer breaks down and the roll cell stops suddenly. Mainly because of diffusion - in this case more by the free convection in order to regenerate the stable layer of density - the initial conditions for an amplified roll cell are regenerated. The breakdown of supply is connected with the slowly diffusion. Therefore relaxation oscillation is mainly observed in liquid-liquid systems with mass transfer.

From the theory we have only a remark to this mechanism from the MIMEG. Because of problems in the computing technique by mass transfer systems, we are only able to develop this model for heat transfer systems.

We calculate the tangential velocity at the surface V_o, the transport of heat by convection $Q_{conv.}$ and by conduction $Q_{diff.}$ during the amplification of a roll cell. V_o is amplified (after the short initial disturbance, which is visible in this curve too) to a maximum and is going to a stationary state with a lower level of V_o. The same behaviour we can see at the convective transport $Q_{conv.}$ respectively at the opposite development of the conductive transport Q_{diff}. That means, that during the amplification of the roll cell instability under conditions of constant temperatures at the boundaries of this open heat transfer system the driving force of this instability $\frac{\Delta\sigma}{\Delta T} \cdot \frac{\Delta T}{\Delta x}$ will be diminished and therefore the intensity of the convection too. Here a steady state can be maintained. In the case of mass transfer in liquid-liquid systems, which lets expect roll cell instability, the driving force can be diminished up to its critical point. Then the roll cell stops suddenly. A new amplification only occurs, if the concentration gradient regenerates by diffusion and exceeds the critical value.

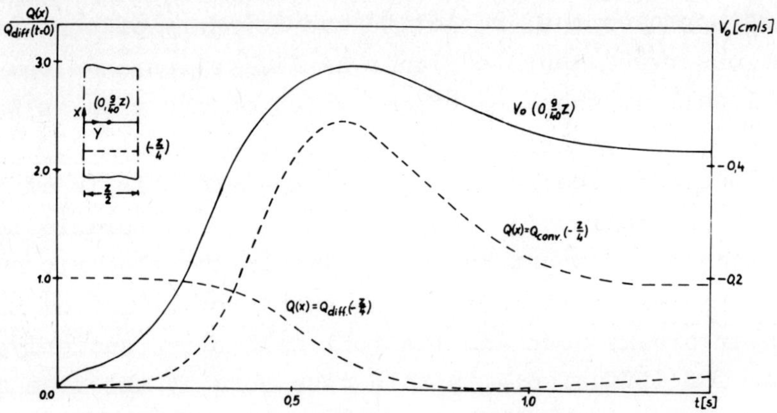

Fig. 17. Tangential velocity V_o and transport of heat by conduction $Q_{diff.}$ and convection $Q_{conv.}$ across the layer $x = -\frac{z}{4}$ during the amplification of roll cells in the time t. MIMEG calculation, fixed temperatures at the boundaries of the open system.

The full spectrum of these roll-cell-like dissipative structures is very roughly screened in the table:

TABLE

Survey about the dissipative structures in connection with the roll cell - Marangoni - instability with decreasing distance from equilibrium

Example of a system:

$$\frac{C_5H_{11}OH}{H_2O}\ \downarrow\ \frac{a}{b}$$

Na-alkylsulfate as diffusing matter

$$e^2 > 1 > r^2 \qquad f > 0 \qquad \frac{\Delta\sigma}{\Delta c} < 0 \qquad \frac{\rho_a}{\rho_b} < 1$$

(f see Sternling-Scriven)

$e^2 = \frac{\nu_a}{\nu_b}$ $\nu_{a,b}$ = kinem. viscosity

$r^2 = \frac{D_a}{D_b}$ $D_{a,b}$ = diffusion coeffic.

$\rho_{a,b}$ = density

The statements are valid also for

$$e^2 > 1 < r^2 \qquad \uparrow\frac{a}{b} \qquad \frac{\Delta\sigma}{\Delta c} < 0 \qquad \left(\frac{\text{gas a}}{\text{liquid b}}\ \text{system}\right)$$

" valid in a weakened form (without or only with very weak relaxation oscillation) for the analogousheat transfer,

" are valid by $\frac{\Delta\sigma}{\Delta c} > 0$, if the direction of transfer is reversed.

Table (contd.)

Great distance from equilibrium

↓ +

increase of the (size Z) of the roll cells and of the wavelength of the relaxation oscillation; decrease of the frequency of the relaxation oscillation resp. of the frequency of the periodic new-amplification

decrease of instability

a. shortperiodic relax. oscillation with small wavelength, stochasticly coexistent in a structure ($^1/10$ up to $^1/20$ s for one spreading, Ø of the spreading cells ~ 0,1 mm)

↓ +

b. less longperiodic relaxation oscillation, similar to a. → + b_1

↓

c. longperiodic relaxation oscillation in transition to each other isolated larger spreading circles (0,5 - 2 s for one spreading, Ø of the spreading circles ~ 1 mm)

↓

d. longperiodic relaxation oscillation similar to c. substructured with very small propagating roll cells (3 - 5 s for one spreading, Ø of the spreading 1 - 2 cm)

↓

e. very longperiodic new-amplification of a structure of propagating small roll cells (5 s - 50 min for one new-amplification, Ø of the small roll cells 0,1 mm)

b_1 large-scaled p-roll cells coexistent in a structure (Ø of the p-roll cells 1 - 5 cm), substructured with propagating roll cells or propagating rollcells or propagating relaxation oscillations.[++]

↓

b_2 large-scaled z-roll cells coexistent in a structure (Ø of the z-roll cells 1 - 5 cm), substructured like b_1[++]

↓

c_1 large-scaled z-single-rollcells (Ø up to 12 cm observed), substructured like b_1[++]

Small distance from equilibrium

↓

+ the arrows represent the transitions, that are possible by increasing instability. There is a different large roundabout way possible over the large-scaled p- or z-roll cells.

++ the large-scaled p- or z-roll cells are seized by a longperiodic new-amplification

Table (contd.)

↓

f. slowly propagating middle small roll cells as a substructure of large scaled more regular roll cells [Ø of the middle small roll cells (two- and three-modal) ~ 2 mm]

↓

g. slowly propagating roll-cylinder (ripples) as a substructure of large scaled less regular roll cells (stripes) (one-modal)

small distance from the equilibrium

From the point of view of the theory of dissipative structures, it is interesting to have information about the critical values of their onset. Contrary to any linear analysis of stability, the GOEM and the MIMEG allow the calculation of the critical driving force. In the system benzophenone-nitrogen with a fixed thickness of the two phases we found for the roll cell instability by heat transfer a good agreement between the theoretical and the experimental value.

Experimental value (Linde and Loeschcke /7/)	8,9	$\frac{\Delta\sigma}{\Delta T}(T_o - T_a{}^o)_{crit} \cdot 100$ dyn·cm^{-1}
MIMEG	16,5	
GOEM	4,0	

The scheme about the conditions of instability shows, that we expect a true oscillation, if the heat transfer is directed from the gas into the liquid at $\frac{\Delta\sigma}{\Delta T} < 0$. From a high degree of instability the oscillation transits from violent, not regular standing three-modal oscillation systems across more regular three- and two-modal standing systems to one-modal standing or propagating oscillation systems at a lower degree of instability. As in the roll cell systems, the roll cell size and the wavelength of oscillation decreases with increasing instability, whereas the frequency of the true (two-parameter) oscillation and of the relaxation (one-parameter) oscillation increase. There is a fundamental difference between roll cells including the relaxation oscillation on the one hand and the true oscillation on the other hand. Contrary to the roll cell system, the oscillation systems behave like a wave at the surface. We observed propagating fronts of the oscillation, which are reflected at a plane or a concave wall like a wave /8/. The standing oscillation can be regarded as a superposition of propagating oscillation systems. We found well synchronized dissipative structures periodically with the space and the time, which have a wavelike interaction with walls of a special geometric form. The oscillation is characterized by a propagating of the moving state and not of the whole boundary layer. (We remember, that the roll cells drift together with the whole boundary layer forming large scaled roll cells.)

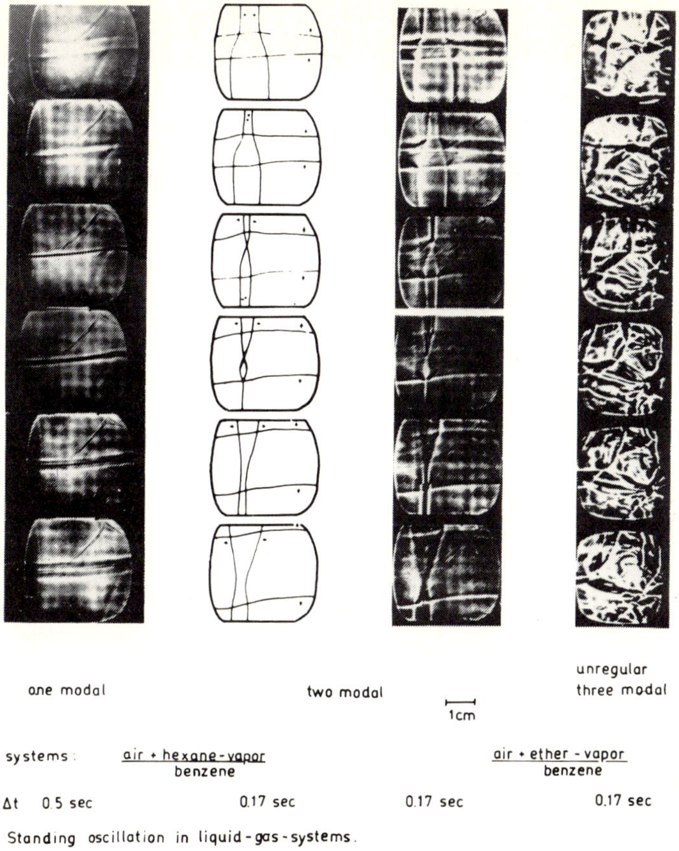

Fig. 18. One-, two- and (irregular) three-modal oscillation.

About the kinematic of the standing oscillation we get an impression from the MIMEG calculation of the tangential velocity in both phases in a small distance to the surface ($\frac{Z}{44}$ Z = wavelength) (Fig. 19). For the oscillation it is typical, that the moving state in the gas is prece-ding in comparison with the moving state in the liquid. The reason is the greater kinematic viscosity of the gas. A real quiescent state is not existing during the oscillation: the old oscillatory convection cell is displaced immediately by a new convection cell with the opposite rotating direction. We see also

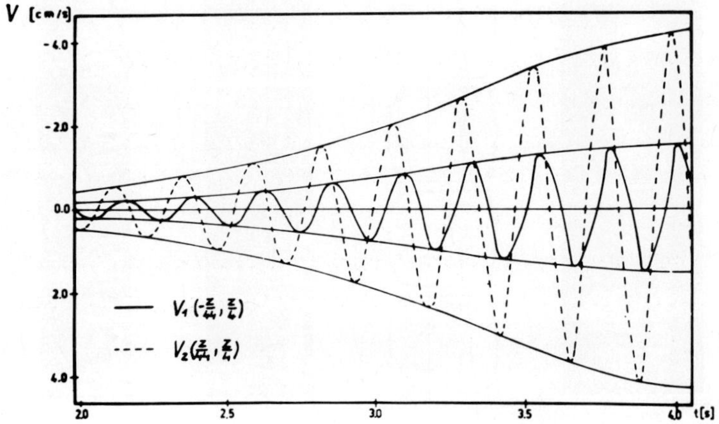

Fig. 19. Amplification of the oscillation and limitation of the amplitudes. The velocity in the gas precedes.
Stationary heat transfer
nitrogen / n-octane ↓

the amplification and later on the development to a steady state. The values of the relative great wavelength in the dimension of 0,5 up to 2 cm, of the frequencies in the dimension from 0,3 to 10 Hz and of the velocity distribution are in good agreement concerning our experimental and numerical results.

3. Mass exchange kinetics in instabilizable systems

From the point of the thermodynamics of open systems, we leave the thermodynamic branch of linear relations between fluxes and forces at the critical point and transit into the nonlinear branch with higher entropy production and higher rates of transport too. In the quasistationary mass exchange kinetics in cells with stirred phases, we expect in the linear branch an exponential exchange kinetics and in the nonlinear branch a nonexponential acceleration of the exchange. With liquid-liquid systems we proofed the mass exchange in the direction, which leads to roll cell instability, as well as the mass exchange direction, which leads to stability /9/. The graph on principle (Fig. 20) shows, that under the condition of a simulated equilibrium by self diffusion we receive an

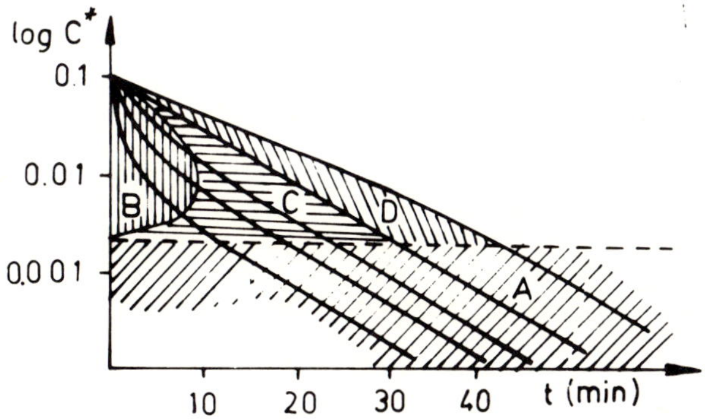

Fig. 20. Mass transfer with Marangoni-instability and Marangoni-stability.

C^* = concentration dimensionless $\frac{n_E - n}{n_E}$

A = true diffusion kinetics

B = self-amplified Marangoni-instability

D = damping of the surface renewal by the Marangoni-stability

C = partly amplified surface renewal of the forced convection

(without forced convection = stable !)

ideal exponential curve. By roll cell instability, we see a striking acceleration of mass exchange, which depends on the difference of the concentration of the employed surface active agents as well as of their surface activity according to the role of $\frac{\Delta\sigma}{\Delta c} \cdot \frac{\Delta c}{\Delta x}$ as the driving force of this instability. But the transition from the nonlinear branch (here the non-exponential exchange) into the linear branch (here the exponential exchange) does not take place with a sharp bend. If we construct the representation of the flux versus the driving force for this stirred exchange system, we also recognize a transition to the nonlinear branch which is not so abrupt (Fig. 21) as expected from the well known analogical represen-

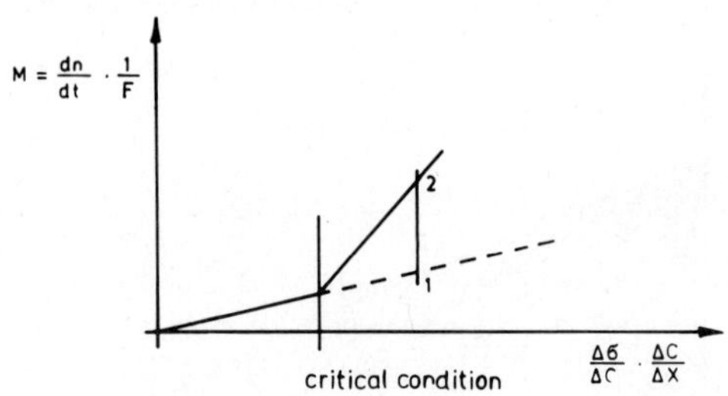

Fig. 21. Mass transfer rates versus the driving force of the Marangoni-instability in a stirred mass exchange cell.

tation of heat transfer without forced convection (Fig. 22). The reason is, that below but near to the critical point we have a simple amplification of the stationary produced disturbances of the forced convection according to $\frac{\Delta\sigma}{\Delta c} \cdot \frac{\Delta c}{\Delta x}$. That is effective as an additional surface renewal. However we have no self-amplification of initial disturbances.

In the case of stability we recognize the opposite effect: the stationary produced disturbances of the forced convection are damped - also according to $\frac{\Delta\sigma}{\Delta c} \cdot \frac{\Delta c}{\Delta x}$ - and the exchange is a little bit slower at the beginning of the exchange experiment. That is the proof of the effect of Marangoni-stability.

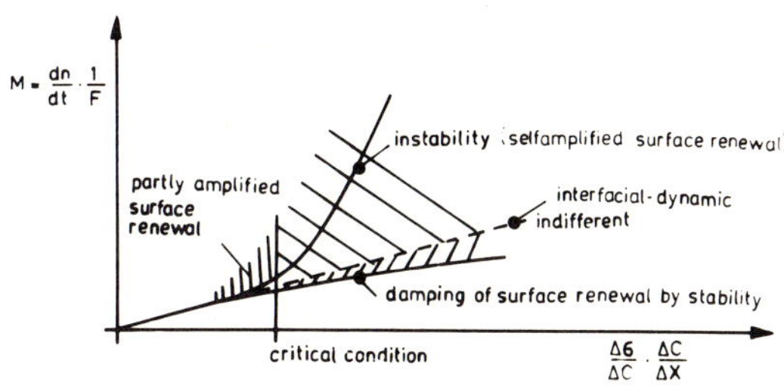

Fig. 22. Mass transfer rates versus the driving force of the Marangoni-instability without forced convection.

A formal approach leads to the equation

$$M = \frac{dn}{dt} \cdot \frac{1}{F} = - K \frac{\Delta C}{\Delta X} = - (K_{lin} + K_{nonlin.}) \cdot \frac{\Delta C}{\Delta X}$$

n = amount of surface active agent, which is transferred across the interphase at the unit of time t

F = exchange area between the phases

ΔC = difference of concentration of the surface active agent

ΔX = thickness of the boundary layer

$K_{lin.} \sim D + A_{\text{turbulent forced convection}}$ describes the exponential part of the mass exchange curve, caused by true diffusion and eddy diffusion of the forced convection (forced surface renewal).

$K_{nonlin.} \sim A_{\text{turbulent simple amplification}} + A_{\text{turbulent self-amplification}}$

describes the nonexponential part of the mass exchange curve. With this $A_{t.\ simple\ a.}$ we can consider on principle the simple positive or negative amplification of disturbances of the

forced convection at the surface.
We estimated in the case of simple positive amplification by an undercritical instability

$$A_{t.\ \text{simple pos. a.}}\ \text{prop.}\left(\frac{\Delta c}{\Delta x}\cdot\frac{\Delta\sigma}{\Delta c}\right)^{1}$$

and in the case of negative amplification (damping) by stability

$$A_{t.\ \text{simple neg. a.}}\ \text{prop.} -\left(\frac{\Delta c}{\Delta x}\cdot\frac{\Delta\sigma}{\Delta c}\right)^{1}.$$

In the case of instability, that means conditions over the critical point and therefore self-amplification of initial disturbances, we estimate

$$A_{t.\ \text{self-amplification}}\ \text{prop.}\left(\frac{\Delta c}{\Delta x}:\frac{\Delta\sigma}{\Delta c}\right)^{2}.$$

The simple positive or negative amplification is an one-step mechanism, represented by the exponent one of the driving force of interfacial dynamics. The self-amplification is a multi-step mechanism with feedback. An iteration of the kind

$$A_{turb.\ N\text{-step of selfamplification}} = \sqrt[2]{\text{const.}\ A_{turb.\ N-1\text{-step of selfampl.}}}\ \frac{\Delta c}{\Delta x}\cdot\frac{\Delta\sigma}{\Delta c}$$

leads directly to

$$A_{turb.\ \text{self-amplification}} = \text{const.}^{2}\left(\frac{\Delta c}{\Delta x}\cdot\frac{\Delta\sigma}{\Delta c}\right)^{2}.$$

This conception can be extended with reflexions about a similar iteration:
the primary disturbance as a shear force L_s leads to a secondary interfacial force L_G, which itself is identical with a secondary shear force and so on with the expression

$$L_G \sim \sqrt{L_S}\ \frac{\Delta c}{\Delta x}\cdot\frac{\Delta\sigma}{\Delta c}.$$

The above mentioned mass exchange kinetic experiments follow this formal approach on principle, as we can see in relation to the effect of instability qualitatively at the following tables. The decrease of $k_{lin.}$ with increasing n is because of the decreasing diffusion coefficients (see also the higher $k_{lin.}$ of the quickly diffusing CH_3CH).

$$C_nH_{2n+1}OSO_3Na \downarrow \frac{C_5H_{11}OH}{H_2O}$$

initial concentration in the upper phase				$K_{lin.} \frac{cm}{min}$	$K_{nonlin.} \frac{cm}{min}$
$\frac{\Delta\sigma}{\Delta c}$	$\frac{\Delta c}{\Delta x}$	n = 10	0,5 %	0,045	0,11
			1,0 %	"	0,44
			1,5 %	"	0,99
	$\frac{\Delta c}{\Delta x}$	n = 14	0,5 %	0,038	0,8
			1,0 %	"	2,4
			1,5 %	"	5,6
		n = 16	1 %	0,033	10

If we compare these experimental results with the results of GOEM, we come to the following conclusion: The roll cell is much more effective in relation to mass transfer than the eddy of the usual surface renewal by the forced convection.

The tangential velocity of the roll cell V_o we can have from the experiment or from the GOEM. [In the heat transfer case, we got the characteristic number N_o from that V_o depends linearly in first approximation

$$N_o = \sqrt{\frac{\tau \cdot \alpha}{\mu_a \cdot 10^2 \cdot 2\pi}} \cdot (K_a \cdot 10^4)^{0,66}$$

α = spatial periodicity number
μ_a = dynamic viscosity of the liquid phase a
K_a = conductivity of temperature of phase a
$\tau = \frac{\Delta\sigma}{\Delta T} \cdot \left(\frac{T_o - T_a}{H_a}\right)$ driving force

$T_{o,a}$ = boundary temperatures of the open system
H_a = thickness of the liquid phase a

The so called Marangoni-number $N_m = \frac{\tau \cdot H_a^2}{K_a \cdot \mu a}$ allowed no correlation between V_o and N_m and is (without the correct coordinate of τ and μ_a) without physical sense.]

If we estimate the additional mass transfer $Z_{rel.}$ from V_o with the penetration theory, there would follow

$Z_{rel.} \sim \sqrt{S}$ S = surface renewal rate

$S \sim V_o$

$V_o \sim \sqrt{\tau}$

$Z_{rel.} \sim \sqrt[4]{\tau}$

The experimental results show for the above systems with instability and forced convection, that

$Z_{rel.} \sim \tau^2$

This extreme contradiction can be discussed, if we keep in mind the complete reversed velocity gradient at the surface, if we compare an eddy, which is forced by a convection from the bulk and a roll cell with a maximum of velocity at the surface. Therefore we have in the latter case extremely smaller and therefore more effective concentration boundary layers in relation to mass transfer.

In the paper of GOEM we have compared the values for Z_{rel} calculated from V_o of the roll cell by the aid of the penetration theory and more physical exactly with the GOEM itself. There we see in special cases, that the GOEM can give values of $Z_{rel,}$ which can be many order of magnitude greater than in the application of the penetration theory.

Also the weak surface active solvent methanol with the initial concentration of 12 % shows on principle the influence of the instability in the same degree like 1 % $C_{10}H_{11}OSO_3Na$.

$$CH_3OH \downarrow \frac{C_5H_{11}OH}{H_2O}$$

12 % CH_3OH inital conc.				$K_{lin.}\ \frac{cm}{min}$	$K_{nonlin.}\ \frac{cm}{min}$
	pure system			0,088	0,34
added surface active agents in the state of equilibrium	$C_nH_{2n+1}OSO_3Na$				
	$\frac{\Delta\sigma}{\Delta c}$	n = 8	0,1%	0,082	0,26
		n = 12	"	0,079	0,18
		n = 16	"	0,059	0,093
		Gardinol	"	0,040	0,075
		Saponine	"	0,035	0,000

If we add the strong surface active agents with increasing surface activity $\frac{\Delta\sigma}{\Delta c}$ in the equilibrium state to the mass exchange system of CH_3OH, we can damp completely the self-amplification: $K_{nonlin.}$ is going to zero. On the other hand also K_{lin} is decreasing: That is the damping influence of the Marangoni-Gibbs-effect on the surface renewal by the forced convection $A_{turb.\ forced\ convection}$. Already here an additional effect is remarkable, if we go to a very high concentration of surface active agents, which are sufficient soluble in both phases: The mass exchange velocity increases again because of the secondary surface renewal by the balance of dynamic ad-

<u>Value of a half exchange τ of $CH_3 \downarrow \frac{C_5H_{11}OH}{H_2O}$ versus the concentration of $C_{16}H_{33}OSO_3Na$ added in the state of equilibrium with the concentration</u> $C_{equ.}$

$C_{equ.}$ [%]	0	0,001	0,01	0,1	0,3	0,6	1	1,5
τ [min]	1,5	1,9	2,3	3,0	3,5	3,5	2,5	2,3

sorption - forced surface convection - desorption.

We can follow these influences of the surface active agents in the equilibrium state also in gas - liquid systems with a stationary transfer of not surface active gases like CO_2 or O_2. In order to get proper experimental conditions for the measurement of the mass transfer coefficients - in this case only K_{lin} - we can direct a jet from the bulk to the plane or gradually deformed but not dispersed surface.

If we show the K_{lin} of CO_2 absorption versus the Re-number of the water jet to a plane surface, we recognize a decrease at the tenth time of K_{lin} with the addition of different surface active agents. Nearly the whole amount of this decrease

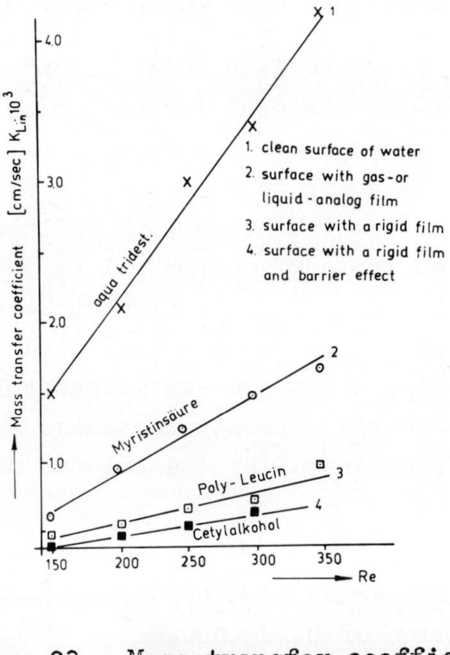

Fig. 23. Mass transfer coefficient of the CO_2 absorption versus Re of the water jet with different films of surface active agents by small deformation of the surface

of K_{lin} is caused by the blockade of surface renewal, only the distance between curve 3 and 4 is a small effect of a direct barrier against the diffusion. At the surface with blocked renewal we have the fundamental balance between shear stress $\mu \frac{\partial v}{\partial x}$ and the gradient of surface tension $\frac{\partial \sigma}{\partial y}$

$$\mu_a \frac{\partial V_a}{\partial x} - \mu_b \frac{\partial V_b}{\partial x} = \frac{\partial \sigma}{\partial y}$$

$$\int \mu \frac{\partial V}{\partial x} \, dy = \int \tau \, dy = \sigma_1 - \sigma_2$$

Already at the curve 2 this balance is completely fulfilled. Between curve 1 and 2 the surface is divided in two parts: one part with renewal and the other part with blocked renewal. Here the balance is not yet fulfilled because of $\int \tau \, dy > \Delta\sigma$, wherein $\Delta\sigma$ is the largest possible difference between the tension of clean water σ_o and the lowest tension σ_{min}, which can be reached with addition of the special surface active agent at all /10/ /11/. The next figure shows, that at a very high level of $\int \tau \, dy$ we got from a trick, using a small area of the surface in order to avoid the dispersion of the liquid - gas system, we can break the balance $\int \tau \, dy = \Delta\sigma$ also at higher concentrations of the surface active agents: K_{lin} will increase again. The crossing of the curves is caused by the effect of secondary surface renewal, which also occurs at a liquid - gas surface in the case of sufficient high concentrations of a soluble surface active agent and a sufficient value of $\int \tau \, dy$.

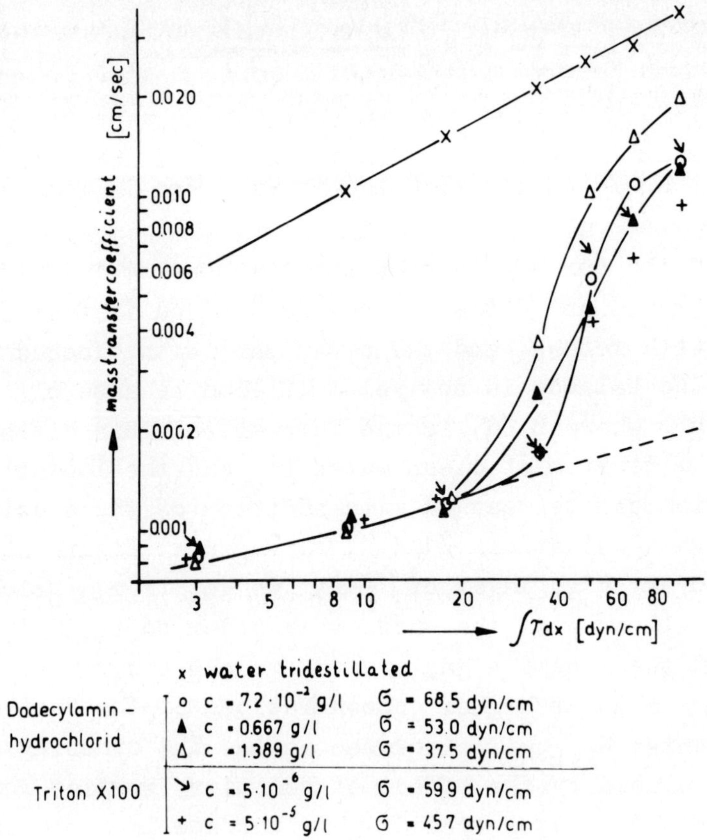

Fig. 24. Mass transfer coefficient of the O_2 desorption from water versus $\int \tau \, dy$

4. A Marangoni-instability analogous to the transition from the laminar flow to turbulence.

The difference in the rate of mass transfer between curve 2 and 3 under the validity of this balance is caused by a special hydrodynamic instability - we call it Marangoni-instability III. This instability needs a sufficient shear stress below the blocked but low viscous surface and is characterized by a dissipative structure of stationary - and at a higher level of instability of oscillatory - shear flow eddies of the surface. This dissipative structure is from its physical origin and its theoretical treatment more analogous to the turbulence.

In order to avoid large scaled additional convections as a disturbance, we need the following onedimensional main-convections at the fluid interphase, which is covered with the adsorptionslayer of the surface active agent to fulfill the balance between shear stress and surface tension gradient :

1. A jet from the liquid phase or from the gas phase is directed vertically to the center of a circular surface /12/.
2. A cylinder is rotating or a band is moving under the surface /13/.
3. Plane Hagen-Poiseuille-flow.
4. Plane Couette-flow.

By the mainconvection produced by a moving band in dependence of an increasing shear stress we can observe:

1. Waves in the plane of the surface.
   This amplitudes of the waves are directed in the direction of the mainconvection.
2. Hairneedlelike countercurrents in the plane of the surface, which can be stationary or instationary.

3. A spatial and temporal periodic structure of oscillation. Fig. 25.

The lineary analysis of stability /14/ for the plane Conette-flow confirms amplificationscoefficients $\hat{\beta} > 0$ for a oscillatory regime $\hat{\beta} = K \cdot U_d$ and a frequency

$$\tilde{\beta} = \pm 2{,}29 \, \frac{\nu}{d^2} \, (K d \, Re)^{1/6}$$

$$Re = \frac{d \cdot U_d}{\nu} \qquad \text{by} \quad \sqrt{K d \, Re} \gg 1$$

$\tilde{\beta} = 2\pi f$ = frequency
ν = kinematic viscosity
d = distance between surface and the moving band
$K = \frac{2\pi}{\lambda}$ = wavenumber
U_d = velocity of the moving band

In the case of the above experiments with $\sqrt{K d \, Re} \approx 100$ the experimental values of the wavenumver were in a sufficient agreement with the theory. This instability occurs also at very small differences in the surface tension ($\Delta\sigma <$ 0,2 dyn/cm) and is observable at natural water-surfaces too.

Concerning the meniscus-instability, which occurs by the propagating of a meniscus relativ to one or to two walls, we refer to our remarks in "Dissipative Strukturen" /16/.

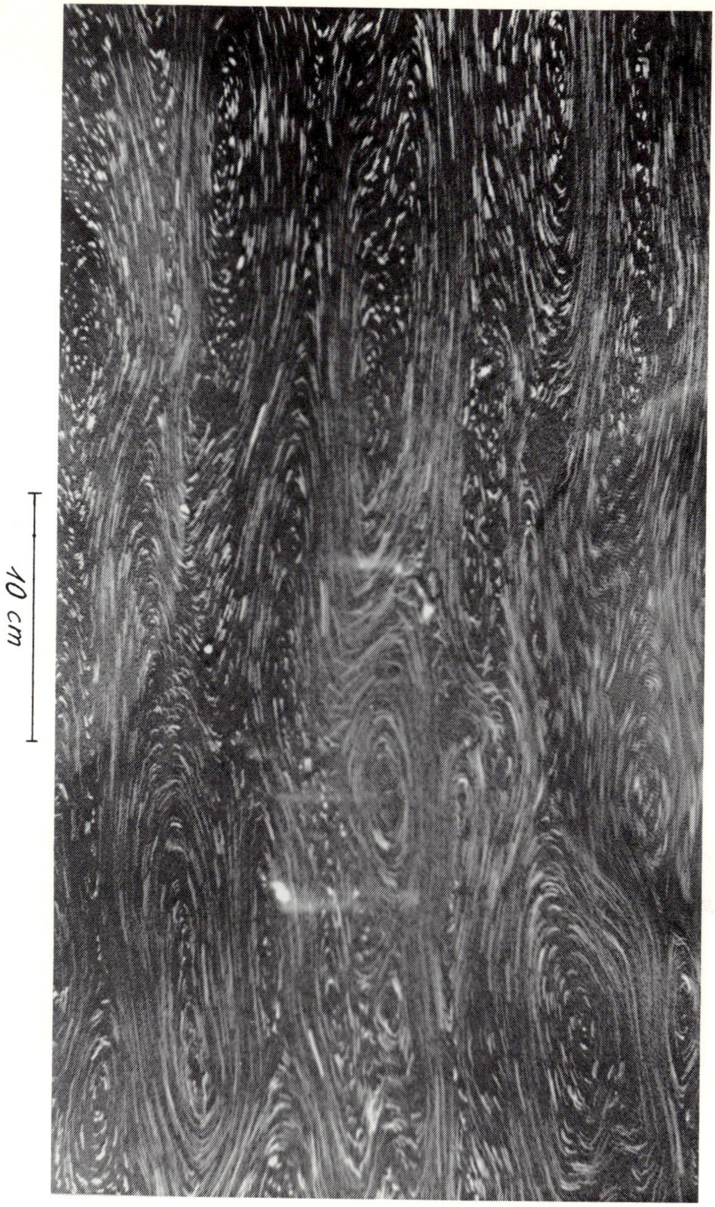

Fig. 25. Turbulence analogous dissipative structure of the Marangoni-instability III.
A moving band with the velocity U = 10 cm/s in a distance H = 1,5 cm to the water surface leads to the one-dimensional shear stress.
Re = 1,5 · 10^3

5. Appendix

The Analytical Interfacial Dynamic Surface Renewal Model (GOEM) /4/ is to complete the Dynamic Surface Renewal Models known until now. The advantage of these known models consists in giving simple mathematical relations between mass- resp. heat transfer and hydrodynamic quantities. The disadvantage of these models is, that they are too general as to give clearly defined explanations concerning hydrodynamic quantities. This is the reason why they cannot help us to explain the influence of mass transfer by the Marangoni-Effect (ME).

The numerical method by H. Wilke /5/ describes the processes of Marangoni-instability caused by ME alone. This same method permits, contrary to the linear theory of stability by Sternling and Scriven /2 /, to calculate approximately the velocities, temperature and the additional heat transfer.

Both methods, GOEM and MIMEG, are based on the following simplifications:

- Only the interfacial tension σ is a linear function of temperature T resp. of concentration c of the transferred mass, the remaining physical parameters being constant.
- A deformation of interface (PHG) in normal direction is not admitted, i.e. the pressure equilibrium at PHG is not considered.
- The existence of spatial periodic solution is presumed. In MIMEG the wave length is taken from Scriven, in GOEM it is governed by the separation of in- and outflow split. The values for check-systems I and III (TS I, TS III) of heat transfer benzophenone, nitrogen are also taken from Scriven /2/.

In MIMEG and GOEM we have the following boundary conditions:

- At the lower and the upper boundary there are fixed boundary temperatures resp. concentrations T_{oa} resp. T_{ob}.
- At PHG ($X = 0$) there has to be the shear stress equilibrium

$$\mu_a \frac{\partial V_a}{\partial X} - \mu_b \frac{\partial V_b}{\partial X} = \frac{d\sigma}{dT} \cdot \frac{\partial T_a}{\partial TX} \quad \text{for } X = 0 \qquad (1)$$

- Continuity of temperature at the PHG resp. a concentration jump corresponding to the equilibrium distributinn coefficient m_{ab} at the interphase

$$T_{-0} = m_{ab} T_{+0} \quad \text{for } X = 0 \text{ with } m_{ab} = 1 \text{ in MIMEG} \qquad (2)$$

- Continuity of mass- resp. heat flow at the PHG

$$\lambda_a \frac{\partial T_a(0,y)}{\partial X} - \lambda_b \frac{\partial T_b(0,y)}{\partial X} = \Sigma \qquad (3)$$

(Σ = source level, in the numerical calculation zero)
(λ = coefficient of thermal conductivity resp. diffusion coefficient)

At the axis of symmetry

- disappearance of velocity component parallel to PHG
- disappearance of derivation

$$\frac{\partial U_a}{\partial y} = \frac{\partial U_b}{\partial y} = 0 \qquad (4)$$

$$\frac{\partial T_a}{\partial y} = \frac{\partial T_b}{\partial y} = 0 \qquad (5)$$

At the lower boundary ($X = H_a$)

$$U(H_a, y) = U_{oa} \sin \alpha y \tag{6}$$

$$V(H_a, y) = 0 \text{ or } \frac{\partial V}{\partial X}(H_a, y) = 0 \tag{7}$$

($\alpha = \frac{2\pi}{S}$ = wave number)

$$U_{oa} \neq 0 \quad \text{in GOEM} \tag{8}$$

$$U_{oa} = 0 \quad \text{in MIMEG} \tag{9}$$

At the upper boundary ($X = H_b$)

$$U(H_b, y) = U_{ob} \sin \alpha y \tag{10}$$

$$V(H_a, y) = 0 \text{ or } \frac{\partial V}{\partial X}(H_a, y) = 0 \tag{11}$$

$$U_{ob} \neq 0 \text{ in GOEM} \tag{12}$$

$$U_{ob} = 0 \text{ in MIMEG} \tag{13}$$

(the same in GOEM)

GOEM looks for stationary solutions of simplified equation of heat transfer resp. equation of diffusion and the equation of motion for incompressible liquids. MIMEG - contrary to GOEM - solves an initial boundary problem, and so one gets additional informationsof temporal development in the process under investigation.

In MIMEG the initial condition (t = 0) is

$$U_a = U_b = V_a = V_b = 0 \tag{14}$$

$$T_a(X_a, y) = T_{oa} - \frac{T_{oa} - T_{ob}}{\frac{\lambda_a}{\lambda_b} H_b + |H_a|} X_a + T_a^S(X_a, y) \tag{15}$$

$$T_b(X_b, y) = T_{ob} - \frac{T_{ob} - T_{oa}}{\frac{\lambda_b}{\lambda_a} |H_a| + H_b} (X_b - H_b) + T_b^S(X_b, y) \tag{16}$$

$T_a^S (X_a, Y)$, $T_b^S (X_b, Y)$ temperature disturbance.

In MIMEG and GOEM there are principally different ways of solving this problem. In MIMEG there is an explicite method of differences, which is based on the SMAC-method by F. H. Harlow and H. A. Amsden /15/ and is adapted to the special case. The numerical stability and exactitude has been verified by numerical experiments.

The results obtained by GOEM and MIMEG have been compared with each other and with experimental results. By this way the range of validity of GOEM can be estimated. For

Reynolds-numbers

$$Re_a = \frac{|V_o| \cdot S}{\mu_a} \leq 5 \qquad \text{liquid phase} \tag{17}$$

$$Re_b = \frac{|V_o| \cdot S}{\mu_b} \leq 500 \qquad \text{gas phase} \tag{18}$$

Péclét-numbers

$$Pe_a = \frac{|V_o| \cdot S}{K_a} \leq 200 \qquad \text{liquid phase} \tag{19}$$

$$Pe_b = \frac{|V_o| \cdot S}{K_b} \leq 0.3 \qquad \text{gas phase} \tag{20}$$

K_a , K_b = temperature conductivity resp. diffusion coefficient of the respective phase.

The results obtained by GOEM represent a sufficient approximation for many cases.

For the MIMEG programme the critical values are in the same order. For MIMEG approximation the exactitude obtained is much higher, but the time of calculation is doubtless much longer.

REFERENCES

1. H. Linde, E. Schwarz and K. Gröger, Chem. Eng. Sci., 1967, 22, 823
2. C. V. Sternling and L. E. Scriven, Amer. Inst. Chem. Engineers J., 1959, 5, 514
3. H. Linde and E. Schwarz, Z. phys. Chem., 1963, 224, 331
4. H. Linde and P. Schwartz, Chem. Techn., 1974, 26, 455, Depot-System Chem. Techn. 1977
5. H. Wilke, Chem. Techn., 1974, 26, 456, Depot-System Chem. Techn. 1977
6. H. Jehring, Nguyen Viet Huyen and E. Horn, J. Electroanal. Chem., 1978, 88, 265

   H. Jehring, H. Linde, G. Wessler and H. J. Krug, Z. phys. Chemie, in preparation
7. H. Linde and K. Loeschcke, Chemie-Ingenieur-Technik, 1966, 39, 65, Mber. Dtsch. Akad. Wiss. Berlin, 1964, 6, 841
8. H. Linde and M. Kunkel, Z. Wärme- und Stoffübertragung, 1969, 2, 60
9. H. Linde and K. Winkler, Z. phys. Chem. , 1965, 230, 207

   H. Linde and D. Thiessen, Z. phys. Chem., 1962, 221, 97 Mber. Dtsch. Akad. Wiss. Berlin, 1962, 4, 710
10. P. Friese, H. Linde and K. Zacharzowsky, Z. phys. Chem., in press, 1978
11. H. Linde and P. Friese, Teor. Osn. Chim. Techn., 1977
12. H. Linde and N. Shulewa, Mber. Dtsch. Akad. Wiss. Berlin, 1970, 12, 883
13. H. Linde and P. Friese, Z. phys. Chem., 1971, 247, 225
14. H. Linde and P. Schwartz, Teor. Osn. Chim. Techn., 1971, 401
15. A. A. Amsden and F. H. Harlow, Los Alamos Sci. Lab., 1971, Report LA-4700
16. E. Kahrig and H. Beßerdich, Fortschritte der experimentellen und theoretischen Biophysik Bd.21, VEB Georg Thieme Leipzig

FORMATION AND THINNING OF LIQUID FILMS

by

I.B. Ivanov*

School of Engineering and Applied Science
Columbia University, New York, New York 10027, USA

and

R.K. Jain

Department of Chemical Engineering
Carnegie-Mellon University, Pittsburgh, Pennsylvania 15213, USA

1. Introduction

When the distance between two bubbles or drops approaching along their line of centres becomes small, their surfaces deform and a draining liquid film appears which either ruptures further on or reaches an equilibrium thickness.[1] The main stages of formation and evolution of the liquid film are shown in Fig. 1.: a) mutual approaching of two slightly deformed bubbles under the action of an external driving force F; b) at small distances the caps of the bubbles are deformed until their curvature changes sign and a cupped-shaped formation, called "dimple," of radius r_d arises; c) the dimple height initially increases but further on it either decreases or the dimple is pushed out and eventually a thinning quasi plane-parallel film of radius R forms (if a long-range positive disjoining pressure Π is present, an equilibrium first film can form); d) when Π is negative, it favours the growth of the corrugations at the film surfaces and at a critical thickness h_{cr} either rupture of the film or formation of thinner black spots occurs; e) the spots increase their radius R^s (or coalesce) until eventually covering the whole initial thicker film; f) at this moment the mechanical equilibrium between the film and the bubble (or the meniscus) is violated and expansion of the black film begins. Finally an equilibrium film, whose lifetime can be unlimited, forms. Analogous is the manner in which the processes occur when a small bubble approaches a rigid surface (wetting film), when two droplets approach each other (emulsion film), etc. The goal of the present paper is to give an outline of the general procedure.

---

*On leave from the University of Sofia, Bulgaria

In all the above processes the main dissipation of energy in the film takes place in a region of radius R_f and maximum thickness $H_f = H(R_f)$. Since at this $\epsilon = (H_f/R_f)^2 \ll 1$, the flow in this region is governed by the lubrication theory equations. In vectorial form they read*:

$$\text{a)} \quad \nabla_s p = \mu\, \partial \vec{v}_s / \partial z^2 \quad ;$$
$$\text{b)} \quad \partial p/\partial z = 0 \quad ; \qquad (1)$$
$$\text{c)} \quad \nabla_s \cdot \vec{v}_s + \partial v_z/\partial z = 0 \quad ,$$

where p is pressure, μ-dynamic viscosity, ∇_s and $\vec{v}_s$-the Hamilton operator and the velocity component in a plane z = const (see Fig. 1a), respectively, v_z is the z-component of the velocity and the dot denotes scalar product. At $r > R_f$ it can be assumed that all quantities have their equilibrium values.[2] For a plane-parallel film (see Fig. 1c) $R_f \equiv R$; in the other cases we shall use the limiting value $R_f \to \infty$, corresponding to $\epsilon \to 0$. The kinematic boundary condition, which has to be used in solving (1), is the condition for a surface to be a bounding one.[3] In lubrication approximation for two different bubbles (or droplets) A and B, it will read (at $z = H^A$ or $z = H^B$):

$$v_z = \frac{\partial H^{A,B}}{\partial t} + \vec{v}_s^{\,A,B} \cdot \nabla_s H^{A,B} \, , \qquad (2)$$

where t is time, $\vec{v}_s^{\,A,B} = \vec{v}_s \Big|_{z=H^{A,B}}$ is the surface velocity and H^A or H^B is the equation of the respective surface (B is the lower bubble). The solution of (1) along with (2) leads to a general differential equation, describing all stages shown in Fig 1:

$$- \frac{\partial H}{\partial t} = \nabla_s \cdot \left[\frac{H}{2} (\vec{v}_s^{\,A} + \vec{v}_s^{\,B}) - \frac{H^3}{12\mu} \nabla_s p \right] \, , \qquad (3)$$

where $H = H^A - H^B$ is the local thickness. For axisymmetric flow (3) reduces to[4]

*Only the axisymmetric case will be treated in any details. Whenever possible, the basic equations will be formulated in vectorial form to account for the possible asymmetry of the flow.

$$-\frac{\partial H}{\partial t} = \nabla_r \left[\frac{H}{2} (v_r^A + v_r^B) - \frac{H^3}{12\mu} \frac{\partial p}{\partial r} \right] , \qquad (4)$$

where $\nabla_r = \frac{1}{r}\frac{\partial}{\partial r} r$ and v_r is the radial velocity.

For the sake of brevity further on we shall mainly consider systems symmetrical with respect to the plane z = 0 and the boundary conditions will be formulated only for the upper surface z=H/2 (see Fig. 1b). Their generalization for non-symmetric systems is straightforward. When solving (3) and (4) we shall use the quasi-steady assumption:[5,6]

$$\frac{\partial H}{\partial t} = \frac{\partial H}{\partial h}\frac{dh}{dt} = - V \frac{\partial H}{\partial h} , \qquad (5)$$

i.e. we shall suppose that H depends on t only implicitly through the thickness $h = H|_{r=0}$ and the velocity of thinning V = - dh/dt at the film axis. The surface velocity depends on the surfactant properties and concentration and on the flow in the droplets (for emulsion films). This problem is considered in detail in Section 3.

When the film surfaces are non-deformable, $\partial H/\partial h = 1$, and (3) or (4) can be integrated to give p. The driving force, F, is then calculated from the force balance[4,5] at the film surfaces*:

$$F = \int_0^{2\pi} \int_0^{R_f} (p - p^1) r d\phi dr , \qquad (6)$$

where p^1 is the pressure in the quiescent liquid at $r \geq R_f$. With deformable interfaces p in (3) and (4) is expressed through the normal dynamic boundary condition giving the jump Δp_{nn} of the normal component p_{nn} (see Fig. 1b) of the stress tensor. When $\Pi = 0$ it reads

$$\Delta p_{nn} \approx p^* - p = \frac{\sigma_o}{2} \nabla_s^2 H , \qquad (7)$$

where p^* is the pressure in the bubble and σ_o is the surface tension of the bulk liquid. Thus H(r,h) is derived, and from (6) and (7) F is calculated. This scheme is easily modified for some particular models considered below. For example, with a plane-parallel film the condition

*It is tacitly assumed that the driving force acting on either bubble is the same.

$$p = p^{\ell} + \Pi \qquad \text{at} \quad r = R$$

has to be used to account for the jump in pressure between the film and the bulk liquid due to the disjoining pressure. For a more detailed discussion of the latter equation as well as of other thermodynamic relations used below, see reference 7. For the theory of the van der Waals' disjoining pressure, see reference 8.

2. FILMS WITH TANGENTIALLY IMMOBILE INTERFACES

The assumption for tangential immobility of the film surfaces is often used because it considerably simplifies the theoretical treatment. However, it is never strictly correct and is only approximately fulfilled at high surfactant concentrations (see Section 3a). The first solutions of Eqs. (1) were obtained for the velocities of thinning V_{Re} and V_T of a film, confined between two solid parallel discs[9] of radii R, and between a solid sphere of radius R_c and an infinite solid plane[10], respectively:

$$\text{a)} \quad V_{Re} = \frac{2h^3 F}{3\pi\mu R^4} \quad ; \qquad \text{b)} \quad V_T = \frac{Fh}{6\pi\mu R_c^2} \quad . \tag{8}$$

2a. Formation and Thinning of Films with Deformable Interfaces

This problem was treated by many authors (see e.g. references 11-13). The variety of their results reveal that the film profile and the velocity of thinning may greatly differ depending on the bubble radius, film thickness, surface velocity, etc. That is why we have confined ourselves to the simpler case of small bubble radii when the dimple is smaller or does not appear at all. The thinning of films with large dimples will be treated in Section 2b using a simplified model.

In order to investigate also the influence of the surface mobility we shall first consider a bubble approaching a solid plane ($H^A = H$, $H^B = 0$, $v_r^B = 0$). For this system, Eq. (4) along with (7) takes the form:

$$\frac{3m^2\mu V}{\sigma_o} \frac{\partial H}{\partial h} = \nabla_r \left[H^3 \frac{\partial}{\partial r} (\Delta_r H) \right] \quad ; \qquad \Delta_r = \nabla_r \frac{\partial}{\partial r} \quad , \tag{9}$$

where m = 1 for freely moving bubble surface and m = 2 for both surfaces tangentially immobile. The velocity of thinning in the latter case is denoted by V_o. Equation (9) was solved[4] by matching the coordinate expansions for great and small values of r. By introducing the dimensionless thickness $\tilde{h} = 4\pi\sigma_o h/F$ the solution can be written in two asymptotic forms, valid for $\tilde{h} > 1$:

$$H = h + r^2/2R_c - \frac{F}{4\pi\sigma_o} \ln\left(1 + \frac{r^2}{2R_c h}\right) \quad ; \tag{10}$$

$$V = 2Fh/3m^2\pi\,\mu R_c^2 \quad , \tag{11}$$

and for $\tilde{h} < 1$:

$$\begin{aligned} H = h + r^2/2R_c - \frac{F}{4\pi\sigma_o} \ln\left(1 + \frac{\pi\sigma_o}{FR_c} r^2\right) - \\ - \frac{r^2}{4R_c} (1 + \pi\sigma_o r^2/FR_c)^{-1} \quad ; \end{aligned} \tag{12}$$

$$V = 8\pi\sigma_o^3 h^3/m^2\,\mu F R_c^2 \quad . \tag{13}$$

The results for $\tilde{h} > 1$ were also derived[5] by the method of parametric expansion with respect to the small parameter $\alpha = F/4\pi\sigma_o R_c$. The same method was applied to a system consisting in two different bubbles A and B of radii R_c^A and R_c^B and surface tensions σ_o^A and σ_o^B with tangentially immobile surfaces (m = 2). By introducing

$$\text{a)}\ \ \bar{R}_c = R_c^A R_c^B/(R_c^A + R_c^B) \quad ; \qquad \text{b)}\ \ \bar{\sigma}_o = \sigma_o^A \sigma_o^B/(\sigma_o^A + \sigma_o^B) \tag{14}$$

the results for the film thickness H and the velocity of thinning V_o can be written in the form:

$$H = h + r^2/2\bar{R}_c - (F/4\pi\bar{\sigma}_o) \ln (1 + r^2/2\bar{R}_c h) \tag{15}$$

$$V_o = Fh/6\pi\,\mu \bar{R}_c^2 \quad . \tag{16}$$

The second terms in (10) and (15) account for the bubble deformation. At a given thickness $h = h_i$ the curvature at $r = 0$ changes its sign and the dimple forms. The thickness of formation h_i and the dimple radius r_d are calculated from the conditions $\partial^2 H/\partial r^2 = 0$ at $r = 0$ and $\partial H/\partial r = 0$, respectively. The dimple radius, r_d, increases very quickly to the equilibrium film radius R. The following equations were derived for h_i and R:

$$a)\ h_i = F/4\pi\bar{\sigma}_o \ ; \qquad b)\ R^2 = F\bar{R}_c/2\pi\bar{\sigma}_o \ . \tag{17}$$

It should be emphasized that (17a) corresponds to $\tilde{h} = 1$, i.e. at $h = h_i$ the parametric expansion fails. That is why Eq. (17a) is an estimate rather than an exact formula. It still gives the same dependence of h_i on R_c ($h_i \sim F \sim R_c^3$) as observed experimentally by McKay and Mason[14], although the experimental values are lower than the theoretical ones. Equation (17b), which is a general expression for the equilibrium radius of contact between two bubbles (with zero contact angles), yields with suitable choice of $\bar{R}_c$ and $\bar{\sigma}_o$ all known particular cases[11,15,16]. Another important conclusion, following from Eqs. (10,12,15), is that the film profile depends only on thermodynamic parameters — indeed, neither of these equations contains μ and m. This is in agreement with Roberts' experimental results.[17]

The case $\tilde{h} < 1$ is very important because it corresponds to the conditions under which the experimental studies of thin films are usually carried out.[1] The analysis of Eq. (12) shows that in a fairly extensive region, lying at (see (17b) with $\bar{R}_c/\bar{\sigma}_o = R_c/\sigma_o$)

$$0 < r < (FR_c/2\pi\sigma_o)^{1/2} = R \tag{18}$$

the film is nearly plane-parallel. On the basis of computer calculations, Buevich and Lipkina[13] arrived to a similar conclusion. The velocity of thinning V_o (see (13) with m = 2) of this part of the film can be written in the form of the Reynolds' equation (8a) if R_c is eliminated by means of (18):

$$V_o = h^3F/2\pi\mu R^4 \ . \tag{19}$$

The coincidence of (19) with (8a) (there is only a small difference in the numerical factors) confirms the applicability of the model of the plane-parallel film, established in many experiments of Scheludko, Exerowa et al.[1]

2b. Thinning of a Wetting Film with Dimple

The problem of the evolution of the dimple was previously treated by Frankel and Mysels.[18] We shall use now a simplified model which will enable us to obtain the results in closed analytical form. We assume that all the energy dissipation takes place in the "barrier" ring with thickness h shown on Fig. 2. The liquid pressures outside the ring, p^{ℓ}, and inside it, p^{i}, are constant. The driving force is given by

$$F = 2\pi \int_0^{R_i} (p^i - p^{\ell}) r\,dr + 2\pi \int_{R_i}^{R} (p - p^{\ell}) r\,dr \ . \tag{20}$$

Then the velocity of thinning, V_o (for tangentially immobile surfaces) is

$$V_o = 2h^3 \Delta P / 3\mu R_{eff}^2 \ ; \tag{21}$$

with

$$R_{eff}^2 = (R^2 + R_i^2) - (R^2 - R_i^2)/\ln (R/R_i) \ , \tag{22}$$

$$\Delta P = P_c \left[\frac{1}{1 - (R_i/R)^2} + \frac{1 + R_c/R_d}{\ln (R_i/R)} \right] - \Pi \ , \tag{23}$$

where ΔP is the driving pressure, $P_c = 2\sigma_o/R_c$ is the capillary pressure of the bubble and $R_d \approx 2R_i^2/(h_d - h)$ is the radius of curvature of the dimple, assumed to be the cap of a sphere. The maximum elevation of the dimple, h_d, above the rigid surface can be calculated from[18] $h_d = 0.285\,(R + R_i)\sqrt{h/R_C}$. With $R_i = 0$, Eq. (21) reduces to Reynolds' equation (8a). Because of the pressure difference $p^i - p^{\ell} = P_c(1 + R_c/R_d)$, under most circumstances V_o will be less than v_{Re}.

2c. Expansion of a Black Spot in a Thicker Film

The black spot, represented in Fig. 1e, was considered as a piston, expanding in r-direction with velocity V_r under the action of a force F_r due both to the difference in tensions $\bar{\Delta} = \Delta - \Delta^s \approx 2\sigma_o - \Delta^s$ (Δ and Δ^s are the film tensions of the thicker film and the black spot, respectively), and to the spot line tension $\varkappa^s$: $F_r = 2\pi R^s(\bar{\Delta} - \varkappa^s/R^s)$. At the same time, the thicker part of the film is thinning with velocity V_o because of the driving force $F = \pi\,[R^2 - (R^s)^2]\Delta P$. The hydrodynamic expressions for these forces were derived, respectively, from $F_r = 2\pi R^s \int_0^h p_{rr}\,dz$ and Eq. (6) with lower limit $r = R^2$ instead of 0. The general expressions for V_o and V_r are given in [19]. An interesting particular case of them is obtained by putting $V_o = 0$ and $\varkappa^s = 0$. Then $V_r = [h/12\mu R^s \ln (R/R^s)]\bar{\Delta}$. This result is in agreement with the proportionality between V_r and $\bar{\Delta}$, found experimentally by Kolarov et al[20]. In the general expressions the influence of a centrifugal field with angular velocity ω was accounted for. For the velocity of thinning of a plane-parallel film without spot ($R^s = 0$), formed in a biconcave meniscus with radius of curvature R_c, we obtained:

$$V_o = (2h^3/3\mu R^2)(\Delta P + \rho\omega^2 R_c^2/4) \quad , \qquad (24)$$

where ρ is the liquid density and $\Delta P = P_c - \Pi$. Due to the term with ω in (24) the velocity V_o can be considerably higher than the velocity of thinning in the absence of a centrifugal field, which is in agreement with Vrij's experiments[21].

2d. Expansion of a Black Film

Let us consider here for simplicity only the case of a film formed in a meniscus which is completely wetting the tube whose radius is R_c. When the expansion of a black film is treated (see Fig. 1f), considerable difficulties arise if the film is a bimolecular membrane* (Newton's film). That is why we assume that the film contains an equilibrium liquid core of constant thickness h_2. We also assume that the contact angle always has its equilibrium value: θ_1 when the meniscus is in contact with a thicker film of radius R_1 and $\theta_2 > \theta_1$ for a thinner film at any radius $R > R_1$. When the spots have covered the whole thicker film the contact angle rises to θ_2, the capillary pressure, $P_c = 2\sigma_o(R_c - R\sin\theta)/(R_c^2 - R^2)$ abruptly drops below its initial value $P_c(R_1,\theta_1)$ and under the action of the driving pressure $\Delta P_c = P_c(R_1,\theta_1) - P_c(R,\theta_2)$ expansion, with velocity $V_r = dR/dt$, occurs. The final equilibrium radius, R_2, is determined by the condition $P_c(R_1,\theta_1) = P_c(R_2,\theta_2)$. For the process under consideration the quasi-steady assumption reads (compare with (5)):

$$\partial H/\partial t = (\partial H/\partial R)(dR/dt) = V_r(\partial H/\partial R) \quad .$$

Hence, the governing equation for a film on a solid substratum will follow from (9) if V_r and $-\partial H/\partial R$ are substituted for V and $\partial H/\partial h$. With m = 1 this equation is also valid for a symmetrical film (as shown in Fig. 1f)

---

*The same problems arise when treating the motion of a three-phase line. They are due to the use of the non-slip boundary condition[22] and can be removed by giving up this condition.

with tangentially immobile surfaces. The integration of the governing equation with the approximations $tg\theta_2 \approx \theta_2$ and $(r - R)\theta_2 >> h_2$ yields:[23]

$$H = \left[\theta_2 - \frac{3m^2 \mu V_r}{\sigma_o} \ln \frac{\theta_2(r-R)}{h_2}\right](r - R) \quad . \tag{25}$$

Equation (25) contains only measurable quantities and can be checked experimentally. For many purposes it is convenient however to use the relation between F and V_r. The respective equations are derived and discussed in reference 23.

3. FILMS WITH TANGENTIALLY MOBILE INTERFACES

The most important effect in the hydrodynamics of real foam, wetting and emulsion films is the mobility of their interfaces. According to Levich[24], the effect of surface mobility is accounted for by the tangential boundary condition (at z=H/2) for the component p_{sn} (see Fig. 1b) of the stress tensor. In the lubrication approximation and with axisymmetric flow this condition reads*

$$\mu \frac{\partial v_r}{\partial z} = \mu^* \left(\frac{\partial v_r^*}{\partial z} + \frac{\partial v_z^*}{\partial r}\right) + \frac{\partial \sigma}{\partial r} + \mu_s \frac{\partial}{\partial r}(\nabla_r v_r) \quad , \tag{26}$$

where μ_s is the sum of the coefficients of surface dilational and shear viscosities and the asterisks denote quantities, referring to the drop (or the bubble). The terms with μ and μ^* account for the viscous friction in the film and in the drop, $\partial\sigma/\partial r$ is the surface force, arising from the perturbation of the surfactant repartition at the surface, and the term with μ_s accounts for the viscous friction inside the surfactant monolayer. In order to determine v_r^* and v_z^* the full Navier-Stokes' equation

$$\frac{\partial \vec{v}^*}{\partial t} + (\vec{v}^* \cdot \nabla)\vec{v}^* = - \frac{1}{\rho^*} + \nabla p^* \quad \nu^* \nabla^2 \vec{v}^* \tag{27}$$

($\nu^* = \mu^*/\rho^*$) must be solved for the drop. The surfactant repartition is governed by the convective diffusion equation

$$\frac{\partial c}{\partial t} + (\vec{v} \cdot \nabla)\, c = D\nabla^2 c \quad , \tag{28}$$

---

*More general formulations and/or discussion of the dynamic tangential boundary condition can be found in reference 25.

where c is concentration and D-diffusion coefficient of the surfactant. In the film, where Peclet's number, $\mathcal{P}e$, is usually small, (28) simplifies to

$$\nabla_s^2 c + \partial^2 c/\partial z^2 = 0 \quad . \tag{29}$$

In the drops $\mathcal{P}e \gg 1$ and the diffusion occurs across a diffusion boundary layer of thickness $\delta_D \ll R$. This allows neglecting the radial derivative in the term $\nabla^2 c^*$ in (28).

The surface convective flux j_{sc} caused by the entrainment of the surfactant by the liquid flow, perturbs its uniform repartition. The system tends to restore the equilibrium by the surface diffusion flux $j_{sd} = -D_s \nabla_s \Gamma$ (D_s is surface diffusivity and Γ-surface concentration) and the normal fluxes j_n from the bulk of the film and the drop onto the film surface. The condition for surfactant conservation is expressed by the balance of these fluxes.[24,26] The normal flux can be represented as composed by two consecutive stages: diffusion flux j_d from the bulk to the subsurface and adsorption flux j_a from the subsurface onto the surface. We were able to show that for the film $j_a/j_d \sim \alpha_a R^2(\partial\Gamma_o/\partial c_o)/Dh$, where the subscript "o" denotes equilibrium value and α_a is defined by[24] $j_a = -\alpha_a (\Gamma - \Gamma_o)$. For the surfactants commonly used $j_a/j_d \gg 1$, i.e. the surfactant transfer is diffusion controlled. Therefore, we shall discuss hereafter only this case. In lubrication approximation Eq. (29) yields $\partial c/\partial z)_{z=H/2} = -\frac{H}{2}\nabla_s^2 c$. Assuming also that $(\Gamma_o - \Gamma)/\Gamma_o \ll 1$, we can write the condition for conservation of a surfactant, soluble only in the film, in the form (at z = H/2):

$$\Gamma_o v_s - \left[D_s(\partial\Gamma_o/\partial c_o) + \frac{DH}{2}\right]\nabla_s c = 0 \quad . \tag{30}$$

In (30) the term $\partial\Gamma/\partial t$, whose relative contribution is $(\Gamma_o - \Gamma)/\Gamma_o \ll 1$, was neglected.

When the surfactant is soluble only in the drops, the contribution of the surface diffusion is* $h\delta_D/R^2 \approx 10^{-4}$ times smaller than in (30) and

*All the estimates here and hereafter are carried out with the typical values (see the notations in the text): $\sigma_o = 50$ dyne.cm^{-1}, $\bar{\Delta} = 0.1$ dyne.cm^{-1}, $P_c = 10^3$ dyne.cm^{-2}, $R = 10^{-2}$ cm, $R_c = 10^{-1}$ cm, $h = 10^{-5}$ cm, $\mu = \mu^* = 10^{-2}$ g.cm^{-1}.s^{-1}, $D = 10^{-5}$ cm^2.s^{-1}, $D_s = 10^{-4}$ cm^2.s^{-1}.

can be disregarded. For this system (30) is replaced by

$$\nabla_s(\Gamma \vec{v}_s) = D^*(\partial c^*/\partial z) \quad . \tag{31}$$

3a. Thinning of Foam and Wetting Films

As far as film thinning is concerned the mobility of the interfaces is by far much more important than their deformation. Indeed, the deformation of the film surfaces alters the velocity of thinning by some 20-30% with respect to V_{Re} (see Section 2a), while surface mobility can increase it by several orders of magnitude.

For a plane-parallel film the method described above leads to (with $\mu_s = 0^*$)[27,28]

$$V/V_{Re} = (1 + \varepsilon^f)/\varepsilon^f \quad , \tag{32}$$

where

$$\varepsilon^f = - \frac{\Gamma_o(\partial\sigma_o/\partial c_o)}{3\mu D[1 + 2D_s(\partial\Gamma_o/\partial c_o)/Dh]} \quad , \tag{33}$$

and V_{Re} is given by (8a). The result for a wetting film without dimple reads

$$V/V_{Re} = (4 + 3\varepsilon^w)/(1 + 3\varepsilon^w) \quad , \tag{34}$$

where

$$\varepsilon^w = - \Gamma_o(\partial\sigma_o/\partial c_o)/3\mu D[1 + D_s(\partial\Gamma_o/\partial c_o)/Dh] \tag{35}$$

Equation (34) holds also for a wetting film with dimple provided that V_o from (21) is substituted for V_{Re} in its left-hand side. It follows from (33-35) that with soluble surfactants $V > V_{Re}$. A very interesting conclusion, suggested by the above results, is the strong influence on V of the surface diffusion, which should be observed at relatively small thickness: $h \leq 2D_s(\partial\Gamma_o/\partial c_o)/D$. In the limit $h \to 0$ (32) and (33) yield:

$$V/V_{Re} = 1 - 6D_s\mu/h\Gamma_o(\partial\sigma_o/\partial\Gamma_o) \quad . \tag{36}$$

Equation (36) was corroborated experimentally in[29]. The same limiting expression is obtained at $h \to 0$ with adsorption controlled surfactant transfer (unpublished results) which means that in this case the mechanism

---

*The relative contribution of the surface viscosity term (see (26)) is of the order of $\mu_s h/\mu R^2$ and can be disregarded with $\mu_s < 0.1$ s.p.

of supplying the surfactant onto the surface is immaterial. The case of tangentially immobile surfaces formally corresponds to ε^f, $\varepsilon^w \to \infty$. For soluble surfactants it is approximately realized at high concentrations and great surface activity.

When the term with the surface viscosity in (26) can be neglected, it is possible to obtain quite general expressions for $v_r^{A,B}$ at $z = H^{A,B}$ even for films with deformable interfaces. Upon substituting these expressions in (4) it takes the form

$$\frac{\partial H}{\partial t} = \nabla_r \left[(1 + Q) \frac{H^3}{12\mu} \frac{\partial p}{\partial r} \right] , \tag{37}$$

where

$$Q = \frac{4 + 3(E^A - E^B)(\Gamma_o^A + \Gamma_o^B)}{(E^A + E^B)(\Gamma_o^A + \Gamma_o^B)} , \tag{38}$$

$$E^{A,B} = \frac{\dfrac{\partial\sigma_o^{A,B}}{\partial c_o}}{3D\mu} \left[1 + \left(D_s^A \frac{\partial\Gamma_o^A}{\partial c_o} + D_s^B \frac{\partial\Gamma_o^B}{\partial c_o} \right) \Big/ DH \right]^{-1} . \tag{39}$$

When the surface diffusion is negligible $E^{A,B}$ reduces to $E_\infty^{A,B} = -(\partial\sigma_o^{A,B}/\partial c_o)3D\mu$ and does not depend on H. Then (37), (5) and (7) lead to

$$\frac{12\mu V}{\sigma_o^A(1+Q_\infty)} \frac{\partial H}{\partial h} = \nabla_r \left[H^3 \frac{\partial}{\partial r} (\Delta_r H^A) \right] \tag{40}$$

The solution of this equation, which is a generalization of Eq. (9), is again Eq. (15), i.e. in this case too the film shape does not depend on the surface mobility. The velocity of thinning, V, at r = 0, is

$$V/V_o = 1 + Q_\infty \tag{41}$$

where V_o is defined by (16). The last equation is a generalization of Taylor's equation (8b). It accounts for the surface mobility, the surface tension and the size of both bubbles. The case of both surfaces tangentially immobile corresponds to Q = 0. The equations for only one surface, B, tangentially immobile can be derived in the same way. They can be obtained also by putting in (38) and (39), $\Gamma_o^B = 0$ and $E^B = E^A - 2/\Gamma_o^A$ which would ensure $v_r^B = 0$. We were not able, however, to find any physical background for this formal transition.

The comparison of (41) with (32) reveals that in this case the relative increase of the velocity of thinning with respect to its value, V_o, for tangentially immobile surfaces is the same as for a plane-parallel film, i.e. the model of the plane-parallel film correctly accounts for the influence of the surface mobility on the velocity of thinning.

Equation (32) fails as $c_o \to 0$, since then $\varepsilon^f \to 0$ and $V \to \infty$. The main reason for this is the failure of the lubrication approximation. Indeed, for tangentially immobile surfaces, Eqs. (1a) and (1b) follow from (27) at small Reynolds' numbers and $\varepsilon = (H_f/R_f)^2 \to 0$, i.e. these equations are zero-order approximations with respect to ε. At this all terms of the order of ε are neglected, e.g. $\partial^2 v_r/\partial r^2$. When the surfaces are mobile, however, the ratio $(\partial^2 v_r/\partial r^2)/(\partial^2 v_r/\partial z^2)$ is of the order of $\varepsilon(V/V_{Re})$. Hence, the radial derivative of v_r can become significant at $V/V_{Re} \gg 1$ even if $\varepsilon \ll 1$. This shortcoming of the theory can be avoided by expanding in series with respect to ε Eqs. (27,28) and the appropriate boundary conditions. By keeping the linear in ε terms, a system of differential equations (which is a generalization of the classical lubrication theory equations) was derived.[30] Applied to a plane-parallel film these generalized equations yield (cf. (32))

$$V/V_{Re} = (1 + 1/\varepsilon^f)/(1 + 4\varepsilon/3\varepsilon^f) \qquad (42)$$

and to the process of mutual approaching in a pure liquid of two bubbles of radii R_c:

$$F = 2\pi\mu R_c V \ln (R_c/h) \quad . \qquad (43)$$

For $\varepsilon^f \to 0$ (i.e. $c_o \to 0$) Eq. (42) yields $V/V_{Re} = 3/4\varepsilon$. Neither this equation nor (43) could be obtained from (1).

3b. Thinning of Emulsion Films

The main difficulty in the hydrodynamic theory of emulsion films arises from the coupling of the flows in the film and in the drops. This effect makes this theory much more involved than the theory of foam films and the first theoretical works in this area appeared only in the last few years.[31,32] In the first part of this section we shall use the model of the plane-parallel film. In the second part the thinning of an emulsion film, formed between two spherical non-deformable drops, will be considered. This is a reasonable model for real emulsions with very small droplets.

We look for a solution of Eq. (27) of the form $v_r^* = Uf(\eta)$ where U is the radial surface velocity and $\eta = (z - h/2)(U/r\ell^*)^{1/2}$. Then (27) leads to the following equation for f:

$$f'''f' - (f'')^2 + f''f^2 + [2ff'' - 3(f')^2]\,\varkappa^e = 0 \qquad (44)$$

where $\varkappa^e(h) \approx 1 - \partial \ln V/\partial \ln h$. The results of the numerical solution of (44) and the values of $a_1 = (df/d\eta)_{\eta=0} < 0$ are presented in [33]. In order to simplify the solution, let us assume $\mu_s = 0$ and consider only two limiting cases: surfactant soluble either in the film (the continuous phase) or in the drops. For the former case we have[34]:

$$\text{a)}\ \ U = Vr/2h(1+\varepsilon^e+\varepsilon^f)\ , \qquad \text{b)}\ \ V/V_{Re} = 1 + 1/(\varepsilon^e+\varepsilon^f) \qquad (45)$$

where the parameter

$$\varepsilon^e = -\,a_1\mu^* h(U/r\ell^*)^{1/2}\Big/6\mu \qquad (46)$$

accounts for the energy dissipation in the drops. For the latter case the solution yields[34] $c^* = c^*(z)$, i.e. $\partial c^*/\partial r = 0$. Therefore, $\partial\sigma/\partial r$ in (26) vanishes and with respect to the film thinning the system behaves as an emulsion system without surfactant. Then V is again given by (45b), but with $\varepsilon^f = 0$.

The numerical estimate of ε^f and ε^e shows that for surfactants of moderate surface activity $\varepsilon^f \approx 1$ and $\varepsilon^e \approx 10^{-5}$. We see, therefore, that the emulsion systems can behave in two wholly different ways, depending on the solubility of the surfactant: first, in the case of a surfactant soluble in the film, the velocity of thinning is somewhat higher but still of the order of Reynolds' velocity ($V \approx V_{Re}/\varepsilon^f$), and is mainly determined by the properties and concentration of the surfactant: second, in the case of a surfactant soluble in the drops the velocity of thinning is much greater than Reynolds' velocity ($V \approx V_{Re}/\varepsilon^e$) and is entirely independent on the presence of surfactants. This effect, which was confirmed experimentally in [35], seems to be related to the Bancroft's rule and its explanation, as given by Davies[36]. Our results indicate that, at least in the case of the droplets' coalescence being preceded by the formation of a thin liquid film, the hydrodynamic factors for direct and reverse emulsions in Davies' theory may differ substantially and cannot be equated, as Davies did. The hydrodynamic factor will be much greater for the droplets formed by the liquid where the surfactant is soluble. This will lead to a faster coalescence of these drops and will favor the formation of the emulsion in which the continuous phase is formed by the liquid where the surfactant is soluble. This conclusion is in accordance with Bancroft's rule.

Equation (45b) for a system without surfactant ($\varepsilon^f = 0$) was also derived[37] using Prandtl's concept of a boundary layer of thickness $\delta \approx 3.5 (h\iota^*/V)^{1/2}$ and v.Kàrmàn-Polhausen's method. The results were very similar to those obtained by solving the general equation (27) with only a small difference in the numerical coefficient in (46). This does not mean, however, that the above theory is restricted by the requirement $\delta/R \ll 1$ for the boundary layer to be present. In fact its validity is larger because it is determined by the condition[33] $\mu^* U/(\Delta P)r \approx (\delta/R)(h/R) \ll 1$. The velocity of thinning (in the absence of surfactants) of a non-symmetric film — thin film between two different drops A and B is[38]

$$V/V_{Re} = 1 + 3(1 + \varepsilon^e_A + \varepsilon^e_B)/(\varepsilon^e_A + \varepsilon^e_B + 4\varepsilon^e_A\varepsilon^e_B) \quad , \tag{47}$$

where $\varepsilon^e_{A,B} = -(a_1^{A,B} h/4\mu)/(U^{A,B}/r\iota^{A,B})^{1/2}$. The last equation is applicable to all systems without surfactant, including one-sided emulsion film (phase B-solid, i.e. $\varepsilon^e_B \to \infty$):

$$V/V_{Re} = 4(1 + \varepsilon^e_A)/(1 + 4\varepsilon^e_A) \quad . \tag{48}$$

Since for all systems of practical interest $\varepsilon^e \ll 1$, a convenient form of (45b) (with $\varepsilon^f = 0$) is

$$V \approx V_{Re}/\varepsilon^e = B(h^5 \Delta P^2/\rho^* \mu^* R^4)^{1/3} \quad , \tag{49}$$

where $B = (-4\sqrt{2}/a_1)^{2/3}$ (for a steady process $B = 2.82$[33]). Note that this equation of thinning is quite different from Reynolds' equation (8a). Moreover, according to (49) V does not depend at all on the film viscosity μ. In the same limit ($\varepsilon^e_A \to 0$) Eq. (48) yields $V/V_{Re} = 4$, i.e. the system behaves as if the drop is a gas bubble (cf. (34) with $\varepsilon^w = 0$). This fundamental difference between symmetric and one-sided emulsion films is due to the fact that in the former case, because of the high mobility of both film surfaces, the energy is dissipated mainly in the drops (in a rather extensive region of thickness $\delta >> h$[33]), while in the latter case, because of the tangential immobility of the solid surface, the velocity U at the surface liquid/liquid has always a small value and the predominant part of the energy is dissipated in the film.

The most general expression for the driving force, moving two non-deformable liquid drops of radii R_c against each other (in the absence of surfactants), was derived by Haber et al[39]. It represents an infinite series of complicated functions and allows only computer analysis. On its bases simple asymptotic expressions for great ($h > R_c$) and small ($h < R_c$) gap widths were derived.[40] In the former case an expansion in series with respect to R_c/h was used and a generalization of the well-known equations of Stokes, Rybzchynski-Hadamard and Lorentz[41] was obtained:

$$F = 6\pi\mu V R_c \left(\frac{2/3 + \bar{\mu}}{1 + \bar{\mu}} \right) \left[1 + \left(\frac{2/3 + \bar{\mu}}{1 + \bar{\mu}} \right) \frac{R_c}{h} \right] \qquad (50)$$

where $\bar{\mu} = \mu^*/\mu$. The latter asymptotes ($h/R_c < 1$) could not be obtained in a straightforward manner using the following approximations:

$$\text{sh}\, x \approx x e^x \quad , \qquad \text{ch}\, x \approx e^x \qquad (51)$$

for any $0 \leq x \leq \infty$. The result for $h/R_c < 1$ thus obtained depends on the parameters $\bar{\mu}$ and $\beta = \bar{\mu}(2h/R_c)^{1/2}$. For the theory of emulsions the case $\beta < 1$ is of special interest — it can be realized either with small drop viscosity ($\bar{\mu} << 1$) or with highly viscous drops ($\bar{\mu} >> 1$) but very small separations ($h << R_c$). In the former case expansions with respect to β and $\bar{\mu}$ were used and Eq. (43) was again derived, while in the latter case the expansions with respect to β and $\bar{\mu}^{-1}$ gave:

$$F = (3\pi^2/8)\, \mu^* V R_c (R_c/2h)^{1/2} \quad . \qquad (52)$$

Note that Eq. (52), like Eq. (49) does not contain the film viscosity μ. Equation (43) was derived also by using the systematic approach of Cox and Brenner[2] and thus the applicability of the approximations (51) was confirmed.

ACKNOWLEDGEMENT

Part of this work was supported by the National Science Foundation, Grant AER-76-81578.

REFERENCES

1. A. Scheludko, Adv. Colloid Interface Sci., 1967, 1, 391.
2. R. Cox and H. Brenner, Chem. Engng. Sci., 1967, 22, 1953.
3. E.B. Dussan, J. Fluid Mechanics, 1976, 77, part 4, 665.
4. D.St. Dimitrov and I.B. Ivanov, J. Colloid Interface Sci., 1978, 64, 97.
5. B.P. Radoev and I.B. Ivanov, Ann. Univ. Sofia, Fac. Chimie, 1971/1972, 66, 631.
6. I.B. Ivanov, B.P. Radoev, E.D. Manev and A.D. Scheludko, Trans. Faraday Soc., 1970, 66, 1262.
7. B.V. Toshev and I.B. Ivanov, Colloid & Polymer Sci., 1975, 253, 558; 1975, 253, 593; G.A. Martynov, I.B. Ivanov and B.V. Toshev, Colloid J., 1976, 38, 474; I.B. Ivanov, B.V. Toshev and B.P. Radoev, in "Wetting, Spreading and Adhesion," p. 37, J.F. Padday, ed., Academic Press, London, 1978.
8. Chr. Vassilieff and I.B. Ivanov, Zeitschrift für Naturforschung, 1976, 31, 1584.
9. O. Reynolds, Phil. Trans. Roy. Soc., London, 1886, A177, 157.
10. Taylor, cited in W. Hardy and I. Bircumshaw, Proc. Roy. Soc., 1952, A108, 12.
11. H.M. Princen, J. Colloid Sci., 1963, 18, 178.
12. O.V. Voynov, Doklady ANSSSR, 1974, 216, 285; V.M. Starov and V.N. Churaev, Colloid J., 1975, 37, 711; 1976, 38, 100.
13. Yu. A. Buevich and E.N. Lipkina, Prikl. Mat. i Teoret. Fiz., 1975, No. 2, 80.
14. G.D.M. McKay and S.G. Mason, Can. J. Chem. Engng., 1963, 41, 203.
15. B.V. Derjaguin and M. Kussakov, Acta Physicochim. USSR, 1939, 10, 25.
16. R.S. Allan, G.E. Charles and S.G. Mason, J. Colloid Sci., 1961, 16, 150; J.C. Lee and Hodgson, Chem. Eng. Sci., 1968, 23, 1375; D.C. Chappelear, J. Colloid Sci., 1961, 16, 186.
17. A. Roberts, J. Phys. D.: Appl. Phys., 1971, 4, 423.
18. S. Frankel and K. Mysels, J. Phys. Chem., 1962, 66, 190.
19. T.T. Traykov, B.P. Radoev and I.B. Ivanov, Ann. Univ. Sofia, Fac. Chimie, 1970/1971, 65, 422.
20. T. Kolarov, A. Scheludko and D. Exerova, Trans. Faraday Soc., 1968, 64, 2364.
21. A. Vrij, personal communication.
22. C. Huh and L.E. Scriven, J. Colloid Interface Sci., 1971, 35, 85; E.B. Dussan V. and S.H. Davis, J. Fluid Mech., 1974, 65, Part 1, 71.
23. I.B. Ivanov, D.St. Dimitrov and B.P. Radeov, J. Colloid Interface Sci., 1978, 63, 166.
24. V.G. Levich, "Physico-chemical Hydrodynamics," Englewood Cliffs, New Jersey, 1962.
25. H.C. Maru, V. Mohan and D.T. Wasan, Engng. Sci., in press; L.E. Scriven, Chem. Engng. Sci., 1960, 12, 98; M.G. Velarde, J.L. Ibañez, T. Sørensen, A. Sanfeld and M. Hennenberg, in "Proceedings of Levich's Conference," in press; M. Hennenberg, T.S. Sørensen and A. Sanfeld, J. Chem. Soc., Faraday Trans. II, 1977, 73, 48.
26. H. Brenner and L.G. Leal, J. Colloid Interface Sci., in press.
27. B.P. Radoev, D.St. Dimitrov and I.B. Ivanov, Colloid & Polymer Sci., 1974, 252, 50.
28. I.B. Ivanov and D.St. Dimitrov, Colloid & Polymer Sci., 1974, 252, 982.
29. E.D. Manev, Chr. Vassilieff and I.B. Ivanov, Colloid & Polymer Sci., 1976, 254, 99.
30. I.B. Ivanov, D.St. Dimitrov and B.P. Radoev, Colloid J., in press.

31. P. Murdoch and D. Leng, Chem. Eng. Sci., 1971, 26, 1881.
32. X.B. Reed, E. Riolo and S. Hartland, Int. J. Multiphase Flow, 1974, 1, 411; 1974, 1, 437.

33. I.B. Ivanov and T.T. Traykov, Intern. J. of Multiphase Flow, 1976, 2, 397.

34. T.T. Traykov and I.B. Ivanov, Intern. J. of Multiphase Flow, 1977, 3, 471.

35. T.T. Traykov, E.D. Manev and I.B. Ivanov, Intern. J. of Multiphase Flow, 1977, 3, 485.

36. J.T. Davies, Proc. 2nd. Int. Congr. Surface Activity, 1957, 1, 426.

37. I.B. Ivanov and T.T. Traykov, Ann. Univ. Sofia, Fac. Chimie, 1971/1972, 66, 641.

38. T.T. Traykov and I.B. Ivanov, Ann. Univ. Sofia, Fac. Chimie, 1974/1975, 69, in press.

39. S. Haber, G. Hetsroni and A. Golan, Int. J. Multiphase Flow, 1973, 1, 57.

40. V.N. Beshkov, B.P. Radoev and I.B. Ivanov, Intern. J. of Multiphase Flow, in press.

41. G.G. Stokes, Trans. Cambr. Phil. Soc., 1851, 9, Pt. II, 8; 258, 181; 213.

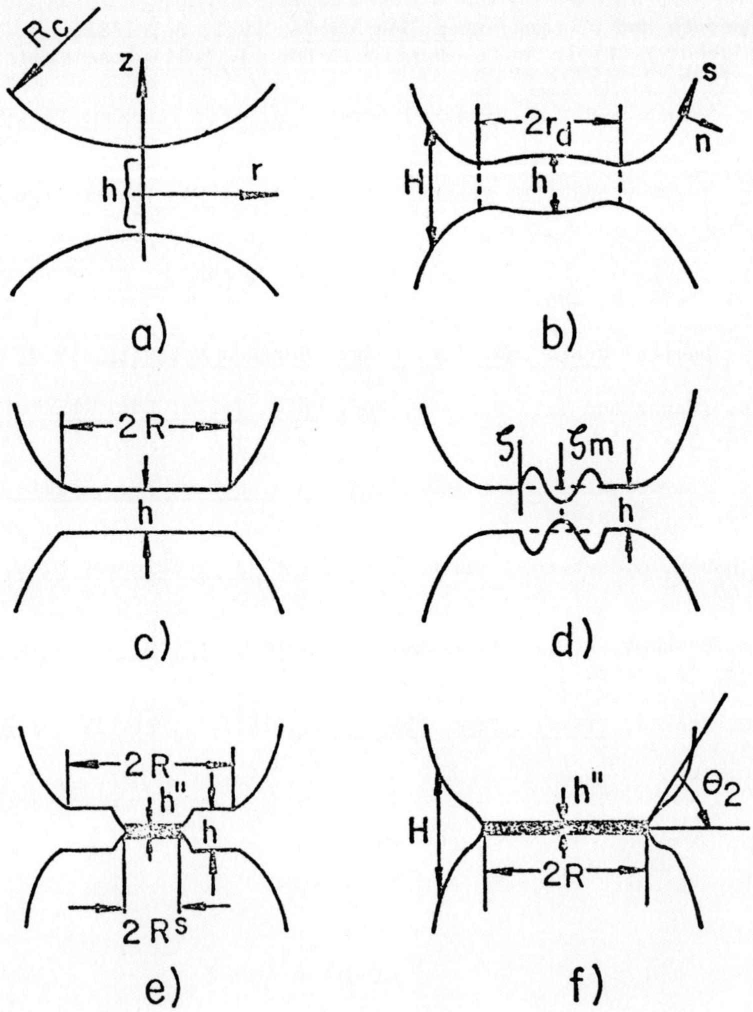

Figure 1. Main Stages in the Evolution of a Thin Film.

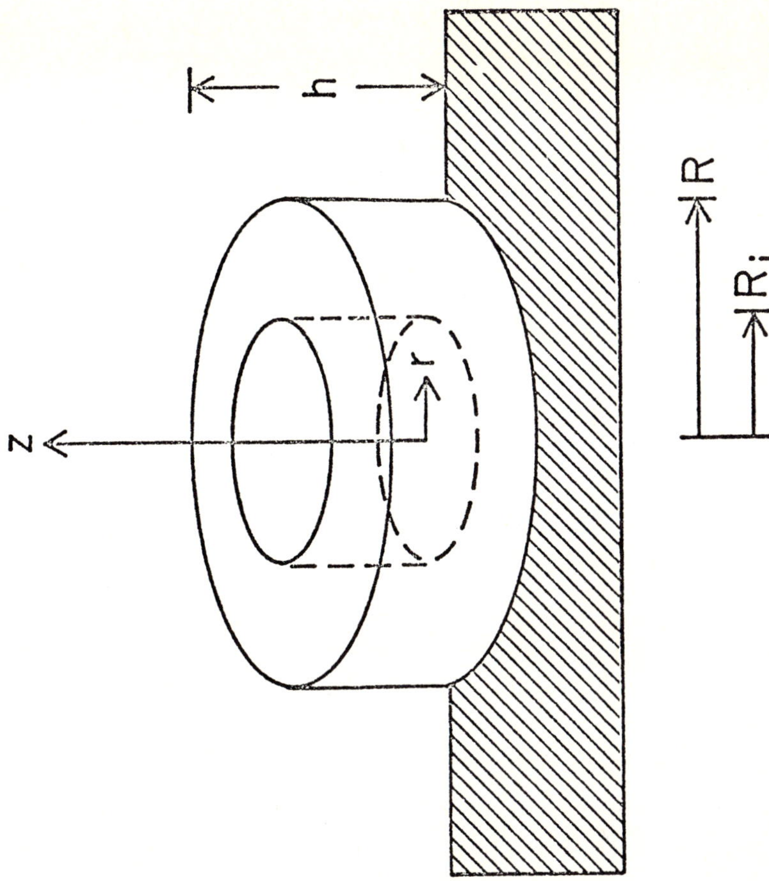
h
z
r
R
R_i

INSTABILITY AND RUPTURE OF THIN LIQUID FILMS

by

Rakesh K. Jain

Department of Chemical Engineering
Carnegie-Mellon University, Pittsburgh, Pennsylvania 15213, USA

Ivan B. Ivanov* and Charles Maldarelli

School of Engineering and Applied Sciences
Columbia University, New York, New York 10027, USA

and

Eli Ruckenstein

Faculty of Engineering and Applied Sciences
SUNY, Buffalo, New York 14214, USA

1. Introduction

Thin liquid films sandwiched between two fluids, or between a fluid and a solid surface are encountered in several industrial and biological systems. In flotation, the thinning and rupture of the liquid film between particle and bubble can be the rate determining step of the process.[1] Films of this type are also important in detergency. When an oil covered surface (e.g. a fabric material) is immersed in aqueous surfactant solution, it is necessary to destabilize the oil film by causing it to thin and rupture to form drops on the surface. These drops may be readily removed by hydrodynamic forces when the system is agitated. In water vapor condensation on a hydrophobic shock tube wall, experiments indicate that a thin film is formed that breaks up into droplets upon reaching a critical thickness of about 100 A.[2] The process of coalescence of bubbles and droplets in foams and emulsions also depends on the rupture of intervening thin liquid film. The formation of microvilli in cancer cells can be explained on the basis of the instability of biological membrane to small perturbations, if the membrane is considered as a thin liquid film sandwiched between the extra- and intra-cellular fluids.[3]

One chapter[4] in this book has dealt with the formation and thinning of thin liquid films. The objective of this review is to provide a unified theoretical framework for analyzing the stability of thin liquid films, to review briefly some of the experimental work done to measure the critical thickness of rupture of these films, and finally, to relate the results

* On leave from the University of Sofia, Bulgaria

obtained with some physicochemical phenomena in industrial and biological systems. While a brief historical perspective of the work done in this area is given here, the major emphasis in the present review is on the recent work on the hydrodynamic stability analysis of these films.

2. Rupture Mechanism

Initially, interest in the stability of thin liquid films arose from studies on the coalescence of particles in unagitated colloidal dispersions. As previously remarked, a common phenomenon in such systems is the formation of thin fluid films between elements of the dispersion. Once created, these films gradually thin and may rupture upon attaining a critical thickness (h_{cr}) of approximately several hundred angstroms. As reviewed by Scheludko,[1] early studies attributed this rupture to the onset of a thermodynamically unstable state during the thinning process. In these studies, the usual approach was to calculate the free energy of the film as a function of its thickness and then to examine the thermodynamic stability from the variation of the free energy with thickness (for an example, see Frenkel[5] for the case of a film on a solid surface.

The thermodynamic arguments, however, did not offer any insight into the kinetic mechanism of rupture. de Vries[6] was perhaps the first to supplement the thermodynamic point of view of film stability with a kinetic treatment. de Vries postulated that a film can rupture if a hole (created by thermal or mechanical fluctuations) forms spontaneously in the film. He calculated that the activation energy required to form such a hole is approximately equal to σh^2 (σ and h are the surface tension and thickness of the film, respectively). The film ruptures when the quantity σh^2 is of the order kT (k denotes Boltzmann's constant and T signifies the temperature). For a film at a temperature of 298 K and with a surface tension of 10 dyne/cm, this mechanism predicts a critical thickness of rupture of 10 Å. However, experimentally measured values of the critical thickness for a film at 298 K and with a surface tension of approximately 10 dyne/cm are of the order of several hundred angstroms. Thus, this mechanism yields a critcal thickness of rupture that is an order of magnitude too small.

In 1960, de Vries[7] modified his "hole" theory in two ways: thermal or mechanical fluctuations can corrugate a deformable interface, and the van der Waals' interaction forces play an important role in the rupture of thin films. Scheludko gave, in 1962, a more detailed thermodynamic treatment to the concept that deformations coupled with interaction forces can lead to film rupture.[8] The existence of corrugation at the film interface has been proven experimentally by light scattering studies.[9] Light incident upon an unperturbed, deformable interface is scattered as well as reflected. The scattering is caused by the unevenness of the surface: A planar interface simply reflects incident light, while a corrugated surface both reflects and scatters it. Scheludko suggested that these natural corrugations, developing on the interfaces of a thin film, increase in amplitude when the growth is accompanied by a decrease in the free energy of the system. Rupture occurs when the enlarging deformations become equal to the thickness of the film.

Scheludko detailed three primary factors that determine the free energy changes of a film subject to growing surface waves: interfacial tension, long range van der Waals interactions and electrical forces. A positive interfacial tension opposes a displacement of the interface because such a displacement increases the surface area and thus raises the free energy of the system.

The long range van der Waals force has an interaction range of the order of 1000 Å. In macroscopic systems, the effect of thickness on this force is usually negligible because characteristic distances are larger than 1000 Å. In thin film systems, however, the role of the thickness is more important because the thickness of the film (at the critical thickness) is of the same order of magnitude. Depending upon the properties of fluids which form the film and bounding media, these forces can be either attractive or reuplsive.[5] When the net force is attractive in nature, it promotes the growth of surface waves.

Electrical forces are significant in the growth of surface waves when ions are present in the liquid from which the film is formed. In this case, double layers are built up at the two interfaces of the film. These double layers oppose the growth of surface waves when their charges have the same sign. This resistance is a consequence of the fact that surface growth is accompanied by a reduction in the thickness of portions of the film and this is energetically unfavorable because it causes an increased overlap of the double layer of the two interfaces and hence an increased repulsion.

Vrij was the first to account for the fact that the growth of surface disturbances occurs in the thinning film.[9] He utilized the corrugation mechanism to develop a theoretical expression for the critical thickness of rupture, suggesting that the thinning and rupture of a film can be viewed as a two stage process.[9,10] In the first stage the uncorrugated film drains from an initial thickness to a thickness h_o. To estimate the time required for this drainage, Vrij considered a circular film thinning under the action of a pressure difference between the film and its border and containing sufficient amounts of surfactant to render the interfaces tangentially immobile. For such films Reynolds' equation relates the dependence of the film thickness (h) on time (t).

$$\frac{d(h^{-2})}{dt} = \frac{4}{3\mu R^2} \Delta p \qquad (1)$$

In Eq. (1), μ is the viscosity of the film, R is the radius and Δp is the pressure difference causing the film to thin.

In the second stage, at a thickness h_o, corrugations develop on the interfaces of the film. Only symmetric varicose disturbances were considered. Using free energy arguments, Vrij has demonstrated that the film is unstable for one of the Fourier components of these disturbances. The film ruptures when the fastest growing component of the subset reaches an amplitude $h_o/2$. The time necessary for this component to cause rupture was estimated by employing a graphical procedure. Vrij identified the critical thickness with the value of h_o that minimized the sum of the drainage and rupture times and introduced a correction for the thinning of the film during the growth of the surface waves. With some additional simplifying assumptions he was also able to derive the following expression for the critical thickness of rupture h or:

$$h_{cr} = 0.267 \left[\frac{A^2 R^2}{f\sigma_o \Delta p} \right]^{1/7} \qquad (2)$$

In Eq. (2), A is the Hamaker constant, σ_o is the surface tension, f is a numerical coefficient approximately equal to six (Δp and R are defined after Eq. (1)). Equation (2) predicts that the critical thickness is independent of the viscosity of the film and varies with the two-sevenths power of the film radius.

The agreement of the elementary calculations of Scheludko and Vrij with experimental observations not only demonstrated the feasibility of the corrugation mechanism but also provided the motivation for a more precise study of the hydrodynamic stability of thin films.

In hydrodynamic stability theory, a geometric perturbation is applied to one or both interfaces of a thin film (Fig. 1). This perturbation produces deformation in the shape of the film as well as initiating fluid motion. If the interfacial perturbations grow in time, the film becomes unstable and may rupture. If the perturbations decay in time, the film is stable with respect to these perturbations. In order to determine the conditions when a film becomes unstable, all the perturbation variables of the system are expressed in the following form:

$$\theta(\vec{R},t) = \hat{\theta}(\vec{R})\, e^{\omega t} \tag{3}$$

Here, θ is any perturbation variable of the system, and it is a function of both position, $\vec{R}$, and time, t. $\hat{\theta}$ is the time independent component of the perturbed quantity, and ω is the complex frequency of motion (also referred to as the angular velocity). The real component of ω is referred to as the growth coefficient (β). If the growth coefficient is positive, the perturbation, θ, will grow exponentially with time, and will cause the film to be unstable. If the growth coefficient is negative, the perturbation will die out eventually. If the growth coefficient is zero, the film is considered marginally stable, and its dynamics will be determined by the imaginary component of ω. The objective of the linear stability analysis is to obtain an analytical expression or numerical values of ω as a function of various physicochemical parameters that characterize the system, by solving the equations of continuity and motion for the system with appropriate boundary conditions.

Two different approaches have been used to account for the long range van der Waals and electrical double layer forces in the hydrodynamic stability analysis: a disjoining pressure procedure and a body force procedure.

In the disjoining pressure approach, the influence of the van der Waals and double layer forces is incorporated in the normal stress balance at the fluid interface of the film. In these balances, the component of the stress tensor which acts in a direction normal to the interface is required to suffer a discontinuity apart from the Laplace jump. On the other hand, in the body force approach, the influence of these forces is incorporated as a body force in the equations of motion. When applied rigorously, both approaches should lead to the same results.

In the next sections, investigations carried out for both non-thinning and thinning liquid films, using these two approaches, are discussed.

3. Non-thinning Films

a. Inviscid Plane-Parallel Film

Felderhof has developed a detailed electrohydrodynamic study of the surface wave dynamics of thin films. The model system he analyzed consisted of an infinite plane-parallel non-thinning film composed of an inviscid fluid and surrounded by vacuum. Dissolved ions were assumed present in the film volume and the gravitational field was neglected. Felderhof obtained analytical expressions for the motion induced by the onset of low amplitude surface waves by using perturbation theory and normal mode analysis to solve the continuity, Euler and Maxwell equations. The Maxwell equations were solved with the assumption that for each instantaneous shape of the film, the ions in the film volume have achieved thermal equilibrium. This quasi-static assumption restricted the treatment to surface waves with a frequency less than the inverse of the time necessary to establish an ionic equilibrium in the film volume and on the surface.

Felderhof studied the effect of one Fourier component of a perturbation of the film and found that there were two independent modes of vibration of the free film: the symmetric mode (squeezing mode) and the antisymmetric mode (stretching or buckling mode). The symmetric mode is a varicose disturbance of the film; the two interfaces are displaced in sinusoids of equal amplitude and wavelength. For this mode, the sinusoids are 180 degrees out of phase. The antisymmetric mode is a sinuous disturbance of the film; as with the symmetric mode, the two interfaces are displaced in sinusoids of equal amplitude and wavelength. For this mode, however, the sinusoids are in phase (Fig. 1). For each mode, there is a characteristic complex frequency. The response of the system to arbitrary interfacial disturbances is composed through a linear combination (over all wavelengths) of the two modes. Dispersion relations for both modes were obtained. From these relations, the conditions for stability can be determined. The dispersion relations were analyzed in the long-wavelength limit. In this limit, his calculations indicated that the stretching mode, in addition to the squeezing mode, can become unstable to surface waves. Felderhof found that the stability of the stretching mode is determined by the sign of a quantity S defined by

$$S = (\sigma_o - \gamma) + \frac{\alpha}{2} - \frac{A}{8\pi h_o^2} \quad . \tag{4}$$

In Eq. (4), σ_o is the bulk surface tension of pure fluid which forms the film, γ is the spreading pressure of ions adsorbed on the surface, α is the electric film tension at constant volume, A is the Hamaker constant and characterizes the London-van der Waals dispersion forces present in the film, and h_o is the unperturbed thickness of the film. Physically, α represents the free energy of formation of the double layers, and it decreases the interfacial tension of film. On the other hand, α represents the effect of surface adsorption of ions on the surface tension. The last term arises from the consideration of London-van der Waals dispersion forces, and leads to a decrease in the film tension. If S is zero, the surface waves are stationary with time. Owing to the neglect in this treatment of viscosity and the consequent absence of viscous dissipation, the surface waves cannot be damped out. If S is negative, the surface waves grow with time. Since the stretching mode does not involve variations in the film thickness, this growth may not be accompanied by rupture. However, the growth does cause a kind of buckling of the film. The nature of this change cannot be ascertained from Felderhof's perturbation study, but it could be evaluated from a non-linear analysis.

b. Viscous Plane Parallel Films (Foam and Emulsion Type)

The stability of an infinite plane parallel non-thinning viscous film was analyzed independently by Lucassen et al.[13,14] and Ruckenstein and Jain.[15] In the former analysis, the thin film was bounded by two different semi-infinite fluids. The semi-infinite phases and the film were assumed to contain surface active agents whose effect on the wave motion was accounted for by assuming that the surfaces were viscoelastic. Diffusional exchange of the surfactant from the film and bulk phases to the interfaces was, however, neglected. These authors used the disjoining pressure approach. The analysis of Ruckenstein and Jain differed from the former in these ways: (i) these authors used the body force approach; (ii) their treatment included a systematic analysis of the effect of surfactants, and (iii) the film was bounded by a vacuum.

While Lucassen and coworkers derived the dispersion equation for a film bounded by two different fluids, they examined in detail the dynamics of a system consisting of a film bounded by two identical semi-infinite phases. For such a system, similar to Felderhof, they found that if gravity is neglected these are two independent modes of vibration of the film; the symmetric mode and the antisymmetric mode. Since only the symmetric mode leads to film rupture due to local thinning of the film, Vrij et al.[14], in a companion paper, examined the conditions for growth of surface waves for the symmetric mode in the long-wavelength limit (i.e. when the wavelength of perturbation is much larger than the unperturbed thickness of the film).

Considering only London-van der Waals dispersion forces, Vrij et al. and Ruckenstein and Jain showed that the growth coefficient, β (the real part of ω), for the symmetric mode is given by a simple expression:

$$\beta = -\sigma_{eff}\, f_1 \tag{5}$$

where

$$\sigma_{eff} = \sigma_o - \frac{A}{\pi h_o^4}\frac{\lambda^2}{4\pi^2} \tag{6}$$

and f_1 is a positive definite function of the system parameters. (The expressions for f_1 are different in references 14 and 15.) In this equation σ_o is the interfacial tension, A is the Hamaker constant, h_o is the unperturbed film thickness, and λ is the wavelength of perturbation. The sign of β, which in turn determines the stability of the system, depends only on the sign of σ_{eff}. An inspection of Eqs. (5) and (6) suggests that it is possible to find the value of a wavelength that corresponds to zero growth rate (condition for marginal stability). This wavelength is referred to as the critical wavelength, λ_c, and is given by:

$$\lambda_c = 2\pi(\pi\sigma_o/A)^{1/2}\, h_o^2 \tag{7}$$

If the wavelength of a perturbation is greater than λ_c, the film will become unstable. For each wavelength greater than λ_c, the value of β is positive and can be calculated from Eq. (5). The wavelength with the largest value of growth coefficient is the most rapidly growing fluctuation. This wavelength dominates the rupture kinetics of the film, and is referred to as the dominant wavelength (λ_d). Furthermore, if one assumes that the linear stability analysis is valid up to the point when the two surfaces come into contact, it is possible to estimate the time constant of rupture (τ_m) as the reciprocal of the growth rate corresponding to the dominant wavelength.

These studies have shown that when the bounding medium is vacuum, a dominant wavelength does not exist in a surfactant free film.[15] However, when large amounts of surfactants are present in the system, the dominant wavelength is given by the following expression:

$$\lambda_d \approx \sqrt{2}\,\lambda_c \tag{8}$$

When the viscosity of the bounding media is non-zero, dominant wavelength exists even for systems without surfactants, and its value is given by

$$\lambda_d \approx \sqrt{3}\,\lambda_c \tag{9}$$

Vrij et al. and Ruckenstein and Jain also examined the effect of the surfactants on the growth coefficient of the symmetric mode in the long wavelength limit. Their detailed numerical calculations demonstrated that the maximum growth rate is essentially a step function of the surface elasticity. For films with a surface elasticity less than a critical value, the maximum growth rate is virtually independent of the elasticity. At the critical value this growth rate is reduced sharply in absolute value. For surface elasticities larger than the critical value, the maximum growth coefficient is again independent of elasticity. Since the time of rupture is approximately the inverse of the maximum growth coefficient, films with elasticities less than the critical value rupture more quickly than those with elasticities greater than this value. Vrij et al. also found that numerical values for this critical surface elasticity are low. They therefore concluded that only small amounts of surfactant need be adsorbed at the interface to render it tangentially immobile. In the limit of large surfactant concentration, the time constant of rupture is given by the following expression:

$$\tau_m = 96\pi^2 \sigma_o \mu\, h_o^5\, A^{-2} \tag{10}$$

The analysis of thin viscous films mentioned above did not include detailed analysis of the double layer forces. Recently, Sche and Fijnaut have extended the electrohydrodynamic analysis of Felderhof to viscous thin films surrounded by a vacuum.[17] The dispersion relations they have obtained are complicated because of the inclusion of these effects, and thus detailed calculations have been undertaken only for the simplified case of a tangentially immobile surface. For this case, the criteria for growth of both the squeezing and stretching modes, in the long-wavelength limit is found to be independent of the viscosity of the film and identical to the criteria established by Felderhof for inviscid flow.

Maldarelli et al.[18] have recently extended the analysis of Felderhof to the general case when the liquid film is bounded by two different viscous phases. In addition to obtaining analytical results for the squeezing and stretching modes for a symmetric film (i.e. film bounded by the same fluid), these authors have studied the role of asymmetry on the growth of perturbations numerically, especially the asymmetry of interfacial tension.

When electrical double layers are absent, the expression for the growth coefficient, β, for the antisymmetric mode is given by the following expression:

$$\beta = -\left(\sigma_o - \frac{A}{8\pi h_o^2}\right) f_2 \tag{11}$$

Here, all the quantities have the usual meaning, and f_2 is a positive definite function of system parameters. Therefore, instability of this type will occur only in low tension systems, i.e. when

$$\sigma_o < \frac{A}{8\pi h_o^2} \tag{12}$$

Biological membranes provide an example of such a system.[3] When a film becomes unstable in this mode, it is possible to estimate the dominant wavelength, λ_d, for systems where $\mu_{media}/\mu_{film} \ll 1$. Its value is:

$$\lambda_d \cong 2\pi h_o/(6\mu_{media}/\mu_{film})^{1/3} \quad . \tag{13}$$

It must be pointed out here that while the value of the quantity (σ_o $-A/8\pi h_o^2$) may be negative during the kinetic process, this quantity must be positive eventually for a system to be stable mechanically and thermodynamically. In addition, reactions at the interface[41] lead to additional terms in Eq. (11) causing the film to be unstable even at positive values of ($\sigma_o - A/8\pi h_o^2$).

When the film is surrounded by different fluids, it is not possible to obtain an analytical expression for the growth coefficient. These authors[18] however have numerically examined such a system in detail, and have found that for a given wavelength and viscosities, the primary factor that determines the ratio of amplitudes of perturbation at two interfaces is the surface tension. In the antisymmetric mode, the amplitude ratio remains close to unity, but in symmetric modes, the amplitude ratio is inversely proportional to the ratio of surface tensions. One would also expect this on the basis of physical arguments, because the larger the value of surface tensions, the more difficult is it for the surface to deform.

c. Spherical Film

Patzer and Homsey,[19] in an analysis similar to the above, have examined the stability of a thin film bounded by two spherically concentric fluid phases. Such a system can be considered as a mathematical model of a soap bubble or a biological cell. The focus of their study was the calculation of the marginally stable thickness of the spherical film as a function of the physiochemical parameters of the system. The base state was assumed to be one of zero flow. Two types of interfaces, tangentially immobile (large surface concentration of surfactant) and completely free (zero surface concentration of surfactant) were considered. Patzer and Homsey found numerically that increasing the surface tension or decreasing the Hamaker constant decreases markedly the marginally stable thickness of the thin film. Their numerical calculations also indicated that the marginal thickness is independent of the type of interface considered (either tangentially immobile or completely free).

d. Film on a Solid Substrate (Wetting Film)

All the aforementioned studies on surface wave dynamics analyzed liquid films that were surrounded by either a viscous fluid or a vacuum. Ruckenstein and Jain[15] and Jain and Ruckenstein[20] have treated the case of a film on a solid substrate. The model system they studied consisted of an infinite plane-parallel non-thinning film situated on top of a solid and bounded from above by a semi-infinite fluid. The surrounding fluid and the film were assumed to contain surface active materials. The lower (film-solid) interface was considered rigid and impenetrable to surfactant. At the upper (film-semi-infinite fluid) interface, the authors formulated a surfactant mass balance to account for the effects on the wave motion of surface diffusion, surface shear viscosity and elasticity and the diffusional exchange of surfactant from the film and the semi-infinite fluid to the interface. Both London dispersion and double layer forces were included in their analysis.

These investigators analyzed the stability of the film with respect to infinitesimal displacements of the upper deformable interface by using perturbation theory and normal mode analysis. Their calculations indicated the existence of only one mode of vibration of the film. This result is a consequence of the assumed rigidity of the lower interface. Explicit expressions for the growth coefficient for the cases of completely free (zero surface concentration of surfactant) and tangentially immobile surfaces were obtained.

In both cases, the growth coefficient, β, is given by the following expression:

$$\beta = - n \sigma_{eff} \times f_3 \; , \tag{14}$$

where:

$$\sigma_{eff} = \sigma_o - \left(\frac{A}{2\pi h_o^4} \right) \frac{\lambda^2}{4\pi^2} \tag{15}$$

and f_3 is a positive definite function of system parameters. Here n=1 for a tangentially immobile surface, and n=4 for a free surface. These authors obtained the following expressions for λ_c and λ_d for systems with and without surfactants:

$$\lambda_c = 2\pi (2\pi\sigma_o/A)^{1/2} h_o^2 \tag{16}$$

and

$$\lambda_d \approx \sqrt{2}\, \lambda_c \tag{17}$$

When the system has no surfactants, the following expression for the time constant of rupture was obtained:

$$\tau_m \approx 48\pi^2 \sigma_o \mu \, h_o^5 \, A^{-2} \tag{18}$$

On the other hand, when surfactants are present in large amounts the time constant increases by a factor of four:

$$\tau_m \approx 192\pi^2 \sigma_o \mu \, h_o^5 \, A^{-2} \tag{19}$$

Their calculations further indicated that for completely free surfaces, the growth coefficient decreases slightly with increasing viscosity of the upper phase, while for tangentially immobile surfaces, the growth coefficent is virtually independent of this viscosity. Jain and Ruckenstein also analyzed numerically the influence of surface elasticity and surface shear viscosity on the growth rate. They obtained a step profile for the behavior of the maximum growth coefficient as a function of the surface elasticity. Their numerical results also demonstrated that for films with reduced surface elasticities, the presence of surface shear viscosity can markedly decrease the maximum growth rate.

4. Thinning Films

a. Foam Films

The studies discussed in Section 3 were devoted to infinite plane-parallel or spherical non-thinning films. However, the films that occur naturally in dispersions are radially bounded structures draining under the influence of capillary suction and negative disjoining pressure. Thus the previous studies, though instructive in detailing the fundamental physiochemical aspects of thin film dynamics, needed to be extended to incorporate the effects of a circular geometry and drainage flow. On the basis of Scheludko's and Vrij's ideas, a systematic analytical approach for thinning films was developed by Ivanov et al.[21] The model system they employed consisted of a plane-parallel circular draining foam film. Symmetric varicose disturbances were superimposed on the thinning and the focus of the analysis was to determine the critical thickness at which the film ruptures as a result of these disturbances.

To obtain analytical expressions for the motion induced by the drainage and surface fluctuations, Ivanov et al. assumed that the velocity vector of this motion can be decomposed into two terms. The first is the velocity due to the drainage of the film. This velocity was obtained by solving the continuity and Navier-Stokes equations in the lubrication approximation for an unperturbed circular draining film. Thus an expression for the velocity of thinning $V = -dh/dt$ was derived. For details see references 4 and 22. The second term is the velocity caused by the surface waves. This velocity was obtained by a quasi-static procedure: The film was assumed to have a fixed thickness and the velocity vector resulting from a symmetrical varicose disturbance, ξ, of the film was calculated as a function of this thickness using the approach employed in the non-thinning films. The drainage was then coupled to the surface growth by assuming that ξ during the thinning is a function of time through the thickness h. Mathematically, the motions due to thinning and the present perturbations can be combined in the following manner:

$$V_\xi = \frac{\partial \xi}{\partial t} = \beta \cdot \xi = \left(\frac{\partial \xi}{\partial h}\right)\left(\frac{dh}{dt}\right) = - V\left(\frac{\partial \xi}{\partial h}\right) \qquad (20)$$

where ξ is the applied perturbation, β is the growth coefficient (real part of the frequency of motion ω), and h is the average film thickness at time t. Due to this assumption, all time-dependent quantities depend on t only via h. Integration of Eq. (20) leads to:

$$\xi(h) = \xi(h_t) \exp\left[-\int_{h_t}^{h} \frac{\beta}{V} dh\right] \qquad (21)$$

where h_t is the transition thickness at which the wave becomes marginally unstable.

The film will rupture when the amplitude of the fastest growing perturbation, ξ_m, is equal to the half thickness of the film at that time, i.e. $h_{cr} = 2|\xi_m|$. Using this condition, and a set of simplifying, but physically consistent, assumptions, Ivanov et al.[21] obtained the following expression for the initial thickness of rupture of foam films with tangentially immobile interfaces:

$$h_{cr}^2 = \left(\frac{4kT}{R^2 \pi_t'}\right)\left[\ln \frac{P_c - \pi_{cr}}{P_c - \pi_t}\right]^{-1} \times$$

$$\exp\left[\frac{R^2 \pi_t'}{4\sigma_o} \ln \frac{P_c - \pi_{cr}}{P_c - \pi_t}\right] \qquad (22)$$

where k is the Boltzmann's constant, T is the absolute temperature, R is the radius of film, P_c is the capillary pressure, σ_o is the interfacial tension, π_t and π_{cr} are the disjoining pressures at the transition thickness, h_t, and the critical thickness, h_{cr}, respectively. The transition thickness, h_t, is obtained by getting the growth coefficient, β, equal to zero in the dispersion equation.

Using this procedure, Ivanov et al. were able to calculate theoretical values of the critical thickness of rupture for films with tangentially immobile surfaces. Their results were in satisfactory agreement with experimentally measured values of critical thicknesses for films with large concentrations of surfactant when theoretical isotherm for disjoining pressure was used[22] (Fig. 21). The authors also studied films that contained small amounts of surfactant. For these films, they utilized a surfactant mass balance at the interface to account for the effect of surface elasticity on the thinning and surface motion. However, surface shear viscosity was neglected. The theoretically calculated critical thicknesses of rupture were found to be independent of the amount of surfactant present, contrary to the experimentally observed decrease in critical thickness with increasing surfactant concentration. In a later study, Ivanov and Dimitrov[23] extended the analysis of Ivanov et al. to include the effects of surface shear viscosity. In this analysis, they demonstrated that the observed decrease in critical thickness with increasing surfactant concentration can be due to surface shear viscosity.

Interesting physical conclusions about thinning films can also be reached, if Eq. (20) is integrated in the form[24]:

$$\xi(h_{cr}) = \xi(h_t) + \int_{h_t}^{h_{cr}} \frac{v_\xi}{V}\, dh \tag{23}$$

It follows from this equation that each factor which effects both β and V to the same extent, will not alter h_{cr}. For example, h_{cr} for films with tangentially immobile interfaces does not depend on μ (see (22)), because both β and V are inversely proportional to μ. On the contrary, if β is decreased more than V, this will result in decreasing h_{cr}, because the growth of perturbations will be hindered and the surfaces will "touch" each other at a smaller mean thickness. This can be observed in the presence of a long-range positive (e.g. electrostatic) disjoining pressure; if it is not large enough to prevent rupture, it will primarily slow down the drainage from the thinnest part of the film without substantially affecting V. If, however, the positive disjoining pressure is a short range one, it will appear and abruptly increase only when the surfaces "touch" each other and thus will not affect h_{cr}. This is the case of formation of black spots — according to the above concept they should appear at the same thickness at which the film would rupture at small surfactant concentrations. These conclusions are in agreement with the experimental results of Manev et al.[25]

b. Emulsion Films

Using the approach discussed above, Ivanov and coworkers have studied the influence of a surfactant on the capillary waves in a symmetrical emulsion system, and have derived expressions for the critical thickness of rupture.[26,27] Using their results, these investigators suggest that the critical thickness of an emulsion film can be reduced by decreasing the value of v_ξ more than that of V. This can be achieved in two different ways: either by increasing the viscosity of drop (because $v_\xi \to 0$ as $\mu_{drop}/\mu_{film} \gg 1$) or by adding surfactants to the drops (because a large surfactant concentration reduces v_ξ considerably, but does not alter the velocity of thinning).

Recently, Dimitrov and Ivanov[28] have obtained the following expression for the rate of thinning of emulsion films with deformable interfaces using the lubrication approximation.

$$-\frac{\partial H}{\partial t} = \nabla_r \left[\frac{H}{2} (v_r^A + v_r^B) - \frac{H^3}{12\mu} \frac{\partial P}{\partial r} \right] \tag{24}$$

where H is the local film thickness, t is time, P is pressure, μ is viscosity, r is the radial position in film, v_r^A and v_r^B are the radial velocities of the upper and lower interfaces, respectively, and $\nabla_r = \frac{1}{r}\frac{\partial}{\partial r} r$. When the interface is tangentially immobile (presumably due to large surfactants concentration), $v_r^{A,B}$ is zero. When the interface is tangentially mobile, $v_r^{A,B}$ is calculated by solving the Navier-Stokes' equation and the diffusion equation for the drop phase and the film along with the appropriate boundary conditions (see the review article by Ivanov and Jain in this book). Note that all hydrodynamic variables, i.e. H, $v_r^{A,B}$ and P in Eq. (24) have two components: one which accounts for the thinning of uncorrugated film, and the other that accounts for the flow induced in the non-thinning film due to corrugations. By neglecting the coupling terms, it is possible to obtain expressions for both v_ξ and V in a straightforward manner. Once the expressions for v_ξ and V are available, the procedure described above can be used to obtain the critical thickness of rupture (unpublished results).

c. Estimation of Coupling Terms

In all the published works of Ivanov and coworkers, the effect of drainage on the wave-motion has been assumed to be negligible. In an attempt to estimate these terms, Gumerman and Homsey[29] also have examined the stability of a plane-parallel circular free draining film subject to surface displacements. Their study, however, was restricted to films containing enough surfactant to render the interfaces tangentially immobile. These authors studied the marginal stability of film and found that hydrodynamic variables of the perturbed state (for example, the pressure and the velocity vector) should, in contrast to the assumptions of Ivanov et al., be dependent upon the drainage. The coupling terms were also estimated by Ivanov.[22]. Based on previous studies suggesting that the length of the wave, dominating the process of rupture, was approximately one tenth of the film radius, he concluded that those terms can be neglected.

We have recently calculated these coupling terms systematically using Eq. (24) for films with tangentially immobile interfaces, and we have derived an analytical expression for the coupling term similar to that found by Gumerman and Homsey. We also found numerically that these terms are negligible for thin films with non-zero surfactant concentrations.

d. Film on a Solid Substrate with Dimple

When a small drop or bubble approaches a solid or liquid interface, the intervening liquid film between the bubble and the interface is usually plane-parallel. However, when the bubble is large, it readily acquires a reverse curvature in its center so that some fluid is entrapped by a thinner ring.[30] The central lens of liquid is usually referred to as "dimple" and the surrounding circular film as "barrier" ring (Fig. 3).

In order to study the role of dimple, we have recently analyzed a model system as shown in Fig. 3: the liquid "barrier" ring of thickness h, and external and internal radii R and R_i formed between a bubble and a solid substrate.[31] The fluid outside the film is considered part of the bulk where the hydrodynamic effects can be neglected. The motion of liquid inside the ring is described by the linearized Navier-Stokes equations. The role of interaction forces present in the film is incorporated by adding the disjoining pressure, Π, in the normal stress balance. The treatment presented here considers the surfactants soluble in the film. Furthermore, the analysis includes the Gibbs-Marangoni effects, and both the surface constitutional dilational viscosity (μ_d) and the surface shear viscosity (μ_s). In the limit $R_i \to 0$, the results reduce to those obtained previously for plane-parallel radially bounded film, and in the limit $R_i \to 0$ and $R \to \infty$, the result for unbounded thin film can be obtained.

It must be pointed out here that in reality the slopes at the boundary of the ring are continuous. In addition the pressure gradients outside the ring may not be zero as assumed here. The model system considered here makes it possible to obtain the exact solution to the problem, as well as retains the essential features of the real system, if the pressures inside and outside the ring p^i and p^ℓ, are calculated by incorporating the geometry of the real system, i.e. the radii of bubble (R_c) and dimple (R_d), respectively (Fig. 3).

Having made these assumptions the problem falls in the realm of hydrodynamic stability theory. The first step is to solve the base case to develop an expression for film thinning. This is followed by the linear stability analysis in which arbitrary perturbations are applied to the upper interface, and conditions under which these perturbations grow to cause the film to rupture are calculated. The growth rate of perturbation gives the time constant of rupture. Finally, using the fact that the film will rupture when the amplitude of perturbation is equal to the thickness of film, the following expression for the critical thickness of rupture is calculated using the approach outlined in Section 4-a:

$$h_{cr}^2 = \frac{2kT}{R_{eff}^2 \Pi_t'} \left[\frac{(R - R_i)^2}{R^2 - R_i^2} \right] \left[\int_{h_{cr}}^{h_t} \frac{\Pi'}{\Delta P} dh \right]^{-1} \times$$

$$\exp \left[\frac{R_{eff}^2 \Pi_t'}{8\sigma_o} \int_{h_{cr}}^{h_t} \frac{\Pi'}{\Delta P} dh \right] \tag{25}$$

with

$$R_{eff}^2 = (R^2 + R_i^2) - (R^2 - R_i^2)/\ln (R/R_i) \tag{26}$$

and

$$\Delta P = P_c \left[\frac{1}{1 - (R_i/R)^2} + \frac{1 + R_c/R_d}{\ln (R_i/R)} \right] - \Pi \tag{27}$$

where ΔP is the driving pressure, $P_c = 2\sigma_o/R_c$ is the capillary pressure of the bubble and $R_d \approx 2R_i^2/(h_d - h)$ is the radius of curvature of the dimple, obtained by assuming the dimple to be the cap of a sphere of radius R_d. The maximum elevation of the dimple, h_d, above the solid surface can be calculated from[32]:

$$h_d = 0.285 (R + R_i) \sqrt{h/R_c} \quad .$$

It must be mentioned here that as $R_i \to 0$, the results for a plane-parallel film can be obtained. In the limit $R_i \to 0$ and $R \to \infty$, the results of Jain and Ruckenstein for a plane-parallel radially unbounded film can be obtained. Since the presence of a dimple does not alter the expression for the growth coefficient, β, the changes in critical thickness are mainly governed by the changes in the velocity of thinning. Velocity of thinning increases when the width of the ring ($R - R_i$) is reduced or the radius of dimple, R_d, is increased. Indeed, the critical thickness of rupture, calculated numerically by solving Eq. (25), decreases as the width of the ring decreases and/or as R_d increases.

5. Applications

a. Industrial Systems

As mentioned in Section 1, quantitative understanding of the dynamics of thin liquid films is of importance in several industrial processes, such as flotation,[1] coalescence of bubbles and droplets in foams and emulsions, detergency, condensation,[2] and power coating. In what follows, the results discussed in previous sections are used to interpret some of the available experimental information.

The most extensive series of experiments to measure the lifetime and critical thickness of rupture of this liquid film have been carried out by Scheludko and his coworkers. In these experiments a radially bounded thin film is formed by either pressing a bubble (or droplet) against a solid substrate, or by sucking out the liquid I from a biconcave meniscus formed by the "drops" or "bubbles" of phase II (Fig. 4). Several variations of these two basic set-ups have been used by various investigators in this area of research;[33] an excellent exposé of the work is given in a comprehensive review by Scheludko.[1] Most of the results obtained from these studies can be explained by the theoretical results reported in Section 4.

Another set of interesting experiments where a thin film is formed on a solid substrate have been conducted by Goldstein[2] in a study of water vapor condensation on a clean hydrophobic shock tube wall. Goldstein reports that in these experiments a continuous film of condensate is formed, which, after reaching a thickness of about 100 Å in about 10μs after the compression begins to break up into many small droplets. In all mechanical systems there are perturbations of various wavelengths; consequently the rupture of the film can be considered in this case to be a consequence of hydrodynamic instability.[15] It is also of interest to mention here that if Hamaker's constant A is taken to be order of 10^{-12} erg and h_o of the order of 50 Å, then the time constant of rupture computed from Eq. (18) with $\sigma_o \approx 70$ dyne/cm and $\mu \approx 16^{-2}$ P is of the order of 10μs, as was found experimentally.[2,15]

Several investigators have measured the coalescence time in foams and emulsions. While these "gross" measurements do not provide accurate information about the time constant of rupture of an individual film, they do provide an order of magnitude. Marucci,[34] for example, has reported that in the presence of surface active impurities the coalescence time in foams is of the order of 10^{-1} sec. One may verify that indeed the value 10^{-1} sec can be obtained from Eq. (10) for an aqueous film about 500 Å thick.

b. Biological Systems

The biological cell provides a different example of a system in which a thin film plays a cruical role.[3] The cell membrane is a highly viscous lamella bounded by two (approximately) spherically concentric liquid phases (i.e. the intra- and extra-cellular fluids). In certain cell mobility processes and in phagocytosis and microvilli formation, the membrane deforms considerably from its approximately spherical configuration. Reasons for the onset of these membrane deformation processes have not been fully elucidated, and it is hoped that a theoretical study of the stability of thin liquid films may provide insight into the origins of such processes.

An inspection of cells with microvilli suggests that microvilli can be considered as the filopodial protusions of the cell membrane.[35] We suggest that intra- or extra-cellular events occurring in living cells cause small fluctuations in the membrane configuration, and under some conditions these fluctuations become unstable in the antisymmetric (or buckling or stretching) mode to form villi.[3] The diameter of microvilli can be estimated from the wavelength of the fastest growing perturbation. Substituting in Eq. (13), $h_o \approx 100$ Å, $\mu_{bulk} \approx 10^{-2}$ P, and $\mu_{film} \approx 10^4$ P, one obtains the value dominant wavelength to be of the order 0.1μm, which is the experimentally measured value of microvilli diameter.[35] Furthermore, Eq. (13) suggests that λ_d is proportional to $\mu_{film}^{1/3}$, i.e. a membrane with a lower viscosity will have a lower value of microvilli diameter, and, hence, a greater number density of microvilli. Indeed, experiments have suggested that a cancer cell membrane is more "fluid" (i.e. less viscous) than a normal cell membrane.[36] This observation in light of the present theoretical framework explains the greater density of microvilli in cancer cells as compared to in normal cells.[37]

The thin film model of a cell membrane can be used to explain the fusion of cells and vesicles, and engulfment of viruses and macromolecules by cells (phagocytosis). In these processes, the cell membrane becomes unstable in the symmetric (or squeezing) mode, and ruptures ultimately so that two cells in contact can fuse. In phagocytosis, the membrane may engulf a particle by bending (stretching mode) as the particle approaches the membrane. Furthermore, in the symmetric mode instability of the membrane thins locally at several spots (Fig. 1). This reduced thickness of membrane may cause the membrane to be more permeable during the onset and progession of instability.

While the present analysis is simple, it describes the available data qualitatively. It can be improved considerably by the inclusion of nonuniform charge density, a more precise analysis of the double-layer interactions, a viscoelastic model of the membrane, a spherical configuration of the cell and chemical reactions.[38,39,40] Inclusion of some of these factors in the stability analysis of thin films is the subject of our current investigations.[41]

ACKNOWLEDGEMENT

Part of this work was supported by the National Science Foundation, Grant AER-76-81578.

REFERENCES

1. A. Scheludko, Adv. Colloid Interface Sci., 1967, 1, 391.
2. R. Goldstein, J. Chem. Phys., 1964, 40, 2763.
3. R.K. Jain, C. Maldarelli and E. Ruckenstein, AIChE Symp. Ser., 1978, 74(182), 120.
4. I.B. Ivanov and R.K. Jain, 1978, in press.
5. J. Frenkel, Kinetic Theory of Liquids, Dover Publ. Inc., New York, 1955.
6. A. de Vries, Rec. Trav. Chim., 1958, 77, 441.
7. A. de Vries, Third Congress of Detergency, 1960, 2, 566.
8. A. Scheludko, Proc. Konink. Ned. Akad. Wet., 1962, B65, 87.
9. A. Vrij, Disc. Faraday Soc., 1966, 42, 23.
10. A. Vrij and J. Overbeek, J. Amer. Chem. Soc., 1968, 90, 3074.
11. S. Chandrasekhar, Hydrodynamic and Hydromagnetic Stability, Oxford University Press, 1961; C. Lin, The Theory of Hydrodynamic Stability, Cambridge University Press, 1958.
12. B. Felderhof, J. Chem. Physics, 1968, 19, 44.
13. J. Lucassen, M. van den Tempel and F. Hesselink, Proc. Koninkl. Ned. Akad. Wet., 1970, B73, 108.
14. A. Vrij, F. Hesselink, J. Lucassen and M. van den Tempel, Proc. Koninkl. Ned. Akad. Wet., 1970, B73, 124.
15. E. Ruckenstein and R.K. Jain, J. of the Chem. Soc., Faraday Trans. II, 1974, 70, 132.
16. V. Levich, Physico-Chemical Hydrodynamics, Prentice-Hall, Englewood Cliffs, N.J., 1967.
17. S. Sche and H. Fijnaut, Dynamics of Thin Free Liquid Films Stabilized with Ionic Surfactants. Paper presented at the Fourth Rolla Conference on Surface Properties of Materials, Rolla, Missouri, 1977.
18. C. Maldarelli, M.S. Thesis, Columbia Univ., New York, 1978; C. Maldarelli, R.K. Jain, I.B. Ivanov and E. Ruckenstein, 1978, to be submitted to J. Colloid and Interface Sci.
19. J. Patzer and G. Homsy, J. Colloid and Interface Sci., 1976, 40, 318.
20. R.K. Jain and E. Ruckenstein, J. Colloid and Interface Sci., 1976, 54, 408.
21. I. Ivanov, B. Radoev, E. Manev and A. Scheludko, Trans. Faraday Soc., 1970, 66, 1262.
22. I. Ivanov, Doctoral Dissertation, University of Sofia, Bulgaria, 1977.
23. I. Ivanov and D. Dimitrov, Colloid and Polymer Sci., 1974, 252, 982.
24. I. Ivanov, B. Radoev, T.T. Traykov, D. Dimitrov, E.D. Manev and C.S. Vassilieff, Proc. Int. Con. Colloid and Surface Sci., E. Wolfram, ed., Vol. 1, p 583, Akademia Kiado, Budapest, 1975.
25. E. Manev, A. Scheludko and D. Exerova, Colloid and Polymer Sci., 1974, 252, 586.
26. B. Radoev and I.B. Ivanov, Ann. Univ. Sofia, Fac. Chimie., 1970/1971, 65, 429.
27. B. Radoev, T. Traykov and I.B. Ivanov, Ann. Univ. Sofia, Fac. Chimie., 1974/1975, 69; I.B. Ivanov, R.K. Jain, P. Somasundaran and T.T. Traynov, in "Solution Chemistry of Surfactants," K.L. Mittal, ed., Plenum Press, to appear.
28. D.S. Dimitrov and I.B. Ivanov, J. Colloid Interface Sci., 1978, 64 97.
29. R. Gumerman and G. Homsey, Chem. Eng. Commun., 1975, 2, 27.

30. B.V. Derjaguin and M. Kussakov, Acta Physicochim., U.R.S.S., 1939, 10, 25; G.A.H. Elton, Proc. Roy. Soc. (London), 1948, A194, 275; L.F. Evans, Ind. Eng. Chem., 1954, 46, 2420; R.S. Allan, G.E. Charles and S.G. Mason, J. Colloid Sci., 1961, 16, 150; G.D.M. MacKay and S.G. Mason, Canadian J. Chem. Eng., 1963, 41, 203; D. Platikanov, J. Phys. Chem., 68, 3619.
31. R.K. Jain, I.B. Ivanov and P. Somasindaran, 1978, to be submitted to J. Chem. Soc., Faraday Trans. II.
32. S. Frankel and K.J. Mysels, J. Phys. Chem., 1962, 66, 190.
33. M.P. Aronson and H.M. Princen, J. Colloid Interface Sci., 1975, 52, 345; T.D. Blake and J.A. Kitchener, J. Chem. Soc. Faraday I., 1972, 68, 1435; T.D. Blake, J. Chem. Soc. Faraday I., 1975, 71, 192; W. Donners, Ph.D. Thesis, University of Utrecht, Netherlands, 1977; S. Hartland, J. Colloid Interface Sci., 1971, 35, 227; T.D. Hodgson and J.C. Lee, J. Colloid Interface Sci., 1969, 30, 94; T.D. Hodgson and D.R. Woods, J. Colloid Interface Sci., 1969, 30, 429; K. Mushiake, T. Imaiznmi and T. Inowe, Eleventh Intl. Mineral Processing Congress, Cagliari, Italy, 1975; A.D. Read and J.A. Kitchener, J. Colloid Interface Sci., 1969, 30, 391; H.J. Schultze, Doctoral Dissertation, Freiberg, 1977; H.J. Schultze, Coll. Polymer Sci., 1977, 253, 730; 254, 438; H. Sonntag, Z. Physik. Chem., 1964, 225, 284; L. Ter Minassian Saraga, J. Chem. Phys., 1955, 52, 181.
34. Marucci, G., Chem. Eng. Sci., 1969, 24, 975.
35. A.B. Pardee, In Vitro, 1971, 7, 95-104; M.M. Burger, Current Topics Cell Reg., 1971, 3, 135-140; M.G.P. Stoker, Proc. Royal Soc., 1972, B181, 1-17; C.A. Pasternak, in Mammalian Cell Membranes, eds. G.A. James and D.M. Robinson, Butterworth, London, Vol. 3, pp 10-15, 1976; K. Porter, D. Prescott and J. Frye, J. Cell Biol., 1973, 57, 815-836; S. Knutton, M.C.B. Summer and C.A. Pasternak, J. Cell. Biol., 1975, 66, 568-576; C.A. Pasternak, J. Theor. Biol., 1976, 58, 365-382; M.C. Willingham and I. Pastan, Proc. Natl. Acad. Sci. USA, 1975, 72, 1263-1267.
36. G.L. Nicolson, Science, 1975, 188, 819-820; G.L. Nicolson, Int. Rev. Cytol., 1974, 39, 89-190.
37. L. Weiss, in Cell Periphery, Metastasis, and Other Contact Phenomena, ed. L. Weiss, North Holland, Amsterdam, p 248, 1967; L. Weiss and J.P. Harlos, in Progress in Surface Science, ed. S.G. Davison (Pergamon, Oxford), Vol. 1, pp 355-405, 1972.
38. M. Hennenberg, T. Sørensen and A. Sanfeld, J. of the Chem. Soc., Faraday Trans. II., 1977, 73, 48; A. Sanfeld and A. Steinchen, Biophysical Chem., 1975, 3, 99; A. Steinchen and A. Sanfeld, Chemical Physics, 1973, 1, 156; M. Velarde, J. Ibañez, T. Sørensen, A. Sanfeld and M. Hennenberg, Proc. Levich Conference, 1978.
39. K. DoLowy, J. Theor. Biol., 1975, 52, 83-97.
40. S. Nir, in Progress in Surface Science, ed. S.G. Davison, Pergamon, Oxford, Vol. 8, pp 1-58, 1977.
41. R.K. Jain, C. Maldarelli, I.B. Ivanov and E. Ruckenstein, 1978, Paper presented at the 71st Annual Meeting of A.I.Ch.E., Miami Beach.

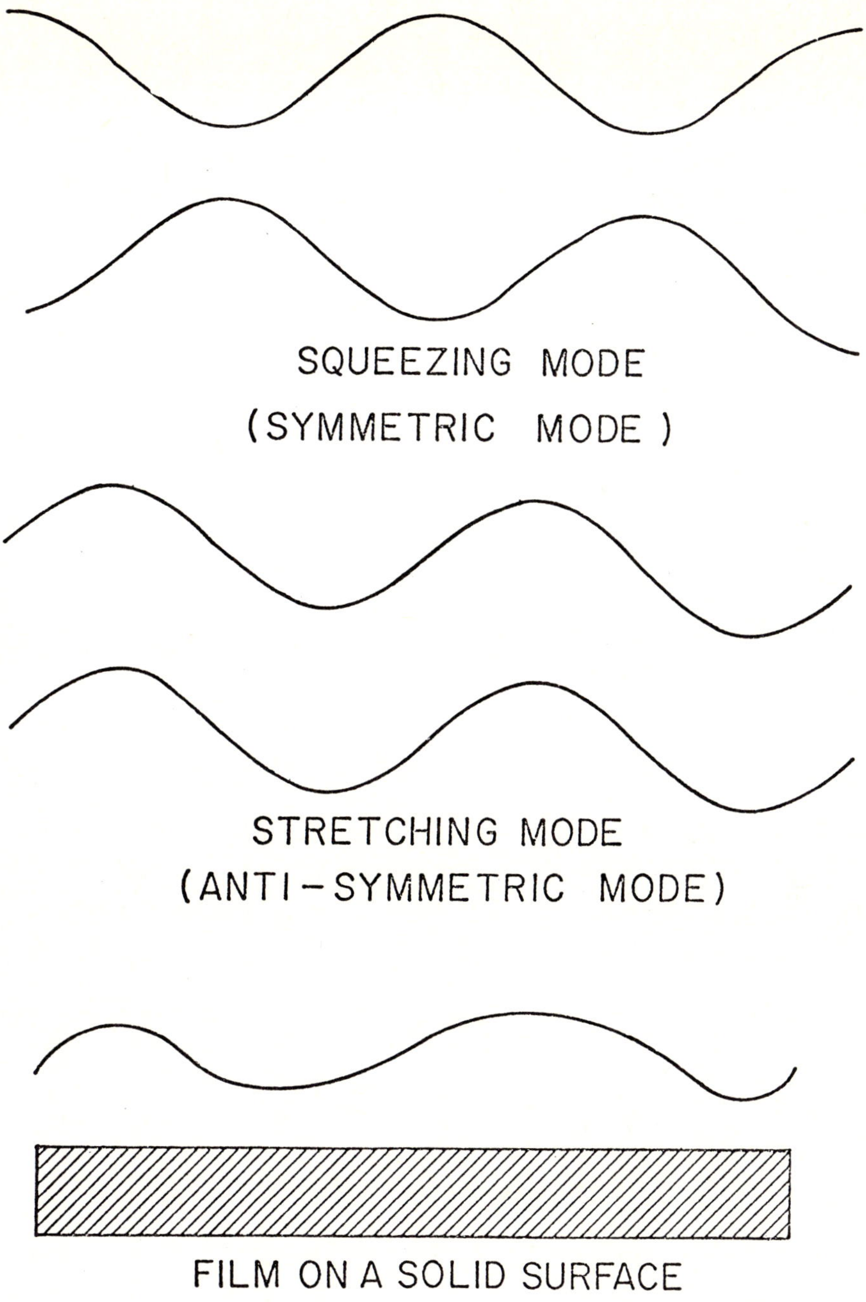

Figure 1. Modes of Instability of a Thin Liquid Film.

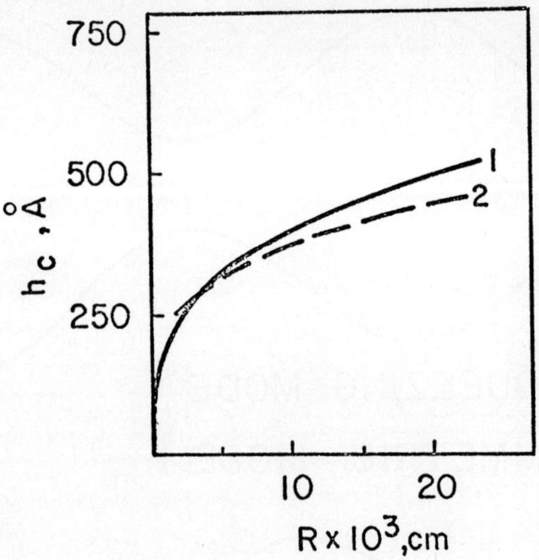

Figure 2. Experimental results (curve 2) and theoretical predictions (curve 1) for the critical thickness of aniline films with tangentially immobile surfaces.

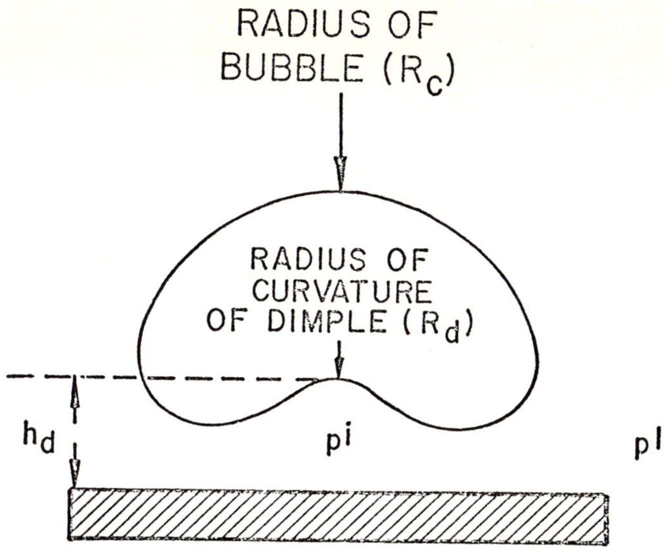

Figure 3(a). Schematic of a "dimple" formed during the approach of a bubble to a solid substrate.

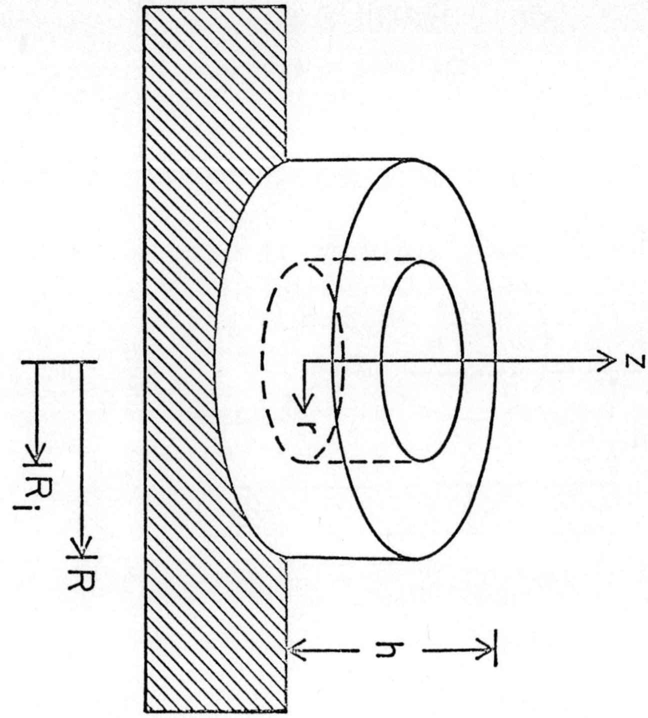

Figure 3(b). Schematic of the ring shaped film and the coordinate system.

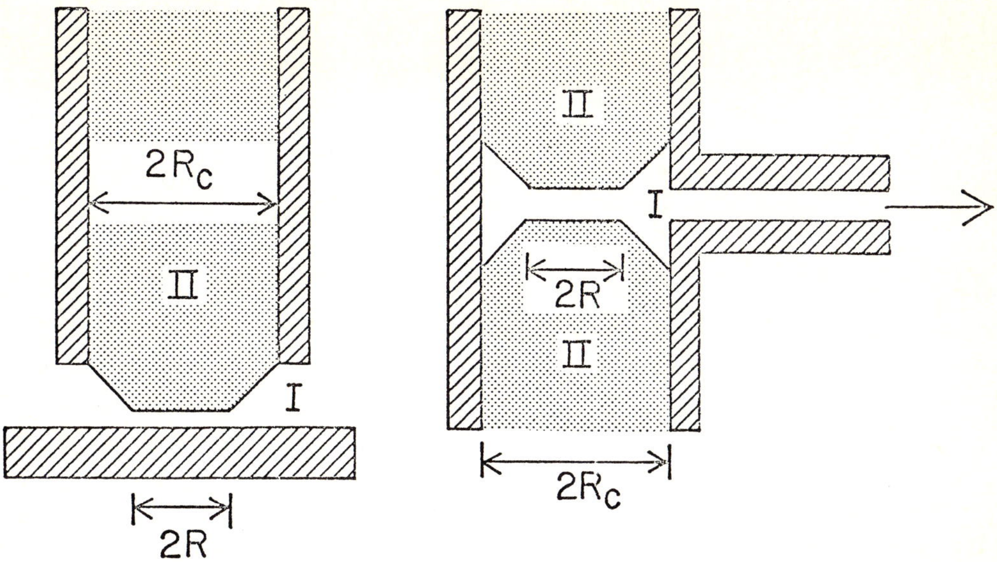

Figure 4. Schematic of experimental set-ups to study the stability of thin liquid films.

MECHANICAL, CHEMICAL, AND ELECTRICAL CONSTRAINTS AND HYDRODYNAMIC INTERFACIAL INSTABILITY.

by

A. Sanfeld, A. Steinchen, M. Hennenberg, P.M. Bisch*, D. Van Lamsweerde-Gallez and W. Dalle-Vedove

Université Libre de Bruxelles, Chimie Physique II, Campus Plaine 1050 Bruxelles, Belgium. /* CNPq - (BRAZIL).

During the last years we developed a hydrodynamic theory of surface motions induced by chemical, mechanical and electrical constraints. The treatment is based on a linear stability analysis of interfaces between two immiscible bulk phases, as was already performed by Chandrasekhar [1] and by Sternling and Scriven [2] [3].

The applied constraints maintain the system out of equilibrium. They are of various types : (a) the slope of the diffusion profiles in the two bulks ;

(b) the affinity of chemical (or electrochemical) surface reactions ;

(c) the affinity of the adsorption-desorption reactions ;

(d) the gravitational forces

(e) the electrical potential difference applied between the two bulks and the surface charge density ;

(f) the dipole orientation in the layer.

The threshold conditions for the onset of longitudinal and transversal surface motion was studied for plane and spherical systems.

For the various situations considered here, the solution of equations for velocities and for electrical potentials in each bulk phase have to satisfy boundary conditions :

classical description can be found, for example, in the excellent book of Chandrasekhar [1] . Solving the Navier-Stokes equation for each volume phase (1) and (2) by expanding the normal (coordinate Z) velocity perturbation in Fourier components, the amplitude v_z may be written as

$$v_z^{(1)} = A^{(1)} e^{kz} + B^{(1)} e^{q^{(1)} z}$$

$$v_z^{(2)} = A^{(2)} e^{-kz} + B^{(2)} e^{-q^{(2)} z} \qquad (1)$$

where k is the wave number and q is a parameter given by

$$q^2 = k^2 + \omega/\nu \qquad (2)$$

with $\nu = \mu/\varrho$ the kinematic viscosity, ω the frequency of the perturbed quantities. The real part of q must be positive or zero in order that the velocity perturbation disappear at infinite distance from the interface.

Assuming a no slip condition at the interface, we have

$$A^{(1)} = -\frac{k+q^{(1)}}{2k} B^{(1)} + \frac{k-q^{(2)}}{2k} B^{(2)}$$

$$A^{(2)} = \frac{k-q^{(1)}}{2k} B^{(1)} - \frac{k+q^{(2)}}{2k} B^{(2)} \qquad (3)$$

A) <u>Transfer of matter due to bulk diffusion for one solute</u>.

The general diffusion-convection perturbed equation in each bulk is

$$\frac{\partial}{\partial t} \delta c + \underset{\sim}{v} \cdot \underset{\sim}{\nabla} c^{\circ} = \mathcal{D} \nabla^2 \delta c \qquad (4)$$

where c° is the one-dimensional concentration distribution with the liquid at rest in our reference state (superscript zero), and $\mathcal{D}$ the diffusion coefficient. We have analyzed two cases in the unperturbed state : first a linear concentration profile solution of eq. (4) and secondly an exponential profile.

no slip of velocities and of electrical potentials at the the interface

Laplace and Marangoni relations

Gauss equation for the electrical field

Surface mass balances

The surface tension is the fundamental coupling parameter between the chemical, the hydrodynamical, and the electrical processes.

The theory may be relevant for the spontaneous onset of interfacial "turbulence" [4] - [9]. The biological membranes constitute a very promising field of application of these methods. It could give informations on cell motions and on primary deformations of cell surfaces induced by chemical or electrical signals during chemotactic motion, phagocytosis and cell fusion [10]-[13]. As a first step in this direction the last part of our paper is then devoted to very crude models of biosurfaces.

This type of instabilities enters the general framework of the concept of "dissipative structures" of Prigogine-Glansdorff (GP) and Nicolis [14] - [16].

An extension of the GP thermodynamic theory of stability for systems in an electromagnetic field and for uncharged surfaces was also developed recently by some of us [17] -[19] but it will not be reported here.

1. Neutral systems.

Let us first consider a flat infinite interface separating the two bulks in a reference steady state at rest. Later on, we will consider also spherical systems.

The Newtonian lower (1) and upper (2) bulks are incompressible in spite of the peculiar repartition of the solute. The general solutions for small perturbations of the linearized hydrodynamic bulk equations are obtained classically. This

Linear profile

In the unperturbed state we have

$$c^{o} = c^{oo} - \beta z \tag{5}$$

Assuming no accumulation of matter at the interface (equality of the fluxes at the surface)

$$\beta^{(1)}\mathcal{D}^{(1)} = \beta^{(2)}\mathcal{D}^{(2)} \tag{6}$$

where both coefficients are positive when diffusion is directed from phase 1 to phase 2.

Let us now assume that, along the perturbed state, the perturbed adsorption is proportional to the perturbed bulk solute concentration at the sublayer. It means

$$\Gamma = K^{(1)}\delta c^{(1)} = K^{(2)}\delta c^{(2)} \tag{7}$$

The linearized perturbed surface mass balance for the solute is

$$\frac{\partial}{\partial t}\delta\Gamma + \Gamma^{o}\,\underset{\sim}{\nabla}_{s}\cdot\underset{\sim}{v}_{s} - \mathcal{D}^{s}\nabla_{s}^{2}\delta\Gamma = \Delta\left[\mathcal{D}\nabla_{n}\delta c\right] \tag{8}$$

Here and in the following we use the abbreviation ΔX for X (liquid 2) - X (liquid 1) when the surface discontinuity is passed. The perturbed linearized surface normal momentum balance is

$$\Gamma^{o}_{tot}\frac{\partial}{\partial t}v_{z} = \Delta\left\{-\delta p + 2\mu\frac{\partial v_{z}}{\partial z} + g\rho\,\delta z\right\} + \sigma^{o}\nabla_{s}^{2}\delta z \tag{9}$$

where δp is the bulk pressure perturbation, δz is the normal deformation of the surface, $g\rho\delta z$ is the gravitational effect, σ^{o} is the unperturbed surface tension and Γ^{o}_{tot} the total adsorption in the reference state.

The perturbed linearized tangential components i of the momentum balance are taken along the interface

$$\Gamma^{o}_{tot}\frac{\partial}{\partial t}v_i^s = \Delta\left\{\mu\left(\frac{\partial v_i}{\partial z}+\frac{\partial v_z}{\partial x_i}\right)\right\}+\frac{\partial\delta\sigma}{\partial x_i}-\eta_d\left(\frac{\partial}{\partial x_i}\frac{\partial v_z}{\partial z}\right)+\eta_s\nabla_s^2 v_i^s \qquad (10)$$

where η_d and η_s are the intrinsic surface dilatational and shear viscosities respectively.

The coupling conditions between hydrodynamical (convection) and chemical effects (mass balance) is given by the surface state equation relating surface tension to adsorption

$$\delta\sigma = -\alpha\delta\Gamma \qquad (11)$$

where α is a positive quantity, and also by the quantity $\Gamma^{o}\nabla_s\cdot v_s$ Remembering that the densities of the bulk liquids 1 and 2 are represented as Heaviside step functions in the distance perpendicular to the surface, there will be no difficulties in defining the position of the surface phase (unique position of the Gibbs dividing surface) [20] . Combining Eq. (1) - (10) it is easy to show the generalized compatibility condition [21]

$$\begin{vmatrix} \omega+k^2\mathcal{D}^s+\sum\frac{\mathcal{D}\cdot r}{K} & -\beta^{(1)}\mathcal{D}^{(1)}\sum\frac{q+k}{q+r} & \Gamma^{o}-\frac{\beta^{(1)}\mathcal{D}^{(1)}}{\omega}\cdot\Delta\left\{\frac{q+k}{q+r}\right\} \\ 0 & -\frac{1}{k}\left[\omega\left(\frac{\Sigma\rho}{k}+\Gamma_t^{o}\right)+\frac{k^2\sigma^{o}-g\Delta\rho}{\omega}\right]-\sum\left\{\mu\left(1+\frac{q}{k}\right)\right\} & +\Delta\left\{\mu\left(1-\frac{q}{k}\right)\right\} \\ \alpha k & +\Delta\left\{\mu\left(1-\frac{q}{k}\right)\right\} & -\frac{1}{k}\left[\Gamma_t^{o}\omega+(\eta_s+\eta_d)k^2+\sum\mu(k+q)\right] \end{vmatrix} = 0 \qquad (12)$$

where $$r^2 = k^2 + \omega/\mathcal{D} \qquad (13)$$

This secular equation was studied to define the critical values of the marginal exchange of stability and overstability.

Exchange of stability.

The critical wavenumber k_{cr} is the positive real solution of

$$\left[\frac{\beta^{(1)}-\beta^{(2)}}{4k^2\Gamma^o}\right]_{cr} = 1 + \frac{1}{\alpha\Gamma^o}\left[2\Sigma\mu + k(\eta_s+\eta_d)\right]\cdot\left[k\mathcal{D}_s + \Sigma\frac{\mathcal{D}}{K}\right] \quad (14)$$

The onset of deformational non oscillating instabilities of a plane liquid-liquid interface due to transfer of surfactant by diffusion is given by the general condition

$$\beta^{(1)}-\beta^{(2)} = \beta^{(1)}\cdot\left[1-\frac{\mathcal{D}^{(1)}}{\mathcal{D}^{(2)}}\right] > 0 \quad (15)$$

In this model, for spontaneous deformations of the interface to occur, the diffusion has to be directed from the liquid phase with the smallest diffusion coefficient to the liquid phase with the greatest. All wavenumbers below a certain critical number k_{cr} will be unstable for a fixed difference of slopes of the concentration profiles. For a fixed wavenumber the difference of sloples has to exceed an instability threshold which increases with the bulk and surface viscosities and the diffusion coefficients, and decreases with increasing of the coupling coefficient between the surface pressure and the surface concentration of the solute.

Furthermore there must be a wavenumber of maximum instability k_{max} for which ω has its maximum value, such that

$$0 < k_{max} < k_{cr} \quad (16)$$

This wavelength is apt to become the dominant, when the perturbations proceed into the non-linear region, and the convection cells in the final "dissipative structure" are likely to have dimensions determined by the value of k_{max}. This is a generalisation of the earlier result of Sternling-Scriven [2] obtained for longitudinal waves and neglecting surface-viscosities, accelerations and diffusion processes.

Both our model and the one studied by Sternling-Scriven are in good qualitative agreement with experimental results obtained by Orell-Westwater [22] for the system ethyleneglycol-ethylacetate with acetic acid as diffusing surfactant [23].

Dispersion curves depend strongly on surface viscosities. More recently another model was used on the basis of pure longitudinal waves. We found ((24) and table 2) also possibilities of non oscillating instability for the case

$$1 < \frac{\mathcal{D}^{(1)}}{\mathcal{D}^{(2)}} < \frac{\nu^{(1)}}{\nu^{(2)}} \qquad (\text{if } \beta^{(1)} > 0 \text{ and } \beta^{(2)} > 0) \tag{17}$$

and

$$1 < \frac{\mathcal{D}^{(2)}}{\mathcal{D}^{(1)}} < \frac{\nu^{(2)}}{\nu^{(1)}} \qquad (\text{if } \beta^{(1)} < 0 \text{ and } \beta^{(2)} < 0) \tag{18}$$

Overstability.

In cases of diffusion from the phase with the higher diffusion coefficient to the phase with the lower, also both models exhibit oscillatory instabilities with critical overstability boundaries [25].

Two different analyses were developed, a numerical approach by Sørensen and al. [26] and a more fundamental analytical development [24] in the case of pure longitudinal waves (see § C)) the general conditions of oscillating instabilities are

$$\frac{\mathcal{D}^{(1)}\nu^{(2)}}{\mathcal{D}^{(2)}\nu^{(1)}} > 1 \qquad (\text{if } \beta^{(1)} > 0 \text{ and } \beta^{(2)} > 0 \text{ transfer } 1 \rightarrow 2) \tag{19}$$

$$\frac{\mathcal{D}^{(2)}\nu^{(1)}}{\mathcal{D}^{(1)}\nu^{(2)}} > 1 \qquad (\text{if } \beta^{(1)} < 0 \text{ and } \beta^{(2)} < 0 \text{ transfer } 2 \rightarrow 1) \tag{20}$$

Exponential profile

A better model for experiments such as those of Orell-Westwater [22] would take into account the finiteness of the diffusion profile. Rigourously an error function type of solution should be used as the reference profile. However, it might be considered that an exponential profile is a qualitatively good approximation. In this case

$$c^{o\,(1)} = c^{\infty\,(1)} + c^{oo\,(1)}\, e^{\beta^{(1)} z}$$

$$c^{o\,(2)} = c^{\infty\,(2)} + c^{oo\,(2)}\, e^{-\beta^{(2)} z} \tag{21}$$

Sørensen and al.[25] [26] obtained a critical condition for the existence of an exchange of stability

$$1 + \frac{1}{\alpha\Gamma^o}\left\{2\,\Sigma\mu + k(\eta_s+\eta_d)\right\}\cdot\left\{k\mathcal{D}^s + \Sigma\frac{\mathcal{D}}{K}\right\}$$

$$= \beta^{(1)}\mathcal{D}^{(1)}\,\frac{c^{oo(1)}}{\Gamma^o}\,\Delta\left\{\frac{1}{\mathcal{D}(2k+\beta)^2}\right\} \tag{22}$$

A computer analysis of Eq. (22) shows that new possibilities arise. Two critical boundaries, one single critical boundary or even no critical boundary for exchange of stabilities are found, with decreasing thickness of the diffusion zone and increasing surface viscosities. The wavelength of maximum instability is of the order of magnitude of the initial cell dimensions given by Orell-Westwater.

B) <u>Adsorption-desorption and surface chemical reactions</u>.

The non equilibrium situations may be characterized by kinetics of adsorption-desorption and by chemical reactions. These processes are considered here as determining steps. The volume phases are instantaneously homogeneous.

The mass balance at the surface is then (instead of Eq. (8))

$$\frac{\partial\delta\Gamma_\gamma}{\partial t} = \sum_\beta G_{\gamma\beta}\,\delta\Gamma_\beta + \mathcal{D}^s_\gamma\,\nabla^2\delta\Gamma_\gamma - \Gamma^o_\gamma\,\underset{\sim}{\nabla}_s\cdot\underset{\approx}{V}_s \tag{23}$$

where

$$G_{\gamma\beta} = \frac{\partial F_\gamma}{\partial\Gamma_\beta} + \frac{\partial J_\gamma}{\partial\Gamma_\beta} \tag{24}$$

is the kinetic matrix element related to the source term F_γ due to chemical surface reactions, and to the adsorption-desorption fluxes J_γ.

Plane interface.

The momentum surface balances (9) and (10) remain unchanged so that combining them with Eq. (23) we obtained the compatibility condition [27]

$$\left|\begin{array}{c:c:c} -G_{\gamma\beta}+\delta_{\gamma\beta}(\omega+k^2\mathcal{D}^s_\gamma) & 0 & -\Gamma_\gamma \\ \hdashline \alpha_\beta\cdot k^2 & \Delta\left[\mu(q-k)\right] & \begin{array}{l} k^2(\eta_s+\eta_d) \\ +\omega\Gamma_{tot} \\ +\sum\{\mu(q+k)\} \end{array} \\ \hdashline 0 & \begin{array}{l} \omega\left(\Gamma_{tot}+\frac{\sum\varrho}{k}\right) \\ +\frac{\sigma k^2-g\Delta\varrho}{\omega} \\ +\sum\{\mu(q+k)\} \end{array} & \Delta\{\mu(q-k)\} \end{array}\right| = 0 \tag{25}$$

For the fluctuating species $G_{\gamma\beta}=G\,;\,\nu_{\gamma\beta}=\nu$ we obtained a marginal state of exchange of stability given by the conditions (see [27] eq. (30))

$$G^{cr} = k^2\mathcal{D}^s + \frac{\alpha\Gamma^\circ k}{k(\eta_s+\eta_d)+2\sum\mu} \tag{26}$$

As intuitively expected we obtained in this neutral situation an exact compensation between the chemical terms G^{cr} and the surface diffusion $\mathcal{D}^s$, the surface activity $\alpha\Gamma^\circ$, the surface and bulk viscosities (η, μ) . A mathematical analysis of the secular equation shows that for small wavelengths, a necessary but not sufficient condition for instability is that the chemical surface reaction of the adsorption-desorption kinetics is unstable by itself which means for example non linear kinetics $(C > 0)$. The effects due to surface diffusion, to surface activity, and to viscosities are stabilizing.

For more than one fluctuating component the coupling between hydrodynamics and out of equilibrium reactions gives rise to new varieties of instabilities. Indeed they may even occur for intrinsic stable kinetics [28] - [30] (see also next paragraph). Furthermore domains of instability might be widely enlarged by the hydrodynamic coupling [29] -[30] . Numerical simul ations have been performed on our theory during the last year. They confirm our theoretical predictions.

First a trimolecular scheme was investigated [31] [32] . Later on a cooperative transformation reaction based on the Changeux and al. [33] mechanism was developed [34][32] . More recently an autocatalytic bimolecular scheme (with also an inhibition step) involving a saturation law of the Langmuir-Hinshelwood type was investigated by Velarde and al. [29][30] .

Analytical analysis for two fluctuating species.

Recently, W. Dalle Vedove and al. [35] have derived the general conditions of instability for two fluctuating species, for zero bulk viscosities, for any reaction scheme. They obtained a third order general dispersion equation :

$$\omega^3 + a_1 \omega^2 + a_2 \omega + a_3 = 0 \tag{27}$$

with

$$a_1 = k^2 \left\{ \frac{\eta_s + \eta_d}{\Gamma_{tot}} + \mathcal{D}_1^s + \mathcal{D}_2^s \right\} - \left\{ C_{11} + C_{22} \right\} \tag{28}$$

$$a_2 = k^4\left[(\mathcal{D}_1^s + \mathcal{D}_2^s)\cdot\left(\frac{\eta_s+\eta_d}{\Gamma_{tot}}\right) + \mathcal{D}_1^s\mathcal{D}_2^s\right]$$

$$+ k^2\left[\frac{\Gamma_1^o\alpha_1}{\Gamma_{tot}} + \frac{\Gamma_2^o\alpha_2}{\Gamma_{tot}} - \left(\frac{\eta_s+\eta_d}{\Gamma_{tot}}\right)(C_{11}+C_{22}) - C_{11}\mathcal{D}_2^s - C_{22}\mathcal{D}_1^s\right]$$

$$+ C_{11}C_{22} - C_{12}C_{21} \tag{29}$$

$$a_3 = k^6\left[\left(\frac{\eta_s+\eta_d}{\Gamma_{tot}}\right)\mathcal{D}_1^s\mathcal{D}_2^s\right]$$

$$+ k^4\left[\frac{\Gamma_1^o\alpha_1}{\Gamma_{tot}}\mathcal{D}_2^s + \frac{\Gamma_2^o\alpha_2}{\Gamma_{tot}}\mathcal{D}_1^s - \left(\frac{\eta_s+\eta_d}{\Gamma_{tot}}\right)(C_{11}\mathcal{D}_2^s + C_{22}\mathcal{D}_1^s)\right]$$

$$+ k^2\left[\left(\frac{\eta_s+\eta_d}{\Gamma_{tot}}\right)\cdot(C_{11}C_{22} - C_{12}C_{21})\right.$$

$$\left. + \frac{\alpha_1}{\Gamma_{tot}}\cdot(\Gamma_2^o C_{12} - \Gamma_1^o C_{22}) + \frac{\alpha_2}{\Gamma_{tot}}(\Gamma_1^o C_{21} - \Gamma_2^o C_{11})\right] \tag{30}$$

Situations for which the chemical matrix elements fulfil the drastic conditions :

$$C_{11} < 0$$

$$C_{22} < 0 \tag{31}$$

$$C_{11}C_{22} > C_{12}C_{21}$$

corresponding to intrinsic chemical stability, have been tested. The analysis of the Routh-Hurwitz criteria for equation (27), giving the stability conditions for the coupled system, is summarized in table 1.

TABLE 1

c_{11}	c_{22}	c_{12}	c_{21}	CONCLUSIONS
<0	<0	0	0	ALWAYS STABLE
		<0	<0	POSSIBLE EXCHANGE OF STABILITY
		>0	>0	POSSIBLE OVERSTABILITY
		<0 >0	>0 <0	POSSIBLE { EXCHANGE OF STABILITY OVERSTABILITY

Unstable situations can be obtained according to the values assumed for the chemical matrix elements $C_{\gamma\beta}$, for the coefficients

$$\alpha_1 = \left(\frac{\partial\sigma}{\partial\Gamma_1}\right)_{T,\Gamma_2} \qquad \alpha_2 = \left(\frac{\partial\sigma}{\partial\Gamma_2}\right)_{T,\Gamma_1}$$

and for the damping parameters : η_s, η_d, $\mathcal{D}_1^s$ and $\mathcal{D}_2^s$.

For a linear kinetics as was proposed by M. Dupeyrat and E. Nakache [37] involving species $C_{16}N^+(CH_3)_3Cl^-$ and picric acid at water-nitrobenzene interface, we have shown that no interfacial motion could occur. Indeed, the chemical kinetics can be represented by the following set of equations

$$[C_{16}N^+(CH_3)_3]_{water} \underset{k_1'}{\overset{k_1}{\rightleftharpoons}} [C_{16}N^+(CH_3)_3]^s \underset{k_2'}{\overset{k_2}{\rightleftharpoons}} [C_{16}N^+(CH_3)_3]_{nitro} \qquad (a)$$

$$[P_i^-]_{water} \underset{k_4'}{\overset{k_4}{\rightleftharpoons}} [P_i^-]^s \underset{k_5'}{\overset{k_5}{\rightleftharpoons}} [P_i^-]_{nitro} \qquad (b)$$

$$[C_{16}N^+(C)_3P_i^-]_{water} \underset{k_6'}{\overset{k_6}{\rightleftharpoons}} [C_{16}N^+(C)_3P_i^-]^s \underset{k_7'}{\overset{k_7}{\rightleftharpoons}} [C_{16}N^+(C)_3P_i^-]_{nitro} \qquad (c)$$

$$[P_i^-]_s + [C_{16}N^+(CH_3)_3]_s \underset{k_3'}{\overset{k_3}{\rightleftharpoons}} [C_{16}N^+(CH_3)_3P_i^-]_s \qquad (d)\ (32)$$

Steps (a)-(c) describe a Langmuir adsorption-desorption mechanism and (d) is a chemical surface reaction. The k_i represents the kinetic constant for the direct step i and k_i' the one for the reverse step. For the two surface active species $C_{16}N^+(CH_3)_3$ and $C_{16}N^+(CH_3)_3Pi^-$, the chemical matrix elements read for an ideal system:

$$C_{11} = -\left\{k_3' + k_6' + k_7 + \frac{1}{N}\left(k_8[C_{16}N(C)_3Pi]_w + k_7'[C_{16}N(C)_3Pi]_n\right)\right\}$$

$$C_{22} = -\left\{k_3[Pi^-]_s + k_1' + k_2 + \frac{1}{N}\left(k_1[C_{16}NC_3^+]_w + k_2'[C_{16}NC_3^+]_n\right)\right\}$$

$$\begin{aligned} C_{12} &= k_3[Pi^-]_s \\ C_{21} &= k_3' \end{aligned} \qquad (33)$$

where N is the saturation adsorption

Using the stationarity conditions, we can easily show that the Routh-Hurwitz criteria for equation (27) are always satisfied, even far from equilibrium. Nevertheless, other mechanisms involving non linear steps could exhibit hydrochemical instabilities far from equilibrium, even for situations satisfying the pure chemical stability conditions. As an example, we suggest the formation of triplets between two cations $C_{16}N^+(CH_3)_3$ and one anion Pi^- at the interface

$$(Pi^-)_s + 2[C_{16}N^+(CH_3)_3]_s \rightleftharpoons [C_{16}N^+C_3Pi^-C_{16}N^+C_3]_s \qquad (34)$$

Another scheme would take into account for non linear solvatation mechanism as proposed by Davies (see p. 337 [37]) (for uranyl nitrate between water and organic solvents). These matters are under consideration.

There is no experimental evidence for unstable chemical reactions per se in the example treated by M. Dupeyrat and E. Nakache. Our purpose is to investigate the occurrence of hydro-chemical instabilities for intrinsicly stable non linear chemical mechanisms.

For systems with non zero bulk viscosities, the secular equation is of higher order and the stability analysis requires numerical methods of computation. We will see later on that electrical effects can also be coupled with chemical mechanisms.

Spherical Interface.

A parallel analysis was performed a few years ago for a spherical interface [38][28]. For one fluctuating specie the state of marginality is determined by see [28] Eq. (63)

$$C^{cr} = \ell(\ell+1)\left[\mathcal{D}^s k^2 + \alpha \Gamma^o/\delta\right] \tag{35}$$

where

$$\delta = \ell(\ell+1)(\eta_s + \eta_d) - 2\eta_s + R(2\ell+1)\Sigma\mu \tag{36}$$

with R the radius of the drop and $\ell \geq 1$ the order of the surface harmonic.

Comparing Eq. (26) to Eq. (27), it is interesting to note that the same formulation describes the neutral state for plane and spherical surfaces.

The first mode $\ell = 0$ corresponding to the pure dilatation of the sphere must be excluded because of the incompressibility condition.

The $\ell = 1$ normal mode corresponds to a net movement of the center of gravity of the drop (movement "in toto"). It is a non uniform translation. Possibilities of "kicking" of the drop may be related to gradients of surface tension along the surface. Chemotactic motion in biological systems could be related to such types of instabilities [11].

For $\ell \geq 2$, we found again all the conclusions obtained for the plane surface.

C) Overstability of pure longitudinal waves.

The approach used to investigate oscillatory marginal stability, for longitudinal waves, is similar to that of Lucassen and vanden Tempel. [31]-[41] The treatment is the same as in the first section, but it is restricted to uncoupled longitudinal and capillary waves. A detailed analysis [24] was performed for large wavelengths. It takes into account the convection in the mass balan ce equations in bulk phases. The complex elasticity modulus, related to the transfer by diffusion, of a soluble surfactant, takes then a more general form. According to the values obtained for this quantity, we can predict oscillatory marginal states, separating stable and unstable regions. We found back, among others, the results previously obtained by Sternling and Scriven [2].

The results obtained are summarized in table 2. The transfer of matter by adsorption-desorption kinetics is now under investigation with the same approach.

2. Charged systems.

Fluid-fluid interfaces can also exhibit electrical charges. Constraints can be imposed, for instance an electrical field, a non equilibrium density of charges, or an electrical current. For example, at water-oil interfaces in the presence of surface active ions, processes of electro-adsorption are observed in certain conditions [42]-[44]. For particular values of the constraints these non-equilibrium systems may be destabilized. Dupeyrat, Guastalla and al. [6][36][45][46] observed spontaneous interfacial motions in the nitrobenzene-water systems when a non-equilibrium concentration of the surface-active ions is imposed. Joos and Verburgh [44] reported on excited longitudinal waves experiments at the nitrobenzene-water interface. They observed less damped waves when ion fluxes cross the interface. These experiments suggest that interfacial instabilities may be related to mass transfer and adsorption kinetics at the interface. Furthermore electrical double layers may also play an important role. Melcher and Zelazo [47][48] considered charged monolayers at liquid-air interfaces with imposed

TABLE 2

NECESSARY CONDITIONS FOR INSTABILITY THROUGH DIFFUSION (transfer from phase a to b with a = 1, 2 and b = 2, 1).

diffusion coefficients and kinematic viscosities	results
$\mathcal{D}^a < \mathcal{D}^b$	unstable states without oscillations - marginal states with exchange of stability
$\dfrac{\mathcal{D}^a \nu^b}{\mathcal{D}^b \nu^a} < 1$ with	unstable states with oscillations
$\dfrac{\mathcal{D}^b}{\mathcal{D}^a} > 1 > \dfrac{\nu_b}{\nu^a}$	marginal states with exchange of stability
$1 < \dfrac{\mathcal{D}^a}{\mathcal{D}^b} < \dfrac{\nu_a}{\nu_b}$	marginal states with exchange of stability
$\dfrac{\mathcal{D}^b}{\mathcal{D}^a} > \dfrac{\nu^b}{\nu^a} > 1$	marginal states with exchange of stability
$\dfrac{\mathcal{D}^a \nu^b}{\mathcal{D}^b \nu^a} > 1$ with	unstables states with oscillations
$\dfrac{\mathcal{D}^a \nu^b}{\mathcal{D}^b \nu^a} > \dfrac{\nu^a}{\nu^b} > 1$	marginal states with overstability
$\dfrac{\mathcal{D}^a}{\mathcal{D}^b} > 1 > \dfrac{\nu^a}{\nu^b}$	marginal states with overstability
$\dfrac{\nu_b}{\nu_a} > \dfrac{\mathcal{D}^b}{\mathcal{D}^a} > 1$	marginal states with overstability

electric field. Hobur and Melcher [49] discussed transport of charges through a varying conductivity interface in miscible liquids.

In this paragraph, we performed a linear stability analysis for a plane charged layer surrounded by diffuse layers in both adjacent fluids. This recent development [50] [51] generalizes the theory of Miller-Scriven [52].

Basic Equations.

The present analysis is restricted to the following hypotheses :

a) validity of the equilibrium macroscopic Boltzmann-Poisson equation both in the steady state and in the perturbed state. It means that the bulk diffusion is very fast, comparing to the other kinetic processes. The transfer of charges through the interface is thus determined by the surface adsorption-desorption step [53].

b) small potential difference across each diffuse layer and very dilute solutions.

Basic Equations.

In a charged fluid the convective motion is coupled to the electrical effects by the ponderomotive forces [54] . Combining the Poisson equation and the equilibrium Boltzmann distribution in the diffuse layers we obtain

$$\nabla^2 \varphi = \frac{4\pi}{\varepsilon} \sum_i z_i \rho_i(\infty) e^{-\frac{n_i \mathcal{F}}{RT}(\varphi - \varphi_\infty)} \qquad (37)$$

where z_i is the charge by unit mass for the ionic species i , n_i is the valence, ε is the dielectric constant and $\varphi(\infty)$ is the electrical potential in the bulk phase.

The convective motion is described by

$$\rho \frac{d\underset{\sim}{v}}{dt} = -\underset{\sim}{\nabla} p + \mu \nabla^2 \underset{\sim}{v} + \underset{\sim}{F} \qquad (38)$$

where $\underset{\sim}{F}$ is the electrical force resulting from the Lorentz and the Kelvin forces, and p is the Kelvin pressure [55] The potential distribution (37) implies that $\underset{\sim}{F}$ derives from a scalar potential.

Steady state.

From Eq. (37) we have in the steady state [56]

$$\bar{\varphi}^{o(1)}(z) = \bar{\varphi}^{o(1)}(0)\, e^{+\kappa^{*(1)} z}$$

$$\bar{\varphi}^{o(2)}(z) = \bar{\varphi}^{o(2)}(0)\, e^{-\kappa^{*(2)} z} \tag{39}$$

with for each phase

$$\bar{\varphi}^{(o)(\ell)} = \varphi^{o(\ell)} - \varphi_{\infty}^{o(\ell)} \tag{40}$$

$$\kappa^{*(\ell)} = \left[\frac{4\pi}{\varepsilon^{(\ell)}} \cdot \sum_i \frac{n_i z_i}{\varphi^*} \rho_i^{o(\ell)}(\infty)\right]^{1/2} \tag{41}$$

$$\varphi^* = \frac{RT}{\mathcal{F}} \tag{42}$$

κ^* is the reciprocal Debye length. For a charged surface without dipoles, we may write the following boundary conditions :

$$\varphi^{o(1)}(0) = \varphi^{o(2)}(0) = \varphi^{oS} \tag{43}$$

and

$$\varepsilon^{(1)}\kappa^{*(1)}\bar{\varphi}^{o(1)} + \varepsilon^{(2)}\kappa^{*(2)}\bar{\varphi}^{o(2)} = -4\pi\sum_i z_i\Gamma_i^o \tag{44}$$

Out of equilibrium, the imposed constraints $\Delta\varphi_{\infty}^{o}$ and $\sum_i z_i\Gamma_i^o$ have arbitrary values.

At true equilibrium these quantities are determined by the equality of electrochemical potentials between the surface and the two bulks.

Perturbed state.

If we assume that the diffuse layers remain at instantaneous equilibrium, the equations for the velocities and for the potential are uncoupled.

From Eq. (37) and (38) we obtain the perturbed velocities given by Eq. (1) and the electrical potential

$$\delta\varphi^{(1)}(z) = C^{(1)} e^{g^{(1)}z}$$

$$\delta\varphi^{(2)}(z) = C^{(2)} e^{-g^{(2)}z} \tag{45}$$

where

$$g = \left[k^2 + \chi^{*2} \right]^{1/2} \tag{46}$$

We assume here also a no slip velocity condition and the continuity of the electrical potential at the surface (Eq. (43)). The Gauss relation (44) is taken into account in the perturbed state. The perturbed linearized normal and tangential surface momentum balances read respectively for one normal mode

$$k^4 \omega \Gamma^o D v^s + k^4 \frac{\sigma^o}{\omega} v^s = \Delta_s\left[\omega \varrho^o D v_z - \mu(D^2+k^2) D v_z + 2\mu k^2 D v_z\right] - \frac{k^2}{4\pi}\left\{\Delta_s\left[\varepsilon \chi^{*2} \bar{\varphi}^o \delta\varphi\right]_{z=0} - \sum_\ell \varepsilon^\ell \chi^{*\ell} \bar{\varphi}^{o\ell} (D\delta\varphi)^\ell\right\} \tag{47}$$

$$\omega \Gamma^o D v^s = k^2 \delta\sigma - (\eta_s + \eta_d) k^2 D v^s + \Delta_s\left[\mu(D^2+k^2) v_z\right] - \frac{k^2}{4\pi} \sum_\ell \varepsilon^\ell \chi^{*\ell} \bar{\varphi}^{o\ell} \delta\varphi^\ell \Big|_{z=0} \tag{48}$$

with

$$D = \frac{d}{dz}$$

The electrical terms in these two relations come from the classical Maxwell tensor [54]. The surface balance for the linearized perturbation reads

$$\omega \delta\Gamma_i = -D_i^s\left\{\kappa^2 \delta\Gamma_i + \frac{n_i \Gamma_i^o}{\varphi^*} \chi^2 \delta\varphi^s\right\} + \Gamma_i^o D v^s + \Delta_s[\delta J_i^z] \tag{49}$$

where the net flux of i through the surface $\Delta_s[\delta J_i^z]$ is given by (see [51] Eq. (17))

$$\Delta_s[\delta J_i^z] = C_i\left(\delta\Gamma_i + \frac{n_i \Gamma_i^o}{\varphi^*} \delta\varphi^s\right) \tag{50}$$

the perturbed state equation is here

$$\delta\sigma = -\sum_i \alpha_i \,\delta\Gamma_i \tag{51}$$

Results.

The present treatment was applied to four simple examples.

1. The surface convection is dominant with regard to diffusion in the surface and to adsorption-desorption. The mass balance equation (49) reduces then to

$$\omega\,\delta\Gamma_i = \Gamma_i^{o}\, D v^{s} \tag{52}$$

For zero bulk viscosities and for long wave perturbations $(\varkappa^*/k \gg 1)$, the secular equation reads

$$\left[\omega^2\left(\Gamma^o + \frac{\Sigma\rho^{o\ell}}{k}\right) + k^2\sigma_T\right]\cdot\left[\omega^2\Gamma^o + \omega^2 k^2(\eta_s+\eta_d) + k^2\Pi_T\right] + k^4\Theta = 0 \tag{53}$$

where

$$\sigma_T = \sigma - \sum_\ell \left(\varepsilon^{(\ell)}/8\pi\right)\varkappa^{*\ell}\left(\bar\varphi^{o\ell}(0)\right)^2 = \sigma_c + \sigma_e \tag{54}$$

$$\Pi_T = \sum_i \alpha_i\Gamma_i^o + \frac{4\pi}{\sum_\ell \varepsilon^{(\ell)}\varkappa^{*(\ell)}}\left(\sum_i z_i\Gamma_i^o\right)^2 \tag{55}$$

$$\Theta = \frac{\Sigma z_i\Gamma_i^o}{\Sigma\varepsilon^{(\ell)}\varkappa^{*(\ell)}}\cdot\frac{1}{4\pi}\Delta_s\left(\varepsilon\bar\varphi^o\right)\Delta_s\left[\varepsilon\varkappa^{*2}(\bar\varphi^o)^2\right] \tag{56}$$

The total surface tension σ_T contains both a chemical contribution σ_c and a negative electrical contribution σ_e due respectively to the short range van der Waals interactions and to the long and middle range electrical interactions (charges and dipoles) in the double layers. see [55] chap. 17 . The electrical contribution depends on the composition of the bulks and of the imposed constraints $\Delta\varphi^o(\infty)$. As already shown by Miller and Scriven [52] , it has a destabilizing character. The positive quantity $4\pi(\Sigma z_i\Gamma_i^o)^2/\Sigma\varepsilon^{(\ell)}\varkappa^{*(\ell)}$ is the electrical contribution to the surface elasticity Π_T (electrical energy of the capacitance of the double layers) and Θ is the electrical coupling between transversal and longitudinal waves. This last quantity is related to the electrical disjoining pressure $\varepsilon^{\ell}(\kappa^{*\ell})^2(\varphi^{o\ell})^2/4\pi$. These new quantities Θ and Π_e are related to the electrical contribution to the Marangoni effect neglected by Miller and Scriven [52] .

For $\Theta = 0$, instabilities may only occur for negative values of σ_T , i.e. if $|\sigma_e| > \sigma_c$ (Rayleigh-Taylor instability induced by electrical effects). This problem could be relevant to explain the transition between emulsion and microemulsions [57] [59]. For $\Theta \neq 0$ the classical Hurwitz criteria for stability reads

$$\sigma_T \Pi_T + \Theta > 0 \tag{57}$$

$$\Theta < 0 \tag{58}$$

Other two conditions are always satisfied.

An exchange of stability may be expected for negative values of Θ , even for $\sigma_T > 0$, if $|\Theta| > \sigma_T \Pi_T$.
Marginal states with overstabilities can be obtained for $\Theta > 0$.

2. The diffusion of the negative species is very fast so that we have

$$\delta\Gamma_- = -\frac{n_- \Gamma_-^o \, \delta\varphi^s}{\varphi^*} \tag{59}$$

The balance for the components (cations and neutral molecules) is assumed to be governed by convective processes

$$\omega \, \delta\Gamma_j = \Gamma_j \mathcal{D} v^s \tag{60}$$

For zero bulk viscosities and neglecting the difference between the electrical disjoining pressure $\Delta_s[\varepsilon \kappa^{*2} \bar{\varphi}^{o2}]$ we obtain

$$\left[\omega^2\left(\Gamma^o + \frac{1}{K}\sum_{\ell} \varrho^{o(\ell)}\right) + k^2\sigma_T\right]\left[\omega^2\Gamma^o + \omega(\eta_s+\eta_d)k^2 + \sum \alpha_j \Gamma_j^o + \frac{z_+\Gamma_+^o}{\bar{K}} \sum z_i \Gamma_i^o\right] = 0 \tag{61}$$

where

$$\bar{K} = \sum_{\ell} \varepsilon^{\ell} \kappa^{*\ell} + 4\pi n_- z_- \Gamma_-^o/\varphi^* \tag{62}$$

An exchange of stability may occur for a net negative surface charge, *i.e.* $\sum z_i\Gamma_i^o < 0$ with $\left|\frac{z_+\Gamma_+^o}{\bar{K}}\sum_i z_i\Gamma_i^o\right| > \sum_j \alpha_j\Gamma_j^o$

3. The mass balance of cation involves an adsorption-desorption kinetics

$$[\omega - (c_+ - k^2\mathcal{D}_+^s)]\delta\Gamma_+ = (c_+ - k^2\mathcal{D}_+^s)\frac{n_+\Gamma_+^o}{\varphi^*}\delta\varphi^s + \Gamma_+^o \mathcal{D}v^s \tag{63}$$

We assume furthermore zero bulk viscosities. Anions are only present in the volume phases, $\lambda \gg \frac{1}{K^*}$, $\Delta_s[\varepsilon K^{*2}(\bar{\varphi}^o)^2] = 0$, secular equation is

$$\left[\omega^2\left(\Gamma^o + \frac{1}{k}\sum \rho^\ell\right) + k^2\sigma_T\right]\left[(\omega - c_+ + k^2\mathcal{D}_+^s)(\omega\Gamma^o + (\eta_s+\eta_d)k^2) + k^2\Pi_T\right] = 0 \tag{64}$$

For $\sigma_T > 0$ no instability due to electrical effects is to be expected. The only destabilizing term is then the chemical contribution $c_+ - k^2\mathcal{D}_+^s$ as shown in the previous sections.

4. Taking into account the bulk fractions, the surface weight due to the gravity, and the dominant surface convective terms in the mass balance for all species, we have

$$\left[\omega^2(\Gamma^o + \sum\rho^{o\ell}/k) + \omega\sum\{\mu^\ell(q^\ell + k)\} + k^2\sigma_T - g\Delta_s[\rho^o]\right]\times$$
$$\left[\omega^2\Gamma^o + \omega(\eta_s + \eta_d) + \omega\sum\mu^\ell(q^\ell + k) + k^2\Pi_T\right] - \left[\omega\Delta_s\{\mu(q-k)\}\right.$$
$$\left.+ k\Delta_s(\Pi_D)\right]\times\left[\omega\Delta_s\{\mu(q-k)\} + kg\Gamma^o - k^3\Theta_E\right] = 0 \tag{65}$$

where

$$\Pi_D^\ell = \varepsilon^\ell(K^{*\ell})^2(\bar{\varphi}^{o\ell})^2/4\pi \tag{66}$$

(electrical disjoining pressure)

$$\Theta_E = \frac{\sum z_i\Gamma_i^o}{\sum \varepsilon^\ell K^{*\ell}}\,\Delta_s[\varepsilon\bar{\varphi}^o] \tag{67}$$

The stability analysis was performed for two situations :

(i) two bulk phases with negligible viscosities $(\mu^{\ell}=0)$
The Hurwitz criteria are

$$k^2(\Delta_s\Pi_D)(g\Gamma^o - k^2\Theta_E) > 0 \qquad (68)$$

$$k^2\Pi_T(k^2\sigma_T - g\Delta_s[\rho^o]) - k^2(\Delta_s[\Pi_D])(g\Gamma^o - k^2\Theta_E) > 0 \qquad (69)$$

The other conditions are always satisfied. Inequalities (68) and (69) were tested for particular experimental values of the parameters (see table 3).

(ii) two bulks with the same kinematic viscosity.
The secular equation was solved numerically for the values reported in table 3. The values in table 3 were chosen in order to fit with the experimental conditions measured in the system studied by Dupeyrat and Nakache [36].

For these realistic values, unstable oscillatory states were obtained. The role of the double layer could thus by itself explain the onset of convection observed experimentally by Dupeyrat and al. [36]. As we have shown previously this phenomenon could be also due to a surface chemical reaction. However both chemical and electrical constraints may also act together, but the exact coupling between the two effects is still to be formulated.

3. Applications to membrane models.

The cell membrane may be described as a phase interface with a surface tension [60] between two immiscible fluids (the intra- and extracellular electrolytic solutions). The first steps of the phagocytosis (deformation of the cell membrane) or of the fusion of two membranes, may be adequately treated as a destabilization of the membrane in response to a chemical or an electrical signal. The membrane model of Singer and Nicolson [61] allows to consider the biomembrane as a fluid mosaic of lipids and proteins. The lipid part, which is a bilayer, plays

TABLE 3

INFLUENCE OF THE DIFFUSE LAYERS ON THE STABILITY

(cm^{-1})	$dyne/cm^2$	dyne	Conclusion for $\mu_n = \mu_w = 0$ unstable oscillating regime for	Conclusion for $\nu_n = \nu_w = 10^{-2}\ \frac{cm^2}{s}$
$2\ 10^5$	97.5	$1.2\ 10^{-8}$	$\lambda < 1.3$ cm	stable
$5\ 10^5$	-554	$7.08\ 10^{-9}$	$\lambda > 1$ cm	stable
$7\ 10^5$	-980	$5.41\ 10^{-9}$	$\lambda > 0.85$cm	stable
$10\ 10^5$	-1.492	$3.9\ 10^{-9}$	$\lambda > 0.75$cm	oscillatory regime unstable for $10.5 > \lambda > 4.2$cm $\lambda_{max} \sim 5.7$
$15\ 10^5$	-2.085	$2.7\ 10^{-9}$	$\lambda > 0.6$cm	oscillatory regime unstable for $14 > \lambda > 3.7$cm $\lambda_{max} \sim 5.5$

$\varepsilon_W = 80$

$\varepsilon_n = 35$

$T = 300°K$

$\varphi_\infty^{W(o)} - \varphi_\infty^{n(o)} = 6\ 10^{-5} V_S$ (18mV)

$z\Gamma^o = 3\ 10^2$ statcoul./cm^2

$\Gamma_T^o = 10^{-12}$ mole/cm^3

$C_n = 0.5\ 10^{-6}$ mole/cm^3

$M_+ = 300$

$\varkappa_n^* = 10^6 cm^{-1}$

$\rho_n - \rho_W = 2 \cdot 10^{-2}$ gr/cm^3

$\sigma_T = 20$ dyne/cm

$\Pi_T = 5$ dyne/cm

an important role in the behaviour of the membrane. Physico-chemical models of the lipid bilayer are extensively studied on an experimental point of view [62] a) phospholipid monolayers at air/water (A/W) or oil-water (O/W) interfaces (the last one may be considered as half a bilayer [63] , b) bilayers either flat (black lipid membrane) or spherical (liposomes)

(i) <u>Layer of macromolecules with transconformation reactions</u>

In a first approach we considered an insoluble macroscopic layer consisting in two electrical double layers and a layer of macromolecules undergoing a transconformation reaction with a change of dipole moment. This could describe the role of proteins in biomembranes. This reaction induces a change in surface tension responsible for the coupling between the chemical process in the monolayer and the hydrodynamic processes.

We assumed the following sequence of reactions to take place in the interface

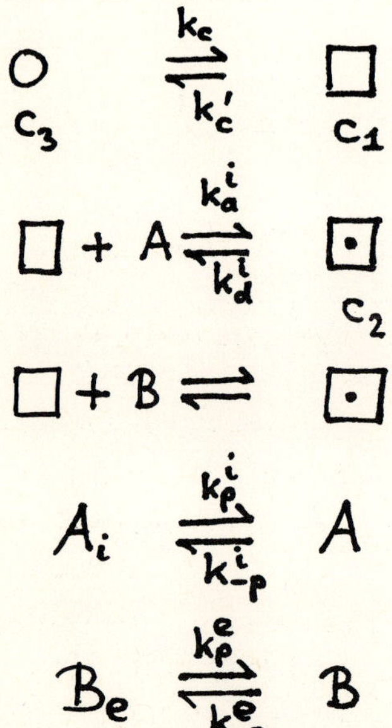

(a) transconformation

(b)

fixation

(c)

(d)

diffusion-migration from the bulk

(e)

where is the concentration (in mass per unit volume) of the macromolecules in the layer ; the subscripts 3, 2, and 1 refer respectively to the molecule in the closed conformation, and to the molecule in the open conformation filled with an ion or free from ion ; A and A_i are the concentration of the ions respectively in the internal double layer and in the internal bulk phase ; B and B_e for the external layer and bulk phase. The transconformations step (a) is associated with a change in the average dipole moment m. The electrical part of the standard affinity ($a-\Delta G$) is then

$$A_{el} = (m_1 - m_3) E \tag{70}$$

The electrical field E is uniform and normal to the interface, throughout the layer.
The transconformation is taken as a cooperative step [64] 6
The pseudo standard chemical affinity is then

$$A_{chem} = A^* + \upsilon (c_1 + c_2) \tag{71}$$

where υ is a positive cooperative constant and A^* the real standard affinity (in the absence of cooperativity). Combining Eq. (70) and (71) we obtained

$$A = A^* + \upsilon(c_1+c_2) + (m_1-m_3)E \tag{72}$$

The pseudo-equilibrium constant k_c'/k_c then reads

$$\frac{k_c'}{k_c} \sim \exp\left\{-\frac{1}{RT}\left[A^* + (m_1-m_3)E\right]\right\}\exp\left\{-\frac{\upsilon}{RT}(c_1+c_2)\right\} \tag{73}$$

From the hydrodynamical and electrical point of view, we consider that the interface region is characterized by a mean value of the convective velocity and by a mean dielectric constant.

Furthermore the kinetic constants may be written in terms of the reduced electrical potential difference $\Delta(\varphi^\circ_\infty/\varphi^*)$ and of the position of a fixation site in the layer p (see [34] eq. 12).

Combining the kinetic equations for the above mechanism with the classical hydrodynamic boundary conditions (3), (9) and (10), a linear perturbation analysis shows that the pure chemical system exhibits multiple steady states with unstable intermediate states. For particular values of the constraint the system undergoes out of equilibrium phase transitions between stable steady states, which can induce the onset of convective motion in the layer. After the instability threshold, the system could reach a mode of maximum instability for a finite initial wavelength of the order of 10^{-5} cm [32]. We have here an example of the Prigogine-Glansdorff dissipative structures [14] in a fluid model of biological membrane.

(ii) Ions and dipoles monolayers with discrete electrical interactions.

Another approach to the biological problem is to study the role of the lipid part of the biomembranes. The monolayer will be considered in the following treatment as a two-dimensional surface between two incompressible fluids. For highly concentrated solutions as in biological solutions, the diffuse layer becomes very thin and the electrical forces are intrinsic to the surface. They can be described by the discrete electrical interaction forces between the ions or dipoles with their appropriate images [66]. This point of view was already adopted by Miller & Scriven [67].

Ions monolayer.

Let us consider the charged surface in contact with an electrolyte solution as composed by the spread monolayer with a compact layer of low dielectric constant ε_c (Fig. 1).

A diffuse layer may exist in the adjacent solution. We assume that the lipids, in the charged layer, are completely ionized $(pH \simeq 7)$.

The plane of the charges determines the inner Helmholtz plane (IHP). We will call "interface" the monolayer of dielectric constant $\varepsilon_m \simeq \varepsilon_c$ together with the compact layer. The width b of the compact layer is small compared to the lattice distance a between the point charged heads of the lipids. The outher Helmholtz plane (OHP) is the limit of the diffuse layer, and determines the plane where the surface tension is defined. Indeed, as the dielectric constant of the solution is high $(\varepsilon_w \simeq 80)$, there will be a dielectric discontinuity at the

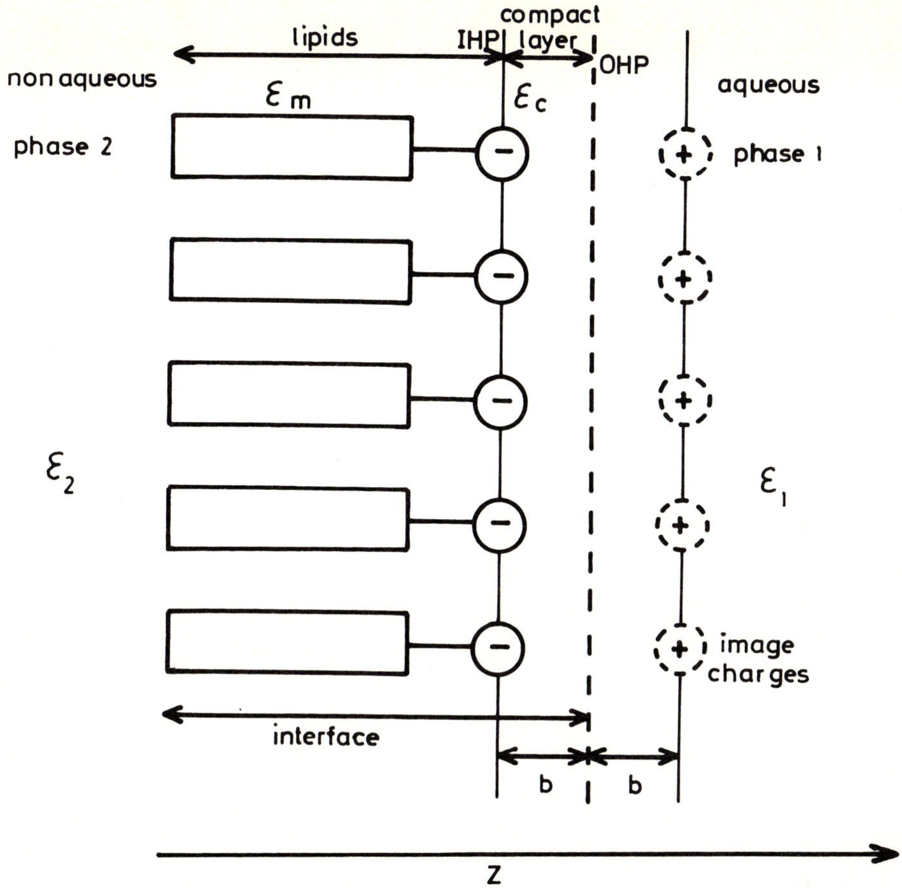

Fig.1. Interface with a spread monolayer of charged lipids.

OHP. The effect of this discontinuity can be determined by introducing fictitious image charges. The deformation of the OHP reads

$$z = \delta z \, f(x, y) \tag{74}$$

with

$$f(x,y) = \exp(i k_x x + i k_y y)$$

and the total lateral displacement of an ion at the IHP reads

$$\zeta_x = -b\left[-\delta z \, \nabla_x f(x,y)\right] + \delta x \, f(x,y) \tag{75}$$

in the x-direction, and similarly in the y direction. The first term represents the lateral displacement due to the fact that the ions are at a distance b from the reference surface, the second term represents the lattice distortion related to the convective motion of the bulk molecules. The wavelength of the perturbation is supposed to be large compared to the lattice distance a : $ka \ll 1$. The perturbed electrical ponderomotive forces at the interface are then given by

$$\delta F_j = z \Gamma^{o} \delta E_j + z \delta\Gamma E_j^{o} \qquad (j = x, y, z) \tag{76}$$

where E_j is the field created by the charges of the hexagonal lattice and by their images and

$$\delta\Gamma = \Gamma^{o} D v_z / \omega \tag{77}$$

is the change in surface density due to the deformation of the surface by convection. With the explicit formulation of δE_j and $z\delta\Gamma$, [66], Eq. (76) reads

$$i k_x \delta F_x + i k_y \delta F_y = A \phi_{xy} k^2 \delta z + A \Psi_{xy} k^2 D v_z$$

$$\delta F_z = A \phi_z k^2 \delta z + A \Psi_z D v_z \tag{78}$$

where A is the electrical energy by unit area including the

infinite summation over the ions lattice and where ϕ or ψ describe electrical interactions in X, Y, Z directions involving the geometrical parameters b and a.

Neglecting the bulk viscosities, the secular equation writes :

$$\left[\omega^2\left(\Gamma^\circ + \sum_\gamma \varrho_\gamma/k\right) + k^2\sigma_T\right]\left[\omega^2\Gamma^\circ + \omega k^2(\eta_s+\eta_d) + k^2\Pi_T\right] + k^2\Theta = 0 \tag{79}$$

with $$\sigma_T = \sigma^\circ - A\phi_z \tag{80}$$

the total surface tension

$$\Pi_T = \alpha\Gamma + A\psi_{xy} \tag{81}$$

the total surface pressure

$$\Theta = A^2\phi_{xy}\psi_z \tag{82}$$

the coupling term between longitudinal and normal modes of deformation.

It is interesting to note here that like in the general continuous system σ_T, Π_T and Θ have also electrical contributions (see Eq. (54), (55) and (56)). Of course in a discrete description these effects are different [68] .

For highly concentrated solutions we expect that all the counter-ions of the solution are situated at the OHP. The solution is taken as a grounded conductor, and the OHP is an equipotential surface with $\varphi^{OHP} = 0$. The image charge q' of each ion q at the IHP reads then

$$q' = -q$$

the explicit quantities A, ϕ_{xy}, ψ_{xy}, ϕ_z and ψ_z are given in table 4

TABLE 4

A	ϕ_{xy}	ψ_{xy}	ϕ_z	ψ_z
$\frac{e^2 n_c^2}{8\pi\varepsilon_c a}\sum_{i,j=1}^{\infty} 1/(m_{ij})^3$	$4(2b/a^2)$	$4(4b^2/a^2)$	$\frac{3}{2}(4b^2/a^2)$	$10(2b/a^2)$

The quantity m_{ij} is the relative distance measured in units of the lattice parameter a. The infinite summation on indexes is evaluated for an hexagonal lattice of charges of density number n_c . The total surface tension $\sigma_T = \sigma_0 - \sigma_e$ is lowered by electrical interactions. A state of marginal stability similar to the state described by Miller and Scriven [67] may be obtained for $\sigma_T = 0$. The total surface pressure $\Pi_T = \alpha\Gamma + \Pi_e$ is enhanced. A new destabilising effect occurs because in this case Θ is always positive. <u>The monolayer then becomes unstable</u>, since the normal and longitudinal modes of deformation will be coupled by electrical effects due to Θ .

For more dilute solutions, however, effects due to diffuse layer seem to be relatively important. Some counter-ions are still situated at the OHP (like a Stern layer), but the others will be distributed in a diffuse layer, following, in a first approximation, the Gouy-Chapman theory [55] . The OHP remains an equipotential surface but with $\varphi^{OHP} = \varphi^{o}$ with φ^{o} the surface Gouy potential. When the ionic strength of the solution increases, the Gouy potential φ^{o} decreases [69] . The boundary conditions for the potential and for the field lead to a new value of the image charge [68]

$$q' = -\left[1 - \beta \frac{(a^2+b^2)^{1/2}}{b}\right] q \tag{83}$$

with

$$\beta = \varepsilon_c \varphi^{G} b/q \tag{84}$$

By summing over all the charges and their images, it appears that the electrical surface tension and pressure are enhanced by the diffuse layer:

$$\sigma_e' = \sigma_e + \beta\sigma_2 \tag{85}$$

$$\Pi_e' = \Pi_e + \beta\Pi_2 \tag{86}$$

On the other hand, the coupling term Θ is not modified by the diffuse layer and the system remains unstable. The constraint seems to be related in such systems to the absence of the counter-ions exchange between the compact layer and the solution which leads to an overestimation of the surface charge density [68] .

Dipoles monolayer.

For lipids with only permanent point dipole of moments there is no double layer and thus no need to distinguish between compact and diffuse layer ans previously done. The dipoles are supposed to lie perpendicular to the reference flat interface (Fig. 2). We may define a dielectric constant in an inner layer with $\varepsilon_d \neq \varepsilon_1$, so that dielectric dipole images will be induced in the solution. The surface tension will be defined at the plane of dielectric discontinuity. The electrical ponderomotive forces are :

$$\delta F_k = \sum_j p_j^\circ \, \delta\left(\frac{\partial E_j}{\partial k}\right) + \sum_j \delta p_j \frac{\partial E^\circ}{\partial k} \qquad (87)$$

where

$$j, k = x, y, z$$

Two situations will be considered [68]

1° The dipoles remain parallel to each other during the deformations. In this case $\delta p_x = \delta p_y = 0$. As for ionized monolayers, the surface tension is lowered by electrical terms due to the repulsive interactions between the dipoles. On the other hand, the term is positive, i.e. the monolayer will be unstable if the dipoles are constrained to remain parallel to each other.

2° The dipoles remain perpendicular to the deformed surface. In this case $\delta p_x = -p\,\delta z \nabla_x f$ and similarly for y the surface tension is higher than in the parallel case, and the term Θ is lowered . A noncentral component of the force between

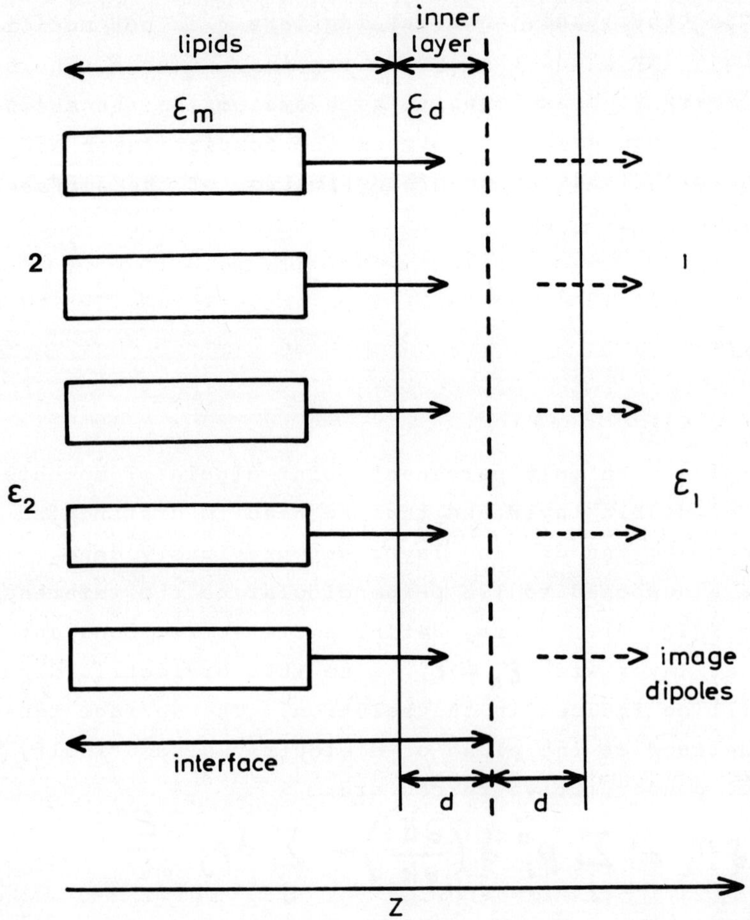

Fig. 2. Interface with a spread monolayer of polar lipids.

two dipoles is primarily responsible for the difference in destabilizing tendency between the parallel and the normal case. The orientation of the dipoles appear thus as a constraint. When the interface is deformed, the polar lipid molecules will orient normally to the perturbed surface. If they are constrained to remain parallel by large interfacial resistance to local rotation, the monolayer will be unstable if bulk viscosities are neglected †. We are now studying the stability of mixed monolayers composed both of charges and dipoles. The treatment will then be extended to bilayers submitted to an external applied potential, to enzymatic reactions [70] , and to adsorption of Ca^{++} which seems directly related to lipid phase transitions [71] .

(†For realistic values of bulk viscos. the layer is stabilised)

All these phenomena have to occur far from equilibrium in order to destabilize the system and to lead to new organization such as dissipative structures [14] in the deformed membrane. Both electrical and chemical processes, coupled with the convection could be relevant for the primary deformation of the cell surface observed during endocytosis [12].

Conclusion.

Threshold conditions are derived for the onset of convection in infinite plane and spherical surfaces undergoing chemical, mechanical and electrical constraints by a linear stability analysis.

We point out the role played by
the diffusion coefficients
the viscosity coefficients
the surface chemical processes
the adsorption-desorption step
the surface activity
the surface charge and the surface electrical potential
the interactions between surface charges and the interactions between surface dipoles in crude models of biosurfaces.

New varieties of instabilities appear due to the coupling between chemical, hydrodynamical, and electrical processes.

These are other examples of spatial and temporal organization entering the general framework of the concept of "dissipative structures" and of "order by fluctuations" developed by Prigogine, Glansdorff, Nicolis [14] [15.]

Acknowledgements.

This work was sponsored by the Danish Government and Technical University of Lyngby (Copenhague), the Fonds National de la Recherche Scientifique (Belgium), the Ministère des Affaires Etrangères de France, the CNPq (Brazil), the Society Solvay and Co and the Belgian Government : Actions de Recherche Concertées convention 76/81-II.3.

References.

1. Chandrasekhar, Hydrodynamic and Hydromagnetic Stability (Oxford University Press, 1961).
2. C.V. Sternling and L.E. Scriven, Amer. Inst. Chem. Engineers J., 1959, 5, 514 ; Fluid Mech., 1964, 19, 321.
3. L.E. Scriven, Chem. Eng. Sci., 1960, 12, 98.
4. J.T. Davies, Turbulence Phenomena (Academic Press 1972).
5. H. Linde, Motion Picture, Diffusions - und Konvektionsanomalien bei Stoffübergan om flüssig-flüssig Phasengrenzen, filed under No. T-HF218 in the Deutsche Zentralinstitut für Lehrmittel, (Kransenstr. 8, Berlin W8, East Germany, 1975).
6. M. Dupeyrat and E. Nakache, in Electrical Phenomena at the Biological Membrane (Ed. E. Roux, Elsevier Publ. Amsterdam, 1977), p. 377 ; VII International Congress Surface Activity Substances (Moscow, 1976) ; C.R. Acad. Sci., 1973, C277, 599.
7. T.K. Sherwood and J.C. Wei, Ind. Eng. Chem., 1957, 49, 1030.
8. D. Thiessen, Z. Phys. Chem., 1966, 232, 27.
9. E. Schartz and H. Linde, Phys. Fluids, 1963, 4, 535.
10. L. Wolpert and D. Gingell, Symposium of the Society of Experimental Biology on "Aspects of Cell Motility" (Cambridge University Press, 1968, chap. XXII).
11. H. Komnick, W. Stocken and K.E. Wohlfarth-Botterman, Int. Rev. Cytol., 1973, 34, 169.
12. G.L. Nicolson, Biochim. Biophys. Acta, 1976, 457, 57 ; ibid. 1976, 458, 1.
13. L. Weiss, in The Chemistry of Biosurfaces (Ed. M.L. Hair, Marcel Dekker, New York, 1972), Vol. 2, Chap. 8.
14. P. Glansdorff and I. Prigogine, Thermodynamic Theory of Structure, Stability and Fluctuations, Wiley-Interscience 1971.

15. G. Nicolis and I. Prigogine, Self-Organization in Non-Equilibrium Sytems, (Wiley, 1977).
16. W. Ebeling, Strukturbildung bei Irreversible Prozessen, Mathematisch-Naturwissenschaftliche Bibliotek Band 60, BSBBG (Teubner Verlags, 1976).
17. A. Sanfeld and A. Steinchen, Bull. Acad. Sci. Belg. 1971, 57, 684.
18. R. Defay, I. Prigogine and A. Sanfeld, J. Colloid Interface Sci., 1977, 58, 498.
19. A. Steinchen and A. Sanfeld in Colloid and Interface Science (Academic Press, New York, 1976), V. II, p. 197.
20. R. Defay, I. Prigogine, A. Bellemans and D.H. Everett, Surface Tension and Adsorption, (Longmans Green, 1966).
21. M. Hennenberg, T.S. Sørensen and A. Sanfeld, J.C.S. Faraday II, 1977, 73n 48.
22. A. Orell and J.W. Westwater, A.I.Ch.E. Journal, 1962, 8, 350.
23. D.A. Haydon, Proc. Roy. Soc., 1958, A243, 483.
24. M. Hennenberg, P.M. Bisch, M. Vignes-Adler and A. Sanfeld, Symposium on Dynamics and Instability of fluid Interface, Fysisk-Kemisk Institut, Technical University of Denmark, May 1978.
25. T.S. Sørensen, M. Hennenberg and A. Sanfeld, J. Colloid Interface Sci., 1977, 61, 62.
26. T.S. Sørensen, F.Y. Hansen, J. Nielsen and M. Hennenberg, J.C.S. Faraday II, 1977, 73, 1589.
27. M. Hennenberg, T.S. Sørensen, A. Steinchen and A. Sanfeld, J. Chimie Phys. 1975, 72, 1204.
28. T.S. Sørensen, M. Hennenberg, A. Steinchen and A. Sanfeld, J. Colloid Interface Sci., 1976, 56, 191.
29. M.G. Velard, J.L. Ibanez, T.S. Sørensen, A. Sanfeld and M. Hennenberg, Proc. Levich Conference, Oxford 1977.
30. J.L. Ibanez and M.G. Velarde, J. Physique, 1977, 38, 1479.
31. A. Steinchen and A. Sanfeld, Chemical Phys., 1967, 1, 156.
32. T.S. Sørensen, M. Hennenberg, A. Steinchen and A. Sanfeld, Progress Colloid Polymer Sci., 1976, 61, 64.
33. R. Blumenthal, J.P. Changeux and R. Lefever, J. Membrane Biol., 1970, 2, 351.
34. A. Sanfeld and A. Steinchen, Biophys. Chem., 1975, 3, 99.
35. W. Dalle-Vedove, P.M. Bisch, A. Steinchen and A. Sanfeld, Private Commun., Brussels, 1978.
36. M. Dupeyrat and E. Nakache, Proc. Levich Conference, Oxford 1977.
37. J.T. Davies and E.K. Rideal, Interfacial Phenomena (Academic Press, 1961), Chap. 7.
38. F. Deyhimi and A. Sanfeld, C.R. Acad. Sci. 1974, D279, 437.
39. E.H. Lucassen-Reynders and J. Lucasse, in Adv. Colloid Interface Sci. (Elsevier-Amsterdam, 1969) 2, p. 347.
40. J. Lucassen and M. van den Tempel, J. Colloid Interface Sci. 1972, 41, 495.
41. M. van den Tempel, J. of Non Newtonian Mechan., 1977, 2, 205.
42. B. D'Epenoux and G. Gavach, J. Colloid Interface Sci., 1976 56, 138.
43. P. Joos and M. Van Bockstaele, J. Phys. Chem. 1976, 80, 1573.

44. P. Joos and Y. Verburgh, in Lecture Notes in Physics (Eds. J.C. Legros and J.K. Platten, Springer-Verlag, 1978, 72, p. 120.
45. H. Colson-Guastalla, M. Dupeyrat, J. Guastalla and J. Michel, J. Chimie Phys., 1970, 67, 1508.
46. Dupeyrat M. and J. Michel, Experientia Suppl. 1971, 18, 269.
47. J.R. Melcher, J. Fluid. Mech., 1973, 60, 417.
48. R.E. Zelaso and J.R. Melcher, Phys. Fluids, 1974, 17, 61.
49. J.F. Hoburg and J.R. Melcher, J. Fluid Mechan., 1976, 73, 333.
50. P.M. Bisch and A. Sanfeld, in Lecture Notes in Physics, (Eds. J.C. Legros and J.K. Platten, Springer Verlag, 1978), 72, p. 112.
51. P.M. Bisch, Proc. Levich Conference, Oxford, 1977.
52. C.A. Miller and L.E. Scriven, J. Colloid Interface Sci., 1970 33, 360 ; ibid. 371.
53. V. Levich, Physico-Chemical Hydrodynamics, (Prentice Hall Englewoods Cliffs, New Jersey, 1967).
54. S.R. de Groot and P. Mazur, Non-Equilibrium Thermodynamics, (North-Holland, 1962) Chap. XIV.
55. A. Sanfled, Thermodynamics of charged and polarized layers, (Wiley, 1968).
56. J.Th. Overbeek, in Colloid Science, (Ed. H.R. Kruyt, Elsevier, Amsterdam), Vol. 1, 1952.
57. A. Sanfeld, A. Steinchen and R. Defay, J. Chimie Phys. 1971 68, 711.
58. A. Sanfeld, A. Steinchen, M. Hennenberg, P.M. Bisch and C. Devillez, Report to the Colloquium "Tertiary Oil Recovery" DGRST Paris 1977.
59. D.O. Shah and R. Schechter, "Improved Oil Recovery by Surfactants and Polymer flooding" (Academic Press, New York, 1977).
60. T. Ueda, M. Muratsugu and Y. Kobatake, Biochim. Biophys. Acta, 1974, 367, 24 ; ibid. 373, 286.
61. S.J. Singer and G.L. Nicolson, Science 1972, 175, 720.
62. P. Mueller, O. Rudin, H. Ti Tien and W.C. Wescott, J. Phys. Chem., 1963, 67, 534.
63. S. Ohki and C.B. Ohki, J. Theor. Biol. , 1976, 62, 389.
64. R. Lefever and J.P. Changeux, C.R. Acad. Sci. Paris, 1972, D275, 591.
65. G. Nicolis and R. Lefever, Membrane, Dissipative Structures and Evolution, in Adv. in Chem. Phys. 1975, 29, 1.
66. D. Van Lamsweerde-Gallez, P.M. Bisch, A. Sanfeld, Bioelectroch. and Bioenergetics, under press 1978.
67. C.A. Miller, L.E. Scriven, J. Colloid Interface Sci., 1974, 46, 477.
68. P.M. Bisch, D. Van Lamsweerde-Gallez and A. Sanfeld, to be submitted for publication J. Colloid Interface Sc.
69. J. Mingins and B.A. Pethica, J.C.S. Faraday Trans. I, 1973, 69, 500.
70. A. Goldbeter and L.A. Segel, Proc. Nat. Acad. Sci. 1977, 74 1543.
71. D. Papahadjopoulos, W.J. Vail, C. Newton, S. Nir, K. Jacobson, G. Poste and E. Lazo, Biochim. Biophys. Acta, 1977, 465, 579.

DYNAMIC EFFECTS OF SURFACTANTS AND MACROMOLECULES ON INTERFACIAL VISCOELASTIC PROPERTIES

by

D. T. Wasan, N. F. Djabbarah, M. K. Vora and S. T. Shah

Department of Chemical Engineering, Illinois Institute of Technology
Chicago, Illinois 60616, USA

1. Introduction

It is now widely recognized that surface-active agents, trace impurities and macromolecules can and often do exert a significant influence on fluid phase separation processes by modifying the interfacial mobility and area (1). For example: the rate of mass transfer may be decreased by as much as an order of magnitude (2), or bubbles and drops may be made to coalesce or not to coalesce (3), or foams and emulsions may be stabilized or broken (4,5), or wavy films may be made to become smooth (6), or thin liquid films may be made to generate waves (7) or rupture into rivulets or drops, or stagnant films may be made to rupture on solid surfaces (8).

Surfactants in solutions tend to accumulate and adsorb at interfaces between their solutions and adjacent phases. The orientation of these molecules, as well as molecular interactions and molecular packing, result in an interfacial behavior different from that in the bulk phases. In some instances, molecular complexing or structural transformations like the formation of a liquid crystalline phase can occur, resulting in peculiar rheological behavior (9). Specifically, such accumulation of surfactants at fluid-fluid interfaces results in interfacial forces that influence the geometry of the phase boundary with consequent effects on interfacial flow and area (10).

An analysis of the dynamic properties is essential in describing behavior of fluid interfaces where interfacial motion is involved (11-14). The characteristic coefficients which describe the response of fluid interfaces to interfacial motion are termed rheological properties. Interfacial rheological behavior of fluid interfaces containing surfactants or macromolecules is often dependent upon shear, area, structure, and age. For simple Newtonian fluid interfaces, the significant rheological properties for interfacial motion are the interfacial shear and dilational viscosities.

As the molecular weight of the adsorbate increases, the interface may exhibit viscoelastic behavior or display rigidity (15). Since a great number of engineering applications involve motion at the fluid-fluid interfaces, a knowledge of the interfacial rheological properties is important for the understanding and treatment of phenomena occurring at fluid interfaces.

The main thrust of our ongoing fundamental study is to design suitable experiments and develop and extend theoretical analyses to lead ultimately to better data and theory on interfacial viscoelastic properties which can then be used to control dynamic rate processes in engineering applications. Partial results of our recent studies can be found elsewhere (16-19).

This paper is concerned with dilute aqueous solutions of polymers and polymer plus surfactant mixtures which exhibit shear and/or aging dependent or independent behavior at both liquid-gas and liquid-liquid interfaces. Also, our most recent measurements of the surface dilational viscoelastic properties, in addition to shear properties, are presented. The relevance of these measurements in such applications as foam and emulsion stability is discussed. In particular, stability of crude oil in surfactant solutions as may be encountered in enhanced oil recovery by surfactant/polymer and caustic flooding processes is highlighted.

2. Measurements of Surface Shear Viscoelastic Properties

The various methods used for surface rheological measurements have been reviewed in a number of recent papers (20,21). In the present study, surface shear viscosity was determined by the so-called deep-channel viscous traction surface viscometric technique (22-25). Basically, it consists of two concentric, stationary vertical cylinders and a rotating flat-bottomed dish containing the fluids (See Figure 1). These cylinders are so placed that they almost touch the bottom of the dish. The dish is rotated, causing the fluids in the channel between the cylinders to rotate. The stationary channel walls tend to shear the fluids, thereby enhancing the effect of the surface. The surface shear viscosity can be calculated from the experimentally determined surface centerline velocities in accordance with appropriate theoretical flow analyses. The surface shear viscosity of a Newtonian surface fluid is given by

$$\frac{\pi\varepsilon}{\eta y_o} = \frac{V^*_{z_c}}{V_{z_c}} - 1 = \frac{4\, V_b}{\pi\, V_{z_c} \cosh\,(\pi D)} - 1 \qquad (1)$$

where η is the bulk viscosity of the fluid, y_o the channel gap, D the dimensionless depth (= $\frac{\text{the depth of the fluid in the channel}}{\text{channel gap}}$), V_{z_c} the surface centerline velocity and V_b the velocity of the floor. The quantity $V^*_{z_c}$ represents the value of V_{z_c} when the surface does not contain the contaminant. Here ε is the surface shear viscosity coefficient.

It was observed that in many systems studied, the centerline surface velocities of surfactant solutions changed continuously with time (26). Figure 2 presents data of the centerline velocity, V_{z_c}, as a function of surface age with ω_o(in R.P.M.) as a parameter for 0.001 wt % Polyox WSR-301 (polyethylene oxide, non-ionic) stabilized by 0.5% isopropanol (non-ionic) in water. These data which were taken at room temperature clearly show the pseudoplastic behavior of the surface viscosity, that is, at the same age, V_{z_c} increases (i.e. ε decreases) as the shear rate (ω_o) is increased. It is interesting to note that a system containing 0.5% isopropanol in water alone displayed negligible surface viscosity ($< 10^{-4}$ g.sec^{-1}). However, upon the addition of only 10 ppm of a high molecular weight polymer there is a rapid build up of a viscous interface. This brings out the importance of aging effects and the complexity of phenomena involving molecular interactions.

Although shear-rate dependent surface viscosities have been recently treated mathematically(25,27,28),these analyses do not take into account the surface aging effects which are generally found to be very pronounced in dilute mixed surfactant solutions (26,29). Furthermore, these systems may also exhibit surface elasticity in addition to shearing viscosity at the gas-liquid interface (30). The latter problem for the non-aging surfactant solutions and insoluble monolayers which exhibit both non-Newtonian steady shear-rate and viscoelastic behavior has recently been discussed by Schechter and Wasan and their collaborators (24,28,30,31).

Figure 3 displays steady state measurements of surface viscosity of an aqueous solution of 0.125% sodium carboxy methyl cellulose (CMC, medium viscosity, anionic) and 0.004% lauryl alcohol (nonionic) at room temperature. This system also exhibited a non-Newtonian surface shear viscosity with a zero shear-rate surface viscosity ε_o of about 0.062 g.sec^{-1}, and an infinite shear-rate surface viscosity ε_∞ of about 0.019 g.sec^{-1}. These values were calculated using the recent shear surface flow analysis of Shah (32) in our laboratory. Although no direct comparison can be made, it is interesting to note that most recently Schechter and his co-workers (33) reported the limiting values of the surface shear visocity as $\varepsilon_o = 0.055$ g.sec^{-1} and $\varepsilon_\infty = 0.01$ g.sec^{-1} for a 0.05 wt % polymeric aqueous solution of polyvinyl alcohol (PVA).

3. Measurements of Shear Viscoelasticity at Liquid-Liquid Interfaces

Reliable measurements of both shear viscosity and elasticity at liquid-liquid interfaces are very limited even though the remarkable stability of emulsions and thin films have been attributed to interfacial shear viscoelasticity (34, 37). Wasan and co-workers (17,19,38) have recently developed a new method using the viscous-traction interfacial viscometer discussed earlier in section 2 to measure shear viscoelastic parameters of liquid-liquid interfaces from displacement data at liquid-gas interfaces alone as shown in Figures 4A and 4B. A check on internal consistency of the method has been made by the comparison of interfacial shear viscoelastic parameters determined from the surface and interfacial displacement experiments (19,38). The utility of this method has been demonstrated by discerning both from a parametric study and from an experimental study the existence of differing displacement patterns depending upon the design and operational features of the liquid-liquid-gas experiment. This method is particularly useful for opaque systems, and we have employed it for the determination of interfacial shear viscosity in crude oil-aqueous surfactant systems (39-41). Since measurements at liquid-gas interfaces are considerably easier to make than those at liquid-liquid interfaces, this technique provides a new avenue of approach to the study of interfacial shear viscoelasticity under various physicochemical conditions in liquid-liquid systems (9, 42). Furthermore, this method is capable of measuring interfacial viscosities from 10^{-4} to 10 surface poise (g.sec^{-1}).

The above method developed in our laboratory was used to obtain interfacial shear viscosity of ethylbenzene and water with and without polyvinyl alcohol (PVA). Polyvinyl alcohol with a molecular weight of 14,8000 and under the trade name Elvanol 52-22 was supplied by Du Pont Company. As previously discussed, PVA is a well studied substance, the concentration of which can be readily determined (43,44). It has negligible solubility in ethylbenzene. The centerline velocity was measured at the gas-liquid surface first with only the oil phase in the channel (See Figure 4A).

It was measured again at the gas-liquid surface with both the oil (light phase) and aqueous phases present (See Figure 4B). The interfacial shear viscosity was calculated from the following equation derived by Wasan and co-workers (17,19):

$$\frac{\pi\hat{\varepsilon}}{\hat{\eta}y_o} = \frac{4V_b}{\pi V_{z_c} \cosh_{(\pi\hat{D})} [\cosh(\pi D) + \pi E \, \mathrm{Sin\,h}(\pi D)]} - 1$$

$$- \frac{\eta}{\hat{\eta}} \left[\frac{\sin h(\pi D) + \pi E \cosh(\pi D)}{\cosh(\pi D) + \pi E \, \mathrm{Sin\,h}(\pi D)}\right] \qquad (2)$$

Where $E = \frac{\varepsilon}{\eta y_o}$ and is given by Equation (1). Here the superscript carat (^) denotes quantities associated with the lower liquid and the liquid-liquid interface. V_{z_c} is obtained from the experiment when the oil phase alone is present in the channel. Using Equations (1) and (2), the value of the interfacial shear viscosity at the liquid-liquid interface can be determined.

The effect of PVA concentration on interfacial shear viscosity is shown in Figure 5. Interfacial shear viscosity increased with increasing concentration of polymer, which varied from 0 to 0.02 wt %. Interfacial shear viscosity corresponding to a 0.02 % PVA concentration was observed to be 0.075 $g.sec^{-1}$ which is large considering that the value of interfacial viscosity at ethylbenzene-water interfaces was negligible ($< 10^{-4}$ $g.sec^{-1}$).

Further work on the measurements of both interfacial viscosity and elasticity is presently continuing in our laboratory.

4. Measurements of Surface Dilational Viscoelastic Properties

Determination of surface dilational properties involves measurement of the response of the surface when subjected to dilational (or compressional) strains. These, in turn, are accompanied by diffusional processes within the surface as well as between the surface and the bulk. Consequently, the theoretical analysis of the experimental data is further complicated by the necessity of coupling the hydrodynamic equations with those governing the conservation of mass at the surface and in the bulk.

A number of methods have been used to measure the dilational viscoelastic properties of fluid interfaces. A critical analysis of various measurement techniques is given elsewhere (21). One of the most promising techniques involves the use of longitudinal waves. This method is more advantageous for the study of surface dilational viscoelasticity because: (a) It is possible to generate purely longitudinal waves by horizontal oscillations of a barrier, thereby making the experimentation simpler and (b) The mathematical analysis involves less stringent limitations.

Lucassen and van den Tempel (45) were the first investigators who attempted the use of longitudinal wave technique for determining surface dilational properties. This technique involves determining the response of the surface of a liquid when subjected to a sinusoidal disturbance of known amplitude and frequency. This was accomplished by measuring the variations in surface tension. This approach gives only the 'apparent' surface dilational viscoelastic coefficients originating from compositional changes at the surface.

We have recently modified the longitudinal wave technique of Lucassen and his co-workers by using an inert tracer particle for detecting the surface wave motion instead of inserting a Wilhelmy plate into the surface (46,47).The use of the latter method is known to interfere with the motion generated. Furthermore, we have employed a linear viscoelastic surface model which includes both the 'intrinsic' or structural and the 'apparent' or compositional surface viscoelastic properties:

$$\tau_{xx} = \sigma_o + (E_{dc} + \Lambda + G)\frac{\partial \zeta}{\partial x} + (\mu_{dc} + \kappa + \varepsilon)\frac{\partial V_x}{\partial x} \tag{3}$$

where τ_{xx} is the tangential component of the surface stress, σ_o the equilibrium surface tension, E_{dc} the surface compositional elasticity, μ_{dc} the compositional viscosity, Λ the intrinsic dilational elasticity, κ the intrinsic dilational viscosity, G the intrinsic shear elasticity and ε the intrinsic shear viscosity of the surface. The surface displacement, ζ, is related to surface velocity, V_x by

$$\zeta = i\omega\, V_x \tag{4}$$

which is valid for small displacements. Here ω is the frequency of wave motion.

Combining Equations (3) and (4) and taking the derivative of Equation (3) with respect x yields

$$\frac{\partial \tau_{xx}}{\partial x} = E^*_n \frac{\partial^2 \zeta}{\partial x^2} \tag{5}$$

where E_n^* is the surface viscoelastic coefficient, defined as-

$$E_n^* = (E_{dc} + \Lambda + G) + i\omega\,(\mu_{dc} + \kappa + \varepsilon) \tag{6}$$

$$= E_n + i\omega\,\mu_n$$

It is clear that the surface viscoelastic coefficient is a complex quantity, the real part of which (E_n) is the sum of all the elastic contributions and the imaginary part $\omega\mu_n$ is the sum of all the viscous contributions of the surface.

For liquids with the ratio of bulk density to bulk viscosity greater than about one hundred ($\frac{\rho}{\eta_b} > 10^2$), we have shown previously that

$$E_n^* = -\frac{i\omega\,\eta_b\,m}{K^2} \tag{7}$$

where K is the complex wave number and is related to the complex constant, m, by

$$m^2 = \frac{i\omega\rho}{\eta_b} + K^2 \tag{8}$$

The value of K can be determined from the longitudinal wave experiments as described below and hence m and E_n^* can be evaluated. Surface dilational parameters can be obtained from Equation (6) provided that surface compositional and shear elasticities are available. The former may be estimated by considering the mass transfer interactions that take place when a system is subjected to dilational strain. The latter can be measured experimentally using an appropriate technique such as the deep channel viscous traction viscometer.

The longitudinal wave apparatus designed and built by Maru (47) in our laboratory is shown schematically in Figure 6. It consists of a 60 x 8 x 1.5 cm teflon trough which contains the solution under study and an adjustable teflon barrier which just touches the surface of the liquid and is supported by an aluminum frame. A ball bearing connects the frame to an aluminum rod (41 cm. long). The other end of the rod is connected to a pin which is fixed eccentrically to a barrel. The barrel is coupled to the shaft of a motor whose circular motion is converted, by means of the eccentric pin, to sinusoidal motion of the barrier.

A tracer particle is placed on the liquid surface at a fixed position away from the barrier. A sinusoidal motion of known amplitude and frequency is imparted to the surface. The response of the surface is determined by observing the tracer particle under a microscope and measuring the amplitude of the particle and the phase lag between the particle and the barrier. Several measurements are obtained for each particle position and several particle positions are considered for each system.

The complex wave number, K, can be determined from the experimentally measured amplitude ratio and phase lag of the tracer particle. This is achieved by a comparison between the experimental wave profile and that predicted by theory.

We have recently reported the surface dilational viscoelasticity data for a single surfactant system such as decanoic acid solutions and stearic acid films (47). Most recently, Djabbarah (48) in our laboratory has extended the analysis to multicomponent systems and used it to determine both the surface shear and dilational viscosities as well as elasticities for aqueous solutions of mixed surfactants of sodium lauryl sulfate - lauryl alcohol. The longitudinal wave experiments were conducted at a barrier frequency of 0.26 radians per second and an amplitude of 0.250 centimeter. The details of the mathematical analysis as well as the experiments are available elsewhere (49).

Djabbarah (50) has also measured the surface excess concentration of the solution of sodium lauryl sulfate and lauryl alcohol using a single bubble foaming apparatus and the surface tension of the mixed surfactant solutions using the Wilhelmy plate method.

Figure 7 depicts the surface tensions of aqueous solution of sodium lauryl sulfate-lauryl alcohol as a function of the bulk concentration of sodium lauryl sulfate at a constant mass ratio of one hundred. The sodium lauryl sulfate used was purified by Soxhelt extraction following by foam fractionation.

Table 1 presents the values of net surface elasticities (E_n), surface viscosities ($\omega\mu_n$) and surface compositional properties as a function of the bulk concentrations of sodium lauryl sulfate. These data clearly indicate that both the net surface viscosity and elasticity increase rapidly with bulk concentration, reach a maximum and then decrease at higher concentrations. A comparison of the values of net (E_n) and compositional elasticities (E_{dc}) indicates that at lower concentrations the major part of surface elasticity is compositional. At higher concentrations, however, the intrinsic elasticity of the surface (Λ + G), becomes appreciable. This leads one to conclude that at lower concentrations the surface behaves as a two dimensional Newtonian fluid and becomes viscoelastic at higher concentrations. This deviation from Newtonian behavior was also observed by Poskanzer and Goodrich (51) for the same mixed surfactant system.

Table 1 also shows the comparison between net and apparent viscosities for the aqueous solution of sodium lauryl sulfate and lauryl alcohol. The intrinsic surface viscosity ($\kappa + \varepsilon$) corresponds to the difference between net and compositional surface viscosities. It is clear that the contribution due to apparent viscosity is essentially negligible and therefore the net surface viscosity ($\kappa + \varepsilon$) is the intrinsic viscosity for this system. This observation was also verified by Djabbarah and Wasan for other concentrations of lauryl alcohol for mixed sodium lauryl sulfate - lauryl alcohol aqueous solutions (49).

Figure 8 shows the total intrinsic surface viscosity as a function of the bulk concentration of sodium lauryl sulfate. This figure also displays values of surface shear viscosity (ε) for this system, which were determined experimentally by Djabbarah (49) using the deep channel viscous traction viscometer. It is interesting to note that surface viscosity increases rapidly with concentration and reaches a maximum and then decreases gradually. The data clearly indicate that surface dilational viscosity (κ) exceeds surface shear viscosity (ε) by more than two orders of magnitude. Recently, Maru and Wasan (47) also reported the similar conclusion ($\kappa \gg \varepsilon$) for surface viscosities of aqueous solutions of decanoic acid as well as for insoluble monolayers of stearic acid.

Figure 9 depicts surface viscosity ($\kappa + \varepsilon$), versus the average area per molecule for the mixed surfactant solutions of sodium lauryl sulfate-lauryl alcohol for a mass ratio of 100. The data show that the surface viscosity is very high (> 10 g. sec^{-1}) for a molecular area less than $30(A^o)^2$/molecule and that the surface viscosity is very low ($< 10^{-2}$g.sec^{-1}) for an molecular area greater than $60(A^o)^2$/molecule, thus pointing out the dependence of surface viscosity on the extent of molecular packing in adsorbed surfactant films.

5. Applications of Interfacial Rheological Properties

Accurate measurements of interfacial rheological properties such as shear and dilational viscosities and elasticities at fluid-fluid interfaces have been attempted during the past twenty years (52). Consequently, it is only in recent years that their role in controlling dynamic rate processes has been recognized in some engineering system, such as foam formation and stability (53,54), emulsion stability and emulsion destabilization by chemical demulsifiers (34,37,55), suspension polymerization (53), Marangoni instability (56,57), interfacial mass transfer (58,59), and oil displacement mechanisms in tertiary oil recovery by chemical flooding processes (60-62).

We are summarizing below some of the applications of interfacial rheological properties which are currently receiving a great deal of attention in our laboratory and which promise to be of much practical interest.

There has been considerable experimental evidence which suggests that interfacial viscoelastic properties may be used as diagnostic tools for assessing the structural changes and transformations which take place at fluid-fluid interfaces (19,37,63,64). In particular, in a series of recent papers, Friberg and Krog and their collaborators (65-67) investigated the presence and location of lyotropic liquid crystals in emulsion systems using optical and electron microscopy and identified the structure of the phases by low-angle X-ray diffractometry. The influence of the presence of the liquid crystalline phase at the oil/water interface on the change in the distance dependence of the van der Waals ineraction was estimated and was found to reduce drastically the van der Waals energy avaiable for promoting coalescence of the droplets, thus pointing out the importance of the interfacial hydrodynamic properties in controlling emulsion stability.

In our experiments with the stability of oil-in-water emulsions we have reported that an increase in interfacial shear viscosity at the oil-water interface was accompanied by an increase in emulsion stability (39,55). We have recently investigated some of the same systems which were studied by Friberg and his co-workers (65) and found that with the appearance of liquid crystals in the interfacial region, the interfacial viscosity increased. Furthermore, in more practical systems involving Salem crude oil and an aqueous solution of petroleum sulfonates containing divalent cations, we observed the presence of liquid crystals at the crude oil-aqueous solution interface. This system also exhibited a very stable crude oil-in-water emulsion as well as a very high interfacial shear viscosity (> 10 $g.sec^{-1}$). It is interesting to note that the interfacial viscosity remained low (< 10^{-2} $g.sec^{-1}$) until the appearance of the liquid crystalline phase and then increased corresponding to the growth of this phase.

In another study on foam stability, Djabbarah (50) in our laboratory has recently correlated the rupture time and the drainage of single bubble lamellae with the surface dilational and shear viscosities for the mixed surfactant system of sodium lauryl sulfate and lauryl alcohol. He has also experimentally determined the surface excess concentrations for this system and found that at high surface concentrations (close molecular packing) where surface viscosities are large, surface viscosities played a dominant role in the thinning process of the lamellae. On the other hand, at low surface concentration, the effect of surface tension gradients or the Marangoni-Gibbs effects become increasingly important. These results are consistent with the recent theoretical prediction of Ivanov and his collaborators (68). Similar conclusions may be noted for emulsion systems. When the interfacial viscosity is high i.e. $(\kappa + \varepsilon) > 10^{-1}$ $g.sec^{-1}$ (surface poise), emulsion will be more stable. On the other hand, when interfacial viscosity is low i.e. $(\kappa + \varepsilon) < 10^{-2}$ $g.sec^{-1}$, interfacial tension gradients or interfacial elasticities may have a greater stabilizing effect.

In a continuing research program on emulsion science and foam stability, we are studying systematically the effects of surfactant association structure, and interfacial structural changes and transformations on interfacial viscoelastic properties, and in turn, the effects of these properties on coalescence rates and stability of foams and emulsions.

6. Conclusions

In summary, it should be stated that interfacial rheological properties assume importance from two points of view: (a) characterizing the interfacial structure and transformation and (b) modelling their effects on multiphase systems.

To assess the importance of interfacial viscoelastic properties in controlling many of the dynamic rate processes, data on interfacial dilational properties in addition to interfacial shear properties are needed. A consistent set of data for the same surfactant or macromolecular system for the most part, is lacking. The interfacial dilational viscoelastic properties are particularly important in many situations where dilational strains are dominant compared with other strains.

It is noted that very few reliable measurements of interfacial viscoelastic properties in liquid-liquid systems are available. Research in this problem area needs to be intensified.

It should also be emphasized that we need to compare values of surface and interfacial rheological properties from different techniques.

ACKNOWLEDGEMENT

Financial support for this study has been provided by the National Science Foundation under Grant NSF-ENG-77-20164 and partly by Department of Energy under contract No. DOE-ET-78-S-02-4944-Aooo.

The experimental data on PVA was obtained by Dr. Nader Aderangi in our laboratory and we are grateful to him.

TABLE 1

Tabulation of Net and Compositional Surface Viscoelastic Properties for Aqueous Solutions of Sodium Lauryl Sulfate-Lauryl Alcohol, $(\frac{SLS}{LOH} = 100)$

SLS Bulk Concentration Mole/cc x 10^6	Net Surface Properties Dyne/cm		Compositional Surface Properties Dyne/cm	
	E_n	$\omega\mu_n$	E_{dc}	$\omega\mu_{dc}$
0.693	5.1863	1.0090	5.3071	0.0015
1.732	8.5527	4.1980	6.6561	0.0013
2.600	13.228	7.2273	6.2808	0.0015
3.468	12.6667	9.4009	6.3327	0.0000
5.200	12.9400	9.8003	2.0478	0.0000
6.935	12.7681	8.600	1.8322	0.0000
8.669	12.0013	8.2006	1.5485	0.0000

REFERENCES

1. J. C. Berg, Recent Developments in Separation Science, N. N. Li, Ed. Vol. II, C. R. C. Press, (1975), p.1.
2. H. Sawistowski and D. R. Jones, Abh. Dent. Akad. Wiss., Berlin, Kl Chem., Acol. Biol. June (1966).
3. R. C. Kintner, Advances in Chemical Engineering, T. B. Drew, et al., Ed., Academic Press, New York (1963).
4. R. Lemlich, Progress in Separation and Purification, E. S. Perry, Ed., Vol. 1, Wiley-Interscience, New York (1968).
5. N. N. Li and A. L. Shiner, Recent Developments in Separation Science, N. N. Li, Ed., C. R. C. Press, Vol. 1, (1972), p. 163.
6. W. B. Krantz and S. L. Goren, Ind. Eng. Chem. Fundamentals, 10, (1971), p. 91.
7. S. Whitaker, Ind. Eng. Chem. Fundamentals, 5, (1966), p. 379.
8. R. K. Jain and E. Ruckenstein, J. Colloid and Interface Sci., 54, (1976), p. 108.
9. S. Srivatsa and D. T. Wasan, paper to be presented at ACS, 53rd Colloid and Surface Science Symposium, University of Missouri-Rolla, Missouri, June, (1978).
10. T. E. Ramabhadran, C. H. Byers and J. C. Friedly, Am. Inst. Chem. Engrs. Journal, 22, (1976), p. 872.
11. L. E. Scriven, Chem. Eng. Sci., 12, (1960), p. 98.
12. J. C. Slattery, Chem. Eng. Sci., 19, (1964), p. 379.
13. M. van den Tempel, J. of Non-Newtonian Fluid Mechanics, 1, (1977), p. 205.
14. F. C. Goodrich, J. Phys. Chem., 66, (1962), p. 1858.
15. J. G. Oldroyd, Proc. Roy. Soc., A232, (1955), p. 567.
16. D. T. Wasan and V. Mohan, Improved Oil Recovery by Surfactant and Polymer Flooding, D. O. Shah and R. S. Schechter, Eds., Academic Press, New York, (1977), p. 177.
17. D. T. Wasan, L. Gupta and M. K. Vora, Am. Inst. Chem. Engrs. Journal, 17, (1971), p. 1287.
18. L. Gupta and D. T. Wasan, Ind. Eng. Chem. Fundamentals, 13, (1974), p. 26.
19. V. Mohan and D. T. Wasan, Colloid and Interface Science-Hydrosols and Rheology, M. Kerker, Ed., Vol. IV, Academic Press, New York, (1976), p. 439.
20. M. Joly, Surface and Colloid Science, Matijevic, Ed., Vol. 5, Wiley-Interscience, New York, (1972), p. 79.
21. H. C. Maru and D. T. Wasan, paper presented at ACS, 51st Colloid and Surface Science Symposium, Buffalo, New York, June (1977).
22. R. A. Burton and Mannheimer, Advances in Chemistry Series, ACS, 3, (1967), p. 315.
23. R. J. Mannheimer and R. S. Schechter, J. Colloid and Interface Science, 32, (1970), p. 195.
24. R. J. Mannheimer and R. S. Schechter, J. Colloid and Interface Science, 32, (1970), p. 225.
25. A. J. Pintar, A. B. Israel and D. T. Wasan, J. Colloid and Interface Science, 37, (1971), p. 52.
26. V. Mohan, L. Gupta and D. T. Wasan, J. Colloid and Interface Science, 57, (1976), p. 496.
27. L. Wei and J. C. Slattery, Colloid and Interface Science, Hydrosols and Rheology, M. Kerker, Ed., Vol. IV, Academic Press, New York, (1976), p. 399.

28. J. W. Gardner, J. V. Addison and R. S. Schechter, Am. Inst. Chem. Engrs. Journal, 24, (1978).
29. R. J. Mannheimer, Am. Inst. Chem. Engrs. Journal, 15, (1969),p. 88.
30. V. Mohan, B. K. Malviya and D. T. Wasan, Canadian Journal of Chemical Engineering, 54, (1976), p. 515.
31. J. W. Gardner and R. S. Schechter, Colloid and Interface Science-Hydrosols and Rheology, Vol. IV, (1976), p. 421.
32. S. T. Shah, M. K. Vora and D. T. Wasan, manuscript in preparation, (1978).
33. J. V. Addison and R. S. Schechter, paper submitted to Am. Inst. Chem. Engrs. Journal, (1978), private communication.
34. T. J. Jones, E. L. Neustadter and K. P. Whittingham, Journal of Canadian Petroleum Technology, April-June, (1978), p. 100.
35. D. M. Grist, E. L. Neustadter and K. P. Whittingham, New Paper No. 7819, submitted to Society of Petroleum Engineers Journal, (1978)
36. J. Boyd, C. Parkinson and P. Sherman, J. Colloid and Interface Science, 41, (1972), p. 359.
37. J. V. Boyd, N. Krog and P. Sherman, Theory and Practice of Emulsion Technology, A. L. Smith, Ed., Academic Press, New York, (1976), p. 123.
38. M. K. Vora and D. T. Wasan, manuscript in preparation, (1978).
39. D. T. Wasan, S. M. Shah, N. Aderangi, M. Chan and J. J. McNamara, Society of Petroleum Engineers Journal, December (1978).
40. D. T. Wasan, J. J. McNamara, S. M. Shah, K. Sampath and N. Aderangi, Society of Rheology Journal, (1978), in press.
41. D. T. Wasan, S. M. Shah, M. Chan, K. Sampath and R. Shah, Symposium on Chemistry of Oil Recovery, Division of Petroleum Chemistry preprint, ACS Meeting, Anaheim, California, March, (1978), p. 705.
42. N. Aderangi and D. T. Wasan, manuscript in preparation, (1978).
43. J. Horacek, Chemickey Prumysl, 12, (1962), p. 385.
44. H. J. Karam, J. Applied Polymer Sci., 18, (1974), p. 1693.
45. J. Lucassen and M. van den Tempel, J. Colloid and Interface Science, 41, (1972), p. 491, and Chem. Eng. Sci., 27, (1972), p. 1283.
46. H. C. Maru, V. Mohan and D. T. Wasan, Part I, Chem. Eng. Sci., (1978), in press.
47. H. C. Maru and D. T. Wasan, Part II, Chem. Eng. Sci., (1978), in press.
48. H. C. Maru, Ph.D. Thesis, Illinois Institute of Technology, Chicago, Illinois, (1975).
49. N. F. Djabbarah and D. T. Wasan, manuscript in preparation,(1978).
50. N. F. Djabbarah and D. T. Wasan, manuscript in preparation,(1978).
51. A. M. Poskanzer and F. C. Goodrich, J. Phys. Chem., 79, (1975), p. 2122.
52. F. C. Goodrich, paper presented at ACS 52nd Symposium on Colloid and Surface Science, Knoxville, Tennessee, June, (1977).
53. B. Kanner and J. E. Glass, Ind. Eng. Chem., 61, 1969, p. 31.
54. D. O. Shah, N. F. Djabbarah and D. T. Wasan, Colloid and Polymer Journal 256 (1978), p. 1002.
55. D. T. Wasan, M. Chan, S. M. Shah, K. Sampath and R. Shah, paper presented at the Engineering Foundation Conference on Theory, Practice and Process Principles for Physical Separations, CEP Symposium Series, (1978), in press
56. L. E. Scriven and C. V. Sterling, J. Fluid Mecahnics,19, (1964), p. 321.
57. M. Hennenberg, T. S. Sorensen and A. Sanfeld, Trans. Faraday Soc., 73, (1977), p. 48.
58. J. T. Davies, and G. R. A. Mayers, Chem. Eng. Sci., 16, (1961), p. 55.
59. R. H. Farley and R. S. Schechter, Canadian Journal of Chemical Engineering, 8, (1963), p. 37.

60. J. C. Slattery, Am. Inst. Chem. Engrs. Journal, 20, (1974), p. 1145.
61. J. C. Slattery, Symposium on Chemistry of Oil Recovery, Division of Petroleum Chemistry Preprint, ACS Meeting, Anaheim, California, March, (1978), p. 683.
62. R. W. Flummerfelt, Annual Report to ERDA on Interfacial Effects in the Recovery of Residual Oil by Displacement, Project No. E-40-1-5075, January, (1977).
63. D. W. Gorpil and F. C. Goodrich, J. Colloid and Interface Science, 62, (1977), p. 142.
64. D. S. Dmitrov, I. Panaiotov, P. Richmond and L. Ter-Minassian-Saraga, J. Colloid and Interface Science, (1978), in press.
65. S. Friberg and L. Rydhag, Kolloid-Zeitschrift und Zeitschrift fur Polymers, 244 (1971), p. 233.
66. S. Friberg, P. O. Jansson and E. Cederberg, J. Colloid and Interface Science, 55, (1976), p. 614.
67. N. Krog, Felte Seifen Anstrichm, 77 (1975), p. 267.
68. I. B. Ivanov, and D. S. Dimitrov, Colloid and Polymer Science, 252, (1974), p. 982.

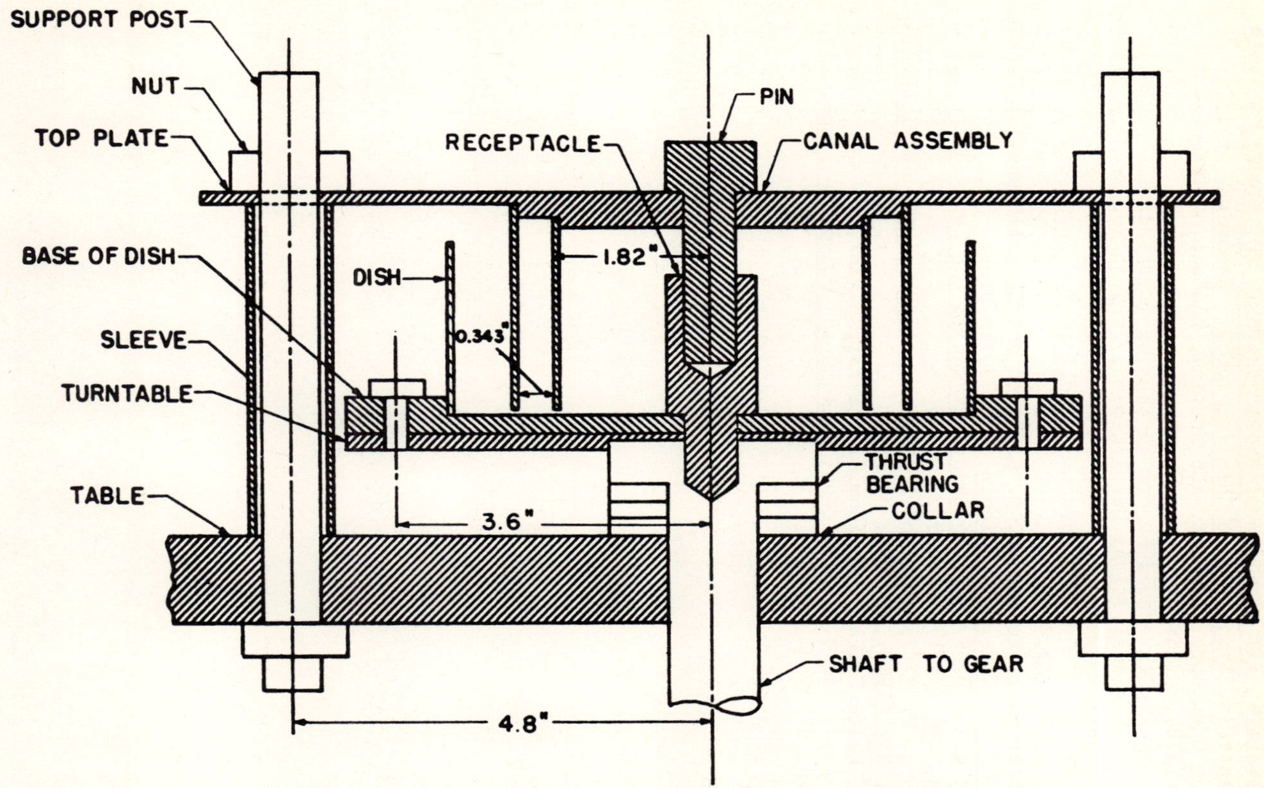

Fig. 1 Cross-Section of the Deep Channel Viscometer used in the Present Work.

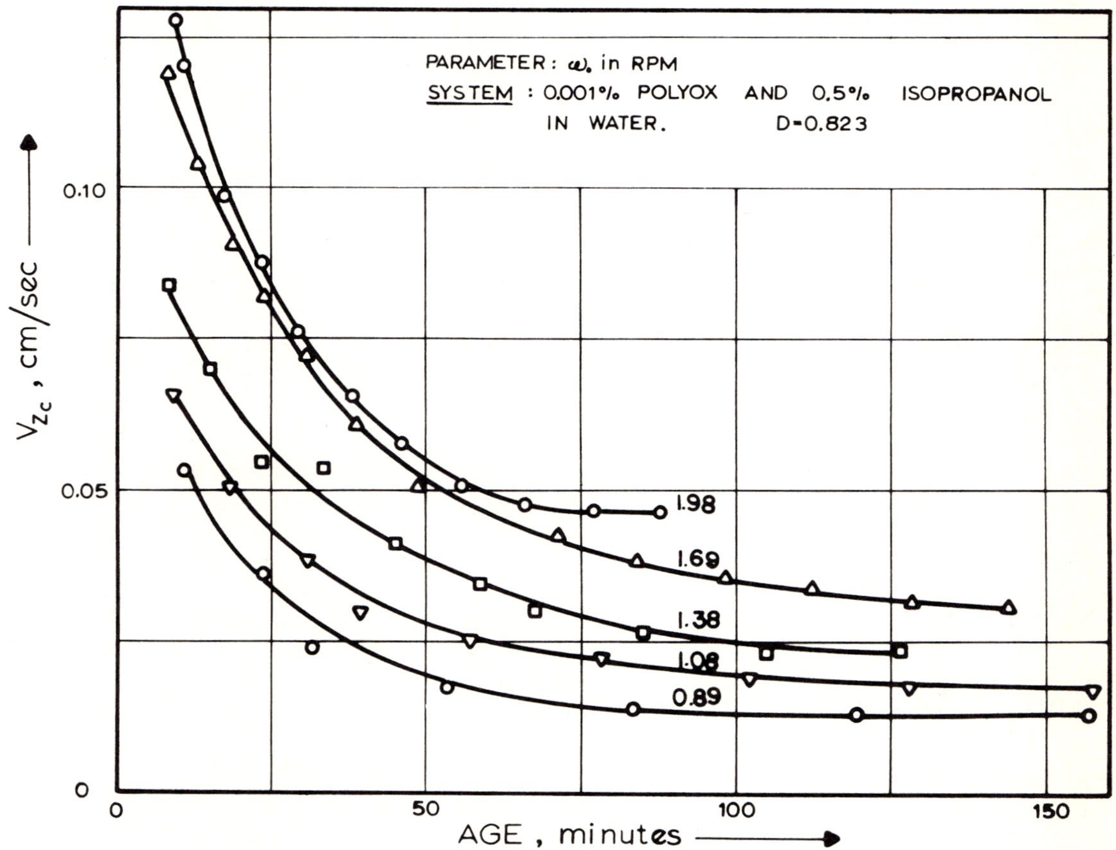

Fig. 2 Time-and Shear-Dependent Centerline Velocity of Aqueous Solution of 0.001% Polyox and 0.5% Isopropanol.

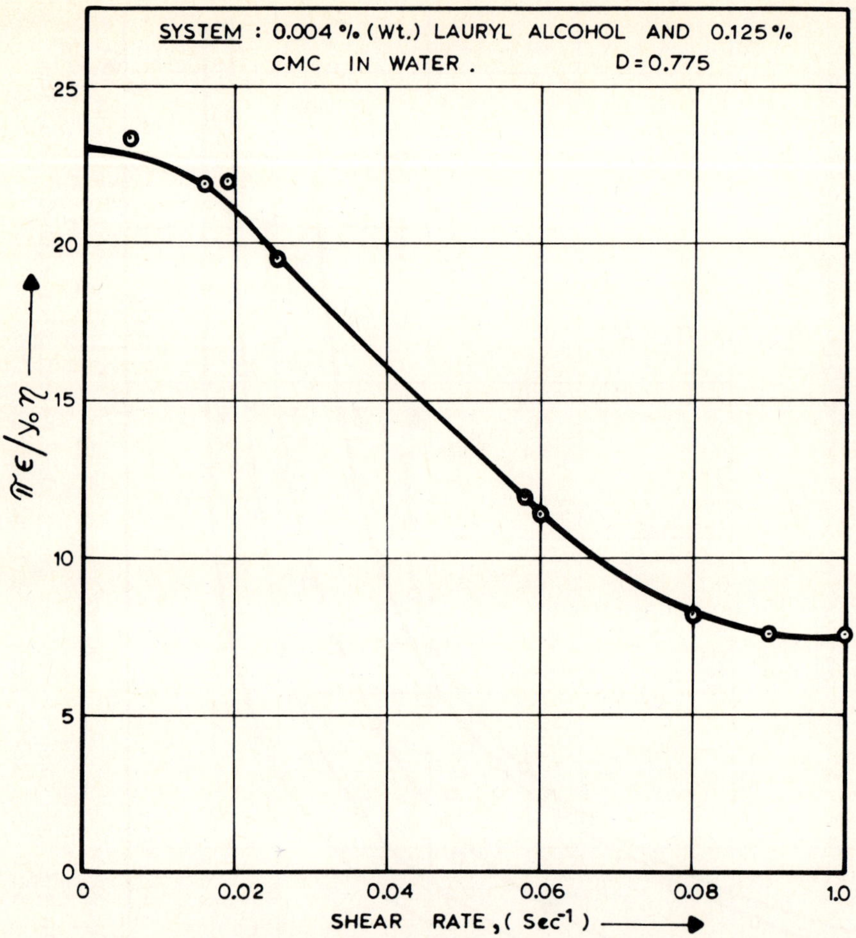

Fig. 3 Non-Newtonian Surface Viscosity of Aqueous Solution of 0.004% Lauryl Alcohol and 0.125% C.M.C.

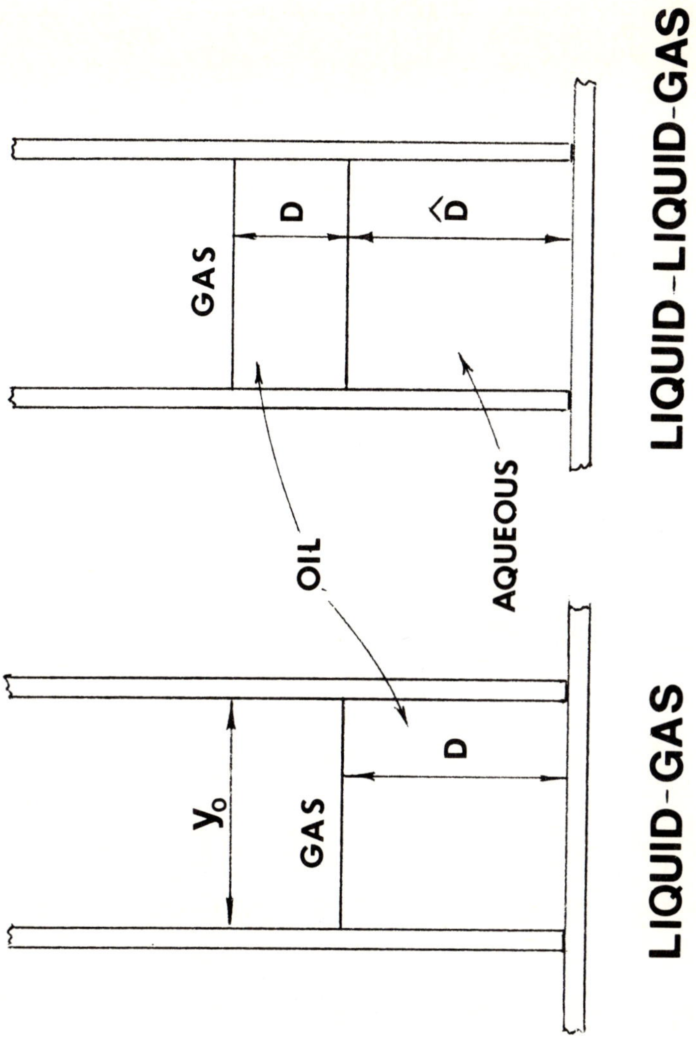

Fig. 4 Liquid-Gas (Fig. 4a) and the Liquid-Liquid- Gas (Fig. 4B) experiments.

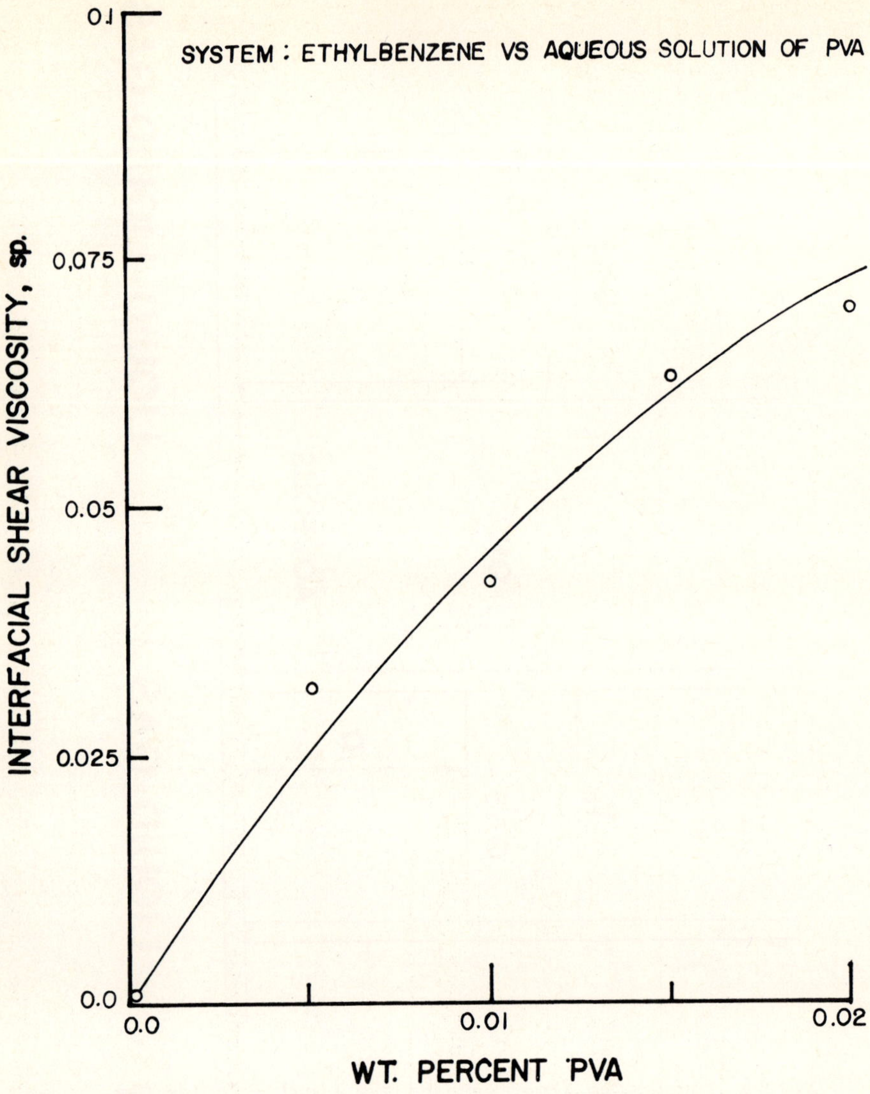

Fig. 5 Interfacial Shear Viscosity of PVA Solutions Versus Ethylbenzene.

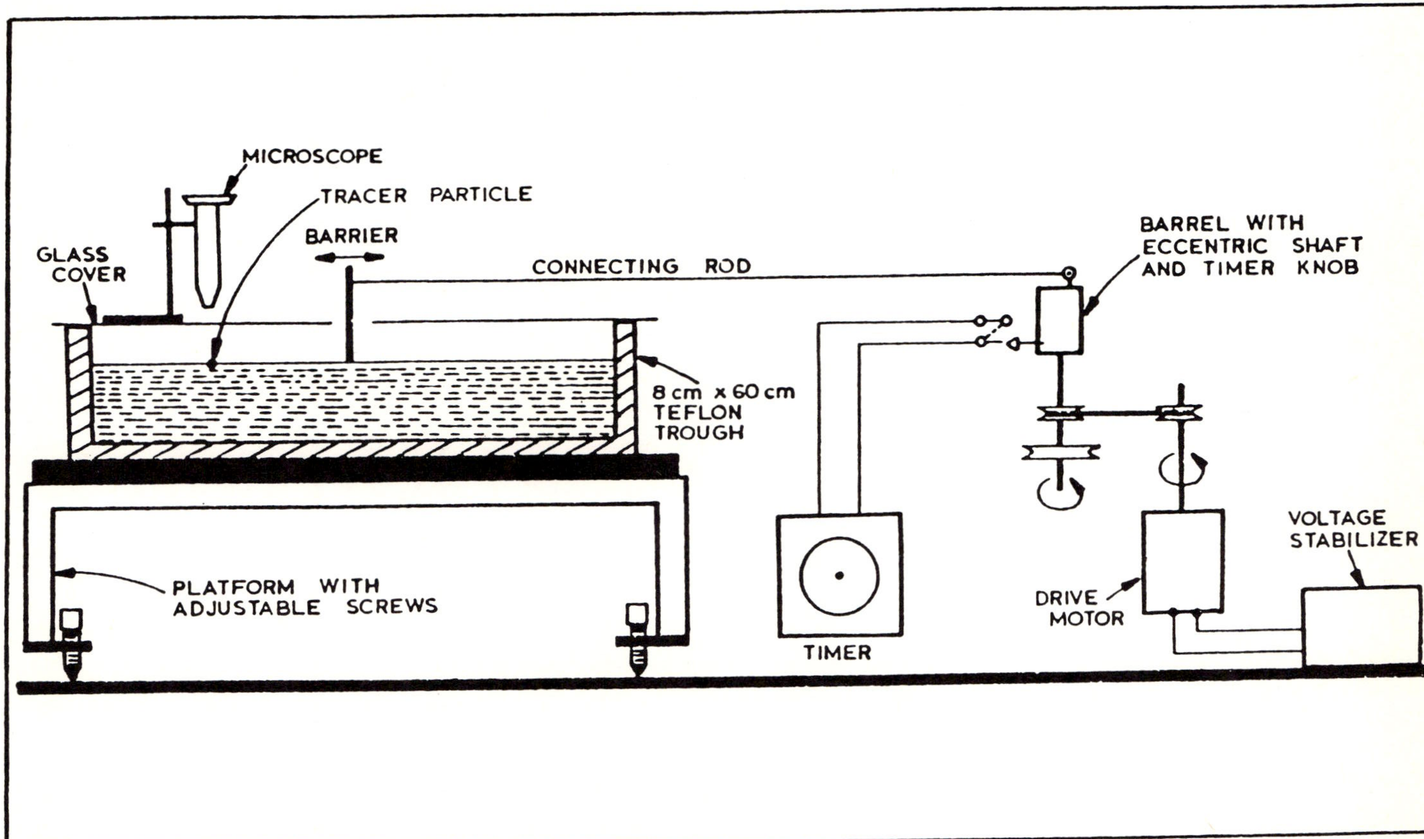

Fig. 6 Schematic Diagram of Apparatus for Longitudinal Wave Measurements.

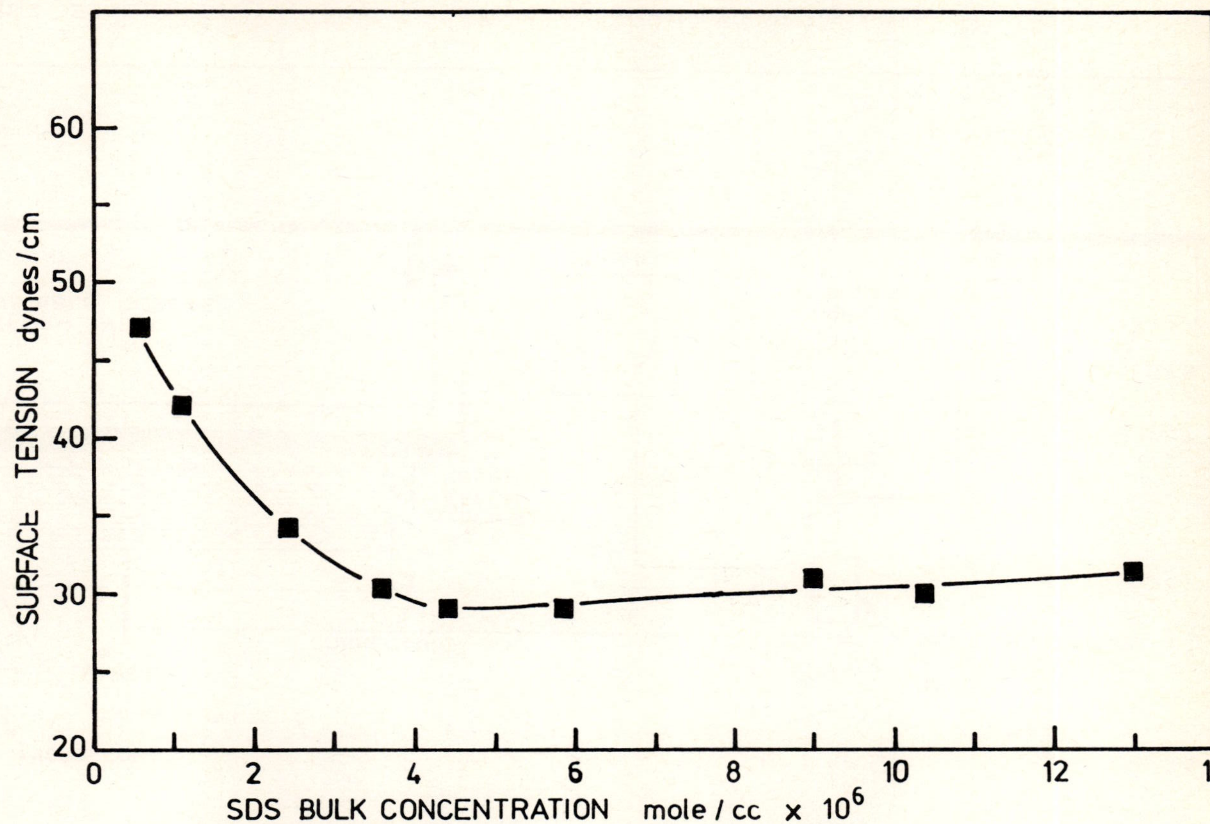

Fig. 7 Surface Tension Versus Bulk Concentration for Aqueous Solutions of Sodium Lauryl Sulfate-Lauryl Alcohol ($\frac{SLS}{LOH}$ = 100).

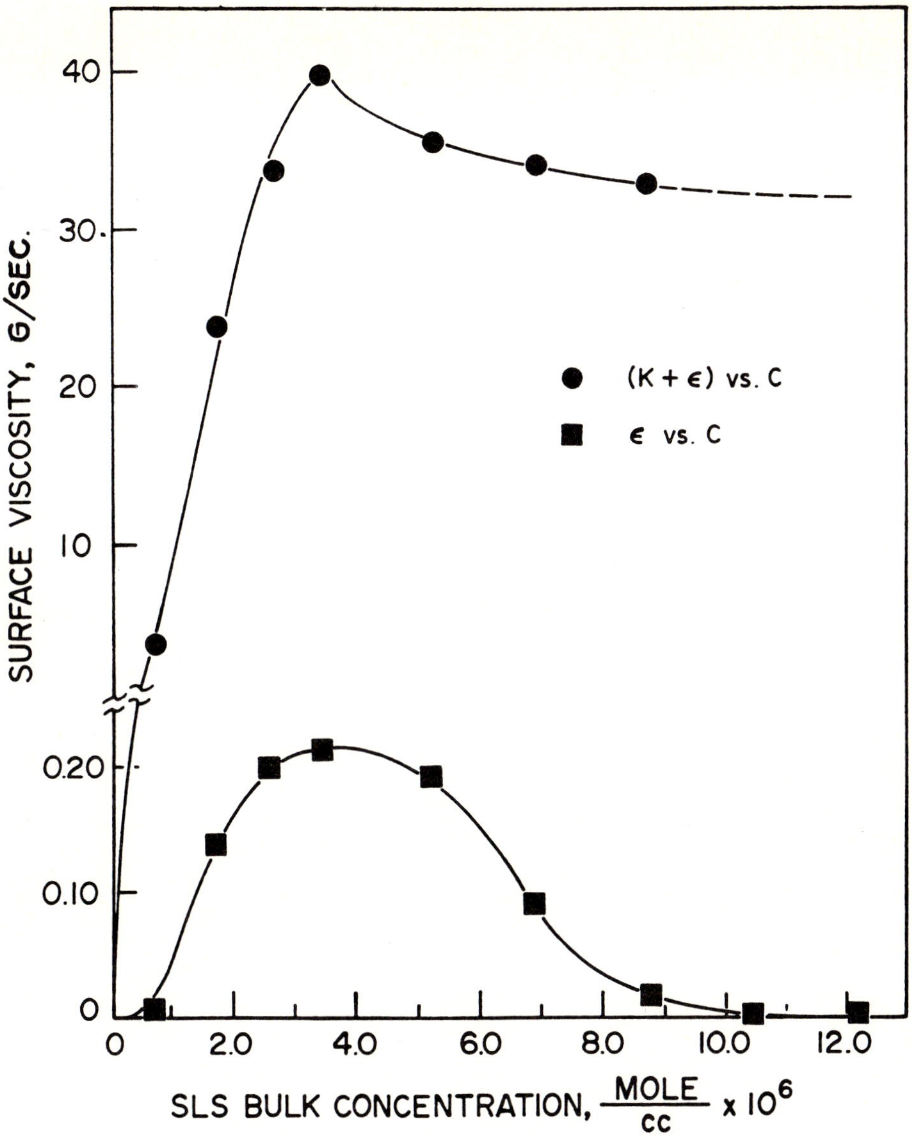

Fig. 8 Surface Viscosity, ($\kappa + \varepsilon$), and Surface Shear Viscosity, ε, as a Function of Sodium Lauryl Sulfate Concentration for Aqueous Solutions of SLS-LOH, $(\frac{SLS}{LOH} = 100)$.

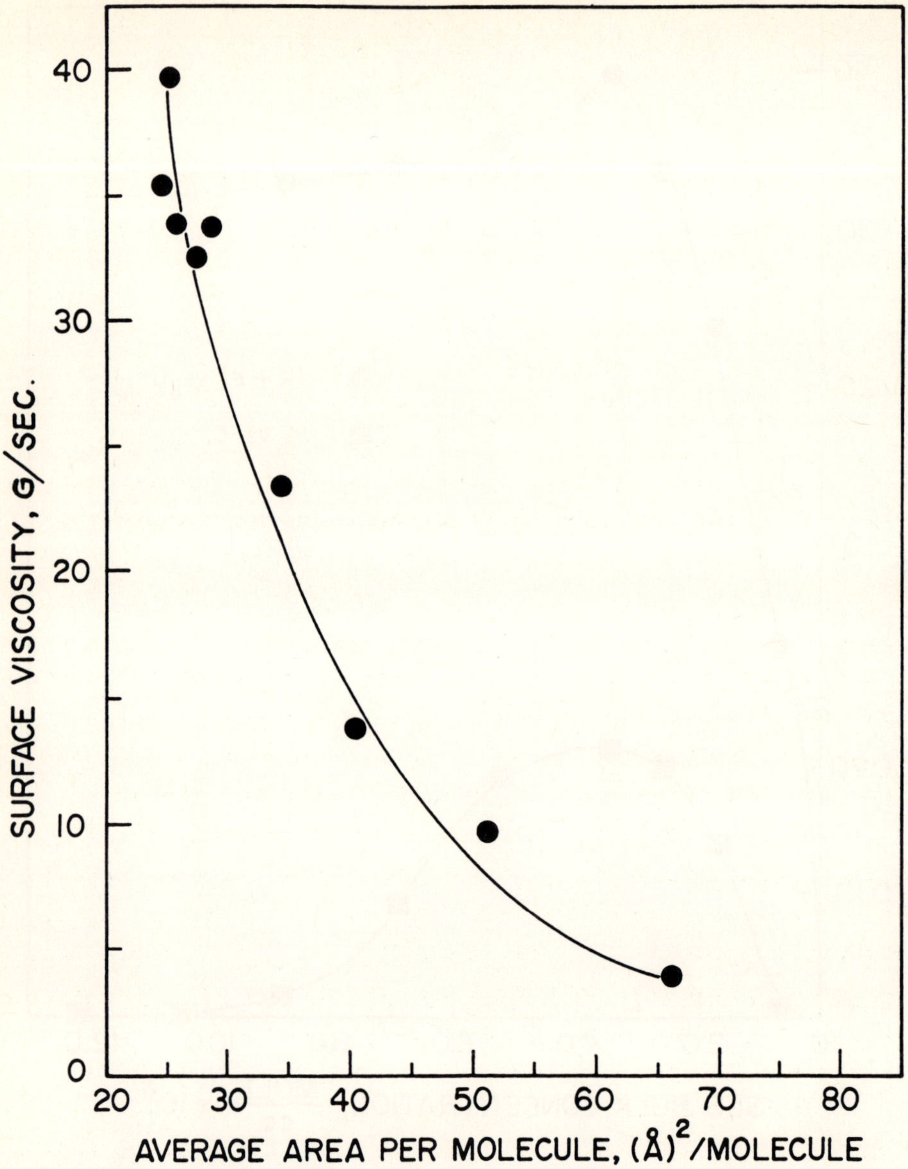

Fig. 9 Surface Viscosity, $(\kappa + \varepsilon)$, Versus the Average Area per Molecule for Aqueous Solutions of SLS-LOH, $(\frac{SLS}{LOH} = 100)$.

INTERFACIAL INSTABILITY AND LONGITUDINAL WAVES IN LIQUID-LIQUID SYSTEMS

by

M. Hennenberg*, P.M. Bisch*,+ M. Vignes-Adler**, A. Sanfeld*

* Université Libre de Bruxelles, ** Laboratoire d'Aerothermique Meudon-France, + CNPq-Brasil.

Introduction

An interface between two liquids may be considered from three complementary points of view : a geometrical, a mechanical and a thermodynamical one. Geometrically the interface is a surface without width or in other words a two dimensional space inbedded in an Euclidian three dimensional one. This feature explains the appearance of mean and total curvature in the complete tensorial description of the interface.

Mechanically it is taken as analogous to a stretched membrane. However to describe the properties of interfaces, one has to take into account the adjacent bulks and the exchange of physical entities such as momentum and matter between the bulks and the surface. Last named has no intrinsic existence but it is essentially the frontier of the volume phases. One of the first mechanical approach taking some of these considerations in a general framework was given by Scriven [1] as a generalization of Boussinesq's pure mathematical analysis. [2] A more detailed version is to be found in Aris' book [3] . Later on, Sternling and Scriven [4] studied the stability of a Newtonian two dimensional fluid.

More generalized balance equations derived from hydrodynamics have been put forward by Slattery [5] . Thermodynamically, theories were developed on the basis of a geometrical surface (Gibbs model), of a mono-and multilayer model [6] . Recently this problem has been studied from the point of view of non equilibrium situations [7] - [12] . A more phenomenological but still mechanical approach was developed by Levich [13] and by van den Tempel [14] . They wanted to distinguish capillary waves induced by normal deformations of the interface from longitudinal waves depending upon the tangential gradient of surface tension. Among the fascinating problems of phenomena far from equilibrium are the interface instabilities induced by mass transfers [4] ; such as chemical reactions [15, 16] . Our purpose in this paper is the study of the surface linear stability when a diffusion process occurs between the two immiscible viscous fluids. Furthermore we assume an instantaneous equilibrium between surface and sublayer. We use as boundary conditions the constitutive properties of the interface between these two fluids as they are expressed in the balance equations of mass and momentum. We consider the reference state to be at rest, the interface between the two bulks being then flat. We neglect all temperature fluctuations.

We give in the first two following sections a general derivation of the compatibility equation of this problem and later on we develop the study of oscillatory disturbances. The non oscillatory regime leading to the exchange of stabilities has been given in a previous paper [17]. Finally we compare our treatment with those developed by previous authors [4],[14],[18].

1. Hydrodynamical perturbed equations at the interface.

Let us call 1 the lower fluid and 2 the upper one. In each bulk (i) the solute is enough dilute so that, we take the total density ρ_i as a constant.

Since we are interested only in dilute solutions, we identify the barycentric velocity to the solvent movement. The incompressibility condition reads :

$$\operatorname{div} \underset{\sim}{v}_i = 0 \tag{1}$$

where $\underset{\sim}{v}_i$ is the velocity of bulk i. The classical Navier-Stokes equation is

$$\rho_i \frac{\partial \underset{\sim}{v}_i}{\partial t} = -\underset{\sim}{\operatorname{grad}}\, p_i + \mu_i \nabla^2 \underset{\sim}{v}_i \tag{2}$$

where μ_i is the constant viscosity of the i bulk. Applying twice the curl operator on both sides of (2) we have

$$\nabla^2 \left(\nu_i \nabla^2 - \frac{\partial}{\partial t} \right) \underset{\sim}{V_i} = 0 \tag{3}$$

where $\nu_i = \mu_i / \varrho_i$ is the kinematic viscosity of bulk i.

We analyse an arbitrary disturbance of a physical quantity (Q) by the standard procedure of expansion into normal modes. Here because of the symmetry of the unperturbed reference state we solve through development of the perturbed quantity δQ in a Fourier integral

$$\delta Q(x,y,z,t) = \int dk\, \delta Q_k(z)\, e^{i(k_x \cdot x + k_y \cdot y) + \underset{\sim}{\omega} t} \tag{4}$$

$$k = \sqrt{k_x^2 + k_y^2} \tag{5}$$

It is sufficient to study one general mode since for the system to be unstable at least one mode should be unstable.

We may use two different approaches :

a)

$$k = k_r - i k_i$$

$$\underset{\sim}{\omega} = i\omega \quad . \tag{6}$$

In this case

$$k = |k| e^{-i\delta} \tag{7}$$

where ω is the frequency, δ is the phase angle, $k_r = 2\pi/\lambda$ and k_i is the damping (>0) or amplification coefficient (<0).

By this expansion we study the instability at a given time with increasing distance from the origin. The marginal oscillating state is defined by $k_i = 0$. This is a peculiarly useful development in case where one analyses the reaction to an oscillating external disturbance. It is mostly this type of fluctuations which we will consider below.

b)

$$k = k_r$$

$$\underset{\sim}{\omega} = \omega_r + i\omega_i \tag{8}$$

Here we study the instability at a given position with increasing time [17]. The real part of $\underset{\sim}{\omega}$ is the damping (<0) or amplification coefficient (>0), while the imaginary part gives the oscillating factor.

The solutions of Eq. (3) are (dropping the index k)
(a) in the upper bulk :

$$v_z^2 = A_2 e^{-kz} + B_2 e^{-q_2 z} \tag{9a}$$

(b) in the lower bulk :

$$v_z^1 = A_1 e^{+kz} + B_1 e^{q_1 z} \tag{9b}$$

with

$$q_j = \sqrt{k^2 + \underset{\sim}{\omega}/\nu_j} \qquad Re(q_j) \geq 0 \quad (j=1,2) \tag{9c}$$

so that the perturbations disappear at infinite distances from the interface.

Equation (9) gives the amplitude of the normal velocities in terms of the four constants A_j, B_j $(j=1,2)$. We have to give four boundary conditions along the interface in order to find the compatibility equation between k and $\underset{\sim}{\omega}$. Two of them are deduced from the hydrodynamic conditions of having no slip at the interface :

$$\underset{\approx}{v}^1\Big|_{z=surface} = \underset{\approx}{v}^2\Big|_{z=surface} = \underset{\approx}{v}^S\Big|_{z=surface} \tag{10}$$

One should keep in mind that strictly speaking the last condition is valid for the velocities of the immiscible solvents taken at the interface and not for the barycentric velocities [7] [9]. But, here again, since the solute is very dilute, we may neglect a sweeping flux [19, 20] proportional to $\left[\frac{(\varrho_{solute})^2}{\varrho_{solvent}} \cdot \underset{\sim}{V}_{solute} \right]$.

Because of Eq. 10 and of the bulk incompressibility (1), one has also

$$\frac{D}{Dz}(V_z^1)\bigg|_{z=\text{surface}} = \frac{D}{Dz}(V_z^2)\bigg|_{z=\text{surface}} \qquad (11)$$

so that one might express the amplitude A_i of the potential part of the velocity in terms of the B_i, amplitude of the rotational part of the velocity as

$$A_1 = -\frac{1}{2k}\left\{(q_2-k)B_2 + (q_1+k)B_1\right\}$$

$$A_2 = -\frac{1}{2k}\left\{(q_2+k)B_2 + (q_1-k)B_1\right\} \qquad (12)$$

2. Surface balance equations.

A) Normal momentum balance.

We linearize the surface momentum and mass balances for small deviations from a plane interface. In this way all the geometric parameters which are linked to the metric of the surface and to its total curvature disappear ([4], [1], [3] and [17]). Only the variation of the mean curvature remains in the normal part of the momentum balance. This generalization of the Laplace equation is

$$\Gamma_{tot}^{o}\frac{\partial V_z}{\partial t} = \Delta\left\{-\delta p + 2\mu\frac{\partial V_z}{\partial z} + g\varrho\,\delta z\right\} + \sigma_o \nabla_s^2 \delta z \qquad (13a)$$

where δp is the bulk pressure perturbation, δz is the normal deformation of the surface, $g\varrho\delta z$ is the gravitational effect. σ_o is the unperturbed surface tension and Γ_{tot}^{o} the total unperturbed adsorption. The operator Δ applied on a function

$$\Delta f = (f^2 - f^1)\Big|_{z=\delta z} \qquad (13b)$$

is the jump calculated in bulk 1 and 2 at the perturbed interface. The last term of the R.H.S. of (13) is the variation of curvature.

Up to a second order term we have

$$\frac{\partial}{\partial t}\delta z = v_z \tag{14}$$

Using a Fourier expansion in normal modes (see Eq.(4) and (9) and Eq. (12), one obtains

$$B_1(q_1-k)\left\{\frac{\varepsilon_D}{2k}\left[\underset{\sim}{\omega}\left(\frac{\rho_1+\rho_2}{k}+\Gamma^{\circ}_{tot}\right)+\frac{k^2\sigma^{\circ}-g\Delta\rho}{\underset{\sim}{\omega}}\right]+\mu_2+\frac{q_1}{k}\mu_1\right\}+B_2(q_2-k)\left\{\frac{\varepsilon_D}{2}\left[\underset{\sim}{\omega}\left(\frac{\rho_1+\rho_2}{k}+\Gamma^{\circ}_{tot}\right)+\frac{k^2\sigma^{\circ}-g\Delta\rho}{\underset{\sim}{\omega}}\right]+\mu_1+\frac{q_2}{k}\mu_2\right\}=0 \tag{15}$$

The dimensionless quantity ε_D is equal to 1 if we take into account the normal deformation of the interface and zero if we restrict ourselves to perturbations leaving the surface undeformed. In this last case $\varepsilon_D=0$, we have to note some important consequences. This hypothesis means

$$v_z\Big|_{z=\text{surface}} = 0 \tag{16}$$

Thus, from Eq.(9) one gets

$$A_i = -B_i \qquad (i=1,2) \tag{17}$$

So that, using the uncompressibility of the two bulks (see (1) and (11)), we write

$$B_1(q_1-k) = -B_2(q_2-k) \tag{18}$$

The equation (15) thus reduces itself to

$$\mu_1(q_1-k) = \mu_2(q_2-k) \tag{19}$$

which implies that there is compatibility between the Laplace condition and the peculiar choice of the velocity (see Eq. (16)). Should we consider (19) as valid and look for frequencies and wavenumbers which are not solution of the classical Rayleigh-Taylor problem [21] we would have then

$$\mu_1 + \frac{\mu_2 q_2}{k} + \tilde{\omega}\left(\frac{\rho_1+\rho_2}{k} + \Gamma^{\circ}_{tot}\right) + \frac{\sigma_0 k^2 - g\Delta\rho}{\tilde{\omega}} \neq 0 \qquad (20)$$

Relation (18) has to be true, there is then no deformation of the surface. We study thus a problem of hydrodynamical instability which is not related to capillary waves. The condition (19) uncouples the Rayleigh-Taylor effect from longitudinal waves.

B) Tangential momentum balances.

The tangential part of the momentum balance depends upon the two dimensional gradient of surface tension (Marangoni effect). The components of the tangential part of this momentum balance are

$$\Gamma^{\circ}_{tot}\frac{\partial}{\partial t}V_i^s = \Delta\left\{\mu\left(\frac{\partial}{\partial z}V_i + \frac{\partial}{\partial x_i}V_z\right)\right\} + \frac{\partial\sigma}{\partial x_i} - \eta_{dil}\left(\frac{\partial}{\partial x_i}\frac{\partial}{\partial z}V_z\right) + \eta_{sh}\nabla_s^2 V_i^s \qquad (21)$$

taken along the interface, where η_{sh} and η_{dil} are the intrinsic surface shear and dilatational viscosities respectively. Combining the incompressibility condition (1) and Eq. (21), we have

$$\Gamma^{\circ}_{tot}\frac{\partial}{\partial t}\frac{\partial}{\partial z}V_z^s = \Delta\left[\left(\frac{\partial^2}{\partial z^2} - \mu\nabla_s^2\right)V_z^s\right] - \nabla_s^2\delta\sigma + \eta_s\nabla_s^2\left(\frac{\partial}{\partial z}V_z^s\right) \qquad (21a)$$

where $\eta_s = \eta_{sh} + \eta_{dil}$ is the sum of the surface intrinsic viscosities. We supposed the interface to be a two dimensional fluid. Only in the case of a plane interface in the reference state do the linearized perturbation problem contain the surface viscosities as one single entity which play a role only in the tangential part of the momentum balance [17].

Neglecting the fluctuations of temperature and pressure in the surface tension, we may write

$$\delta\sigma = -\alpha\,\delta\Gamma \tag{22}$$

where Γ is the surface adsorption of the surfactant, and α is a positive quantity.

After expansion in normal modes and using the hydrodynamical boundary conditions (12) and the equation of state (22), the tangential part of the momentum balance becomes :

$$-B_1(q_1-k)\left\{q_1\mu_1 + k\mu_2 + \frac{k^2}{2}\eta_s + \frac{\omega}{2}\Gamma^{o}_{tot}\right\} + B_2(q_2-k)\left\{q_2\mu_2 + k\mu_1 + \frac{k^2}{2}\eta_s + \frac{\omega}{2}\Gamma^{o}_{tot}\right\} - k^2\alpha\,\delta\Gamma = 0 \tag{23}$$

C) Interfacial mass balance.

As we stated before, we have a soluble surfactant diffusing accross the two bulks so that the mass balance at the interface is

$$\varepsilon_J\left(\frac{\partial}{\partial t}\Gamma + \frac{\dot{a}\Gamma}{2a} + \mathrm{div}_s(\Gamma\underset{\sim}{v}^s) + \mathrm{div}_s J_s^d\right) = \Delta J^d\cdot\underset{\sim}{n} \tag{24}$$

where J_i^d is the bulk diffusion flux from bulk i ; Γ the adsorbed matter,

J_s^d is the surface diffusion flux ;

$\underset{\sim}{n}$ is the outward directed unit normal,

a is the surface metric [3]

ε_J is equal to 1 if there is adsorption of matter at the interface and 0 if there is no adsorption.

This last will be the case for the reference stationary state.

In order to solve this last equation (24) one has to calculate the perturbed diffusion fluxes and one has to give the relation between $\delta\Gamma$ and the variation of the mass concentration of the solute δc_i in each bulk.

Let us assume in the reference state

$$d\Gamma^{\circ} = K_i \, dc_i^{\circ} \Big|_{z=0} \tag{25}$$

and in the perturbed state

$$\delta\Gamma = K_i \left[\delta c_i + \varepsilon_D \frac{d}{dz} c_i^{\circ} \, \delta z \right] \tag{26}$$

the balance equation for each bulk is

$$\frac{\partial}{\partial t} \delta c_i = \varepsilon_G V_z \frac{dc_i^{\circ}}{dz} + \mathcal{D}_i \nabla^2 \delta c_i \tag{27}$$

where ε_G is equal to 1 if one takes into account the convection and zero otherwise, $\mathcal{D}_i$ is the bulk diffusion coefficient of the solute.

In the unperturbed reference state we assume a linear concentration profile solution of the one dimensional Fickian law

$$c_i^{\circ} = c_i^{\circ\circ} - \beta_i z \tag{28}$$

and no accumulation of matter along the interface which means

$$\beta_1 \mathcal{D}_1 = \beta_2 \mathcal{D}_2 \tag{29}$$

The coefficients β_i are positive quantities if transfer is from 1 to 2 and negative if transfer is the other way round.

Now in the perturbed state, the diffusion flux is

$$J_i^d = - \left[\mathcal{D}_i \frac{d}{dz} \delta c_i + \varepsilon_D \cdot \mathcal{D}_i \cdot \left(\frac{d^2 c_i^{\circ}}{dz^2} \right)_{z=0} \delta z \right] \tag{30}$$

The last term of the R.H.S. of (30) is zero because of Eq. (28). By expanding in normal modes Eq. (27) and replacing the velocities by their respective definitions (9), using also the conditions (12) one has

$$\delta c_1 = \mathcal{C}^{(1)} \exp r_1 z + \varepsilon_c \left[\mathcal{C}_1^{(1)} \exp kz + \mathcal{C}_2^{(1)} \exp q_1 z \right]$$

$$\delta c_2 = \mathcal{C}^{(2)} \exp -r_2 z + \varepsilon_c \left[\mathcal{C}_1^{(2)} \exp -kz + \mathcal{C}_2^{(2)} \exp -q_2 z \right] \tag{31}$$

with $r_i = \sqrt{k^2 + \underset{\sim}{\omega}/D_i}$ and $\mathrm{Re}(r_i) \geq 0$

$$\mathcal{C}_1^{(i)} = \frac{-\beta_i A_i}{D_i (k^2 - r_i^2)} \quad ; \quad \mathcal{C}_2^{(i)} = \frac{-\beta_i B_i}{D_i (q_i^2 - r_i^2)} \tag{32}$$

Using Eq. (26), we express the amplitudes $\mathcal{C}^{(1)}$ and $\mathcal{C}^{(2)}$ in terms of $\mathcal{C}_j^{(i)}$ and of the amplitude of $\delta\Gamma$, $\delta\tilde{\Gamma}$. Combining Eq. (24) and (30) with a surface Fickian law, we obtain the surface mass balance equation

$$\left\{ \varepsilon_J \left[\underset{\sim}{\omega} + k^2 D_s \right] + \sum \frac{r_j D_j}{K_j} \right\} \delta\tilde{\Gamma} + B_1 (q_1 - k) \left[-\varepsilon_J \frac{\Gamma^o}{2} + \frac{\beta_1 D_1}{\underset{\sim}{\omega}} \left(\frac{r_1 + r_2}{2} \right) (\varepsilon_G - \varepsilon_D) - \frac{\beta_1 D_1}{\underset{\sim}{\omega}} \cdot \frac{q_1 + k}{q_1 + r_1} \varepsilon_G \right] + B_2 (q_2 - k) \cdot \left[\varepsilon_J \frac{\Gamma^o}{2} + \frac{\beta_1 D_1}{\underset{\sim}{\omega}} \left(\frac{r_1 + r_2}{2} \right) (\varepsilon_G - \varepsilon_D) - \frac{\beta_1 D_1}{\underset{\sim}{\omega}} \frac{q_2 + k}{q_2 + r_2} \varepsilon_G \right] = 0 \tag{33}$$

D) Compatibility equation.

The equations (15), (23), (33) are an homogeneous linear algebraic system of three equations in the three unknown $\delta\tilde{\Gamma}$, $B_1(q_1 - k)$, $B_2(q_2 - k)$. In order to have a non trivial solution, the determinant of the coefficients has to be zero. This is the compatibility condition of the problem. Some simple manipulations of the columns lead us to the following form of the compatibility condition : (see eq. (34) on next page).

This determinant is the product of two different stability effects together with a coupling term. Indeed, it is easy to see that the second term of the second line is the well-known Rayleigh-Taylor problem of the stability of an adverse gradient of density in the gravity field, the quantity $\frac{k^2 \sigma^o}{\underset{\sim}{\omega}}$ is related to the Laplace condition and thus to the normal deformation of the surface. The Rayleigh-Taylor instability will not be discussed furthermore.[21]

$$\begin{vmatrix} \begin{array}{l} \varepsilon_J\left[\underset{\sim}{\omega}+k^2 D_s\right] \\ +\sum \frac{D_i r_i}{K_i} \end{array} & \begin{array}{l} -\frac{\beta_1 D_1}{\underset{\sim}{\omega}}\left[\left(\frac{q_1+k}{q_1+r_1}+\frac{q_2+k}{q_2+r_2}\right)\varepsilon_G \right. \\ \left. \quad +\frac{r_1+r_2}{k}(\varepsilon_D-\varepsilon_G)\right] \end{array} & \begin{array}{l} \varepsilon_J\Gamma^\circ \\ +\frac{\beta_1 D_1}{\underset{\sim}{\omega}}\left[\frac{q_1+k}{q_1+r_1} \right. \\ \left. -\frac{q_2+k}{q_2+r_2}\right]\varepsilon_G \end{array} \\ 0 & \begin{array}{l} -\frac{\varepsilon_D}{k}\left[\underset{\sim}{\omega}\left\{\frac{\rho_1+\rho_2}{k}+\Gamma^\circ_{tot}\right\} \right. \\ \left. +\frac{k^2\sigma^\circ-g\Delta\rho}{\underset{\sim}{\omega}}\right]- \end{array} & \begin{array}{l} \varepsilon_D\left[\mu_1\left(\frac{q_1}{k}-1\right) \right. \\ \left. -\mu_2\left(\frac{q_2}{k}-1\right)\right] \end{array} \\ \alpha k & \begin{array}{l} \sum\left\{\mu_i\left(1+\frac{q_i}{k}\right)\right\} \\ \varepsilon_D\left[\mu_1\left(\frac{q_1}{k}-1\right)-\mu_2\left(\frac{q_2}{k}-1\right)\right] \end{array} & \begin{array}{l} -\frac{1}{k}\left[\Gamma^\circ_{tot}\underset{\sim}{\omega}+\eta_s k^2 \right. \\ \left. +\sum\{\mu_i(k+q_i)\}\right] \end{array} \end{vmatrix} = 0 \qquad (34)$$

The second effect is the Marangoni process due to the variation of the surface tension along the interface and to the movement of the adjacent vulks it induces. Both problems are coupled through the last term of the second line of (34) that is

$$\varepsilon_D\left[\mu_1\left(\frac{q_1}{k}-1\right)-\mu_2\left(\frac{q_2}{k}-1\right)\right]$$

If we do not consider any normal deformation of the interface, we uncouple completely the two problems and we can better focuse our attention on the Marangoni effect. We shall do so and put the determinant in the following form :

$$\alpha k^2\left[\frac{\varepsilon_J\Gamma^\circ+\varepsilon_G(\beta_1 D_1/\underset{\sim}{\omega})\cdot\left(\frac{q_1+k}{q_1+r_1}-\frac{q_2+k}{q_2+r_2}\right)}{\varepsilon_J(\underset{\sim}{\omega}+k^2 D_s)+\sum\left(\frac{D_i r_i}{K_i}\right)}\right]+\Gamma_{tot}\underset{\sim}{\omega} + k^2\eta_s+\sum\{\mu_i(q_i+k)\} = 0 \qquad (35)$$

This is a generalization of the Marangoni problem as studied by Sternling and Scriven to cases where one takes into account the variation of the adsorption at the interface [4].

3. Surface elasticity and longitudinal waves.

We interest ourselves to motion induced by the horizontal variation of the surface tension along a compressible surface since

$$\frac{\delta A_e}{A_e} = \frac{1}{\underset{\sim}{\omega}} \operatorname{div}_s \underset{\sim}{V}_s = -\frac{1}{\underset{\sim}{\omega}} \frac{d}{dz} V_z$$
$$= -\frac{1}{\underset{\sim}{\omega}} B_1 (q_1 - k) \exp\left[i(k_x x + k_y y) + \underset{\sim}{\omega} t\right] \tag{36}$$

we have variation of area A_e either through compression or through dilatation but because of surface tension there will be a resistance against surface deformation. We introduce then a surface elasticity modulus

$$\varepsilon = \frac{\delta \sigma}{\delta(\ln A_e)} \tag{37}$$

The elasticity coefficient should reflect the different behaviour between soluble and insoluble surfactant. Indeed, the diffusional exchanges between the bulks and the surface may induce relaxation processes [14], [22] [23] [24.] The most general definition is then

$$\frac{\delta \sigma}{\delta \ln A_e} = |\varepsilon| \exp i\theta = \varepsilon_r + i\,\varepsilon_i \tag{38}$$

Let us now justify the validity of the definition (38). Combining Eq. (21a), (36) and (37), we have

$$k^2 \left[\eta_s + \frac{\varepsilon}{\underset{\sim}{\omega}}\right] \frac{DV_z}{Dz} = \Delta\left[\mu\left(\frac{D^2}{Dz^2} + k^2\right) V_z\right] - \Gamma^0_{tot} \underset{\sim}{\omega} \frac{DV_z}{Dz} \tag{39}$$

It is important to note that we have not used the state equation (22) to derive the last equation. Equations (9), (17), (18) and (30) give then the compatibility condition of our problem

$$k^2\left[\eta_s + \frac{\epsilon}{\tilde{\omega}}\right] + \sum_\alpha \mu_\alpha (q_\alpha + k) + \Gamma^\circ_{tot}\,\tilde{\omega} = 0 \tag{40}$$

and ϵ hides the physical nature of the processes leading to the various exchanges between bulks and surface . Previously we have developed completely this problem by assuming that exchanges of matter by diffusion were at the core of new features. Thus the comparison of Eq. (40) and (35) gives explicitly the elasticity coefficient ϵ in terms of these exchanges

$$\epsilon = \left\{ \frac{\epsilon_J \Gamma^\circ + \frac{\beta_1 D_1}{\tilde{\omega}}\left[\frac{q_1+k}{q_1+r_1} - \frac{q_2+k}{q_2+r_2}\right]\epsilon_G}{\epsilon_J(\tilde{\omega} + k^2 D_s) + \sum \frac{D_i r_i}{K_i}} \right\} \alpha\,\tilde{\omega} \tag{41}$$

and using a restrictive definition of the classical Gibbs elasticity

$$\epsilon_0 = \alpha\Gamma^\circ = \left.\frac{d\sigma}{d\ln\Gamma^\circ}\right|_{\Gamma=\Gamma^\circ} \tag{42}$$

we obtain

$$\epsilon = \epsilon_0\,\tilde{\omega}\left\{ \frac{\epsilon_J + \frac{\beta_1 D_1}{\tilde{\omega}\Gamma^\circ}\left(\frac{q_1+k}{q_1+r_1} - \frac{q_2+k}{q_2+r_2}\right)\epsilon_G}{\epsilon_J(\tilde{\omega} + k^2 D_S) + \sum \frac{D_i r_i}{K_i}} \right\} \tag{43}$$

The hypothesis (38) is thus justified because generally k or $\tilde{\omega}$ are complex quantities (see Eq.(6),(7) and (8)). The phase angle

$$\frac{\epsilon}{\epsilon_0} = \frac{\epsilon_J + \frac{\beta_1 D_1}{\tilde{\omega}\Gamma^\circ}\left(\frac{q_1+k}{q_1+r_1} - \frac{q_2+k}{q_2+r_2}\right)}{\epsilon_J(\tilde{\omega} + k^2 D_s) + \sum_i \frac{D_i r_i}{K_i}} \tag{44}$$

is generally not a multiple of 2π, especially since we are looking at the response of the system to an oscillatory disturbance. The elasticity modulus is a function of neither the total intrinsic surface viscosity η_s nor the surface acceleration $\Gamma^{\circ}_{tot}\tilde{\omega}$. Last quantity gives only a small contribution to the surface stability [17]. Furthermore very scarce experimental values of η_s are available [25], [26]. The explicit derivation of equation (43) does not depend upon the chosen strain-stress model. It relies only upon the exchange of matter between bulks and surface and upon the state equation. The dynamic quantity ϵ is defined for out of equilibrium situations assuming the equilibrium between layer and sublayer.

Like in the classical theory of viscoelasticity for three dimensional phenomena, ϵ_r is a storage modulus and ϵ_i a loss modulus. We have thus a kind of viscoelastic behaviour of the interface [14] [27]. The comparison should not be pushed too far since this new feature is entirely due to exchanges between bulks and surface. There are thus two kinds of dissipative processes at the interface. The first one is linked to the intrinsic viscosity and does not depend upon those exchanges. The other is solely due to the exchanges of matter between bulks and surface and lead to an important change in the rheological properties of the interface.

4. Long wavelengths approximations.

Let us look to oscillatory disturbances so that we will use Eq. (7).

$$k = |k| e^{-i\delta} \quad ; \quad \tilde{\omega} = i\omega$$

Since wavelength induced by an oscillatory disturbance will be of the dimension of the apparatus they will be much greater than the penetration depth so that we can restrict ourselves to consider cases where

$$|r_j^2| > |q_j^2| \gg |k^2|$$

$$q_j \cong (1+i)\sqrt{\frac{\omega}{2\nu_j}} \qquad (j=1,2) \qquad r_j \cong (1+i)\sqrt{\frac{\omega}{2D_j}} \tag{45}$$

The problem becomes very parallel to the well known hydrodynamic

Stokes second problem [28],[29]. It is the flow about an infinite plate which oscillates longitudinally. We have a no-slip condition expressed by Eq. (10), (11), (12). We may then introduce a momentum penetration depth

$$\delta_j^{\nu} = \sqrt{\frac{\nu_j}{\omega}} \qquad (j = 1,2) \tag{46}$$

beyond which the exchange of momentum has no importance. Within this very thin layer of width δ_j^{γ} the oscillatory movement of the interface is taken into account. Beyond this layer, the bulk viscosity forces have restored the stationary reference state. It has some of the characters of the boundary layer theory originally developed to explain the behaviour of real, viscous fluids near a wall. It is a sound comparison since both fundamental conditions of the boundary layer theory are present here: shearing bulk stresses due to viscosisty, and a no-slip condition with zero normal velocity at the interface. Together they help to explain the decrease from the velocity near the interface to zero in the bulks. We know that near a surface there exist a very thin layer in which concentration changes abruptly [30] so that mass transfer through diffusion or convection are of the same size. A mass penetration depth may also be defined in the vicinity of the surface. The approximations (45) are thus proportional to the inverse of the penetration depth of mass and momentum respectively. Since, in general

$$\mathcal{D}_j \ll \nu_j \qquad (j = 1,2) \tag{47}$$

the mass penetration depth is smaller than the momentum penetration depth. Taking into account the approximation (45) the uncoupling condition (19) reads

$$\frac{\mu_1}{\mu_2} = \frac{\rho_2}{\rho_1} \tag{48}$$

In figure 1, we plot ρ_1/ρ_2 as a function of μ_1/μ_2. We obtain a branch of equilateral hyperbola symmetric around the first bissectrix which in this representation is the locus of all the states where both bulks have the same kinematic viscosity.

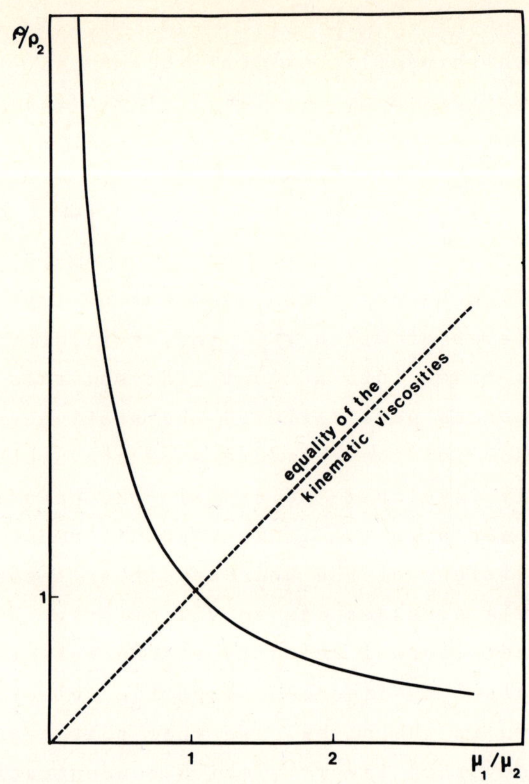

Fig. 1. Uncoupling condition between the Rayleigh-Taylor problem and the Marangoni effect for large wavelengths.

Both curves intersect each other in only one point where the densities are equal, and the viscosities are equal. Experimental evidences [24], however, show that to uncouple longitudinal and capillary waves it is sufficient that μ_1/μ_2 and ρ_2/ρ_1 are of the same order of magnitude. Replacing r_j and q_j in Eq. (40) and (41) by their approximation (46), and using (38), one has

$$|\varepsilon|\,|k|^2 = \sum_j \sqrt{\mu_j \rho_j}\;\omega^{3/2} \tag{49a}$$

$$\delta = \frac{\pi}{8} + \frac{\Theta}{2} \tag{49b}$$

We are looking to space damping or amplification of the disturbance in the third quadrant:

$$0 < \delta < \frac{\pi}{2} \qquad \text{or} \qquad -\frac{\pi}{8} < \frac{\Theta}{2} < \frac{3\pi}{8}$$

stable solution

$$-\frac{\pi}{2} < \delta < 0 \qquad \text{or} \qquad -\frac{3\pi}{8} < \frac{\Theta}{2} < -\frac{\pi}{8}$$

unstable solution (50)

so that the marginal state with overstability corresponds to

$$\delta = 0 \qquad \Theta = -\frac{\pi}{4} \qquad tg\Theta = -1 \tag{51}$$

We have the general definition of the surface elasticity modulus

$$\mathcal{E} = \mathcal{E}_0 \left\{ \frac{\mathcal{E}_J - i\mathcal{E}_G \Lambda}{\mathcal{E}_J + \tau - i\tau} \right\} \tag{52a}$$

$$tg\,\Theta = \frac{\mathcal{E}_J \tau - \Lambda(\mathcal{E}_J + \tau)\mathcal{E}_G}{\mathcal{E}_J(\Lambda + \tau) + \mathcal{E}_G \tau \Lambda} \tag{52b}$$

The dimensionless quantities τ and Λ are defined by

$$\tau = \sqrt{\frac{1}{\omega}}\,\underset{\sim}{\alpha} = \frac{1}{K_1}\sqrt{\frac{D_1}{2\omega}} + \frac{1}{K_2}\sqrt{\frac{D_2}{2\omega}} \tag{52c}$$

$$\Lambda = \frac{\underset{\sim}{\gamma}}{\omega} = \frac{\beta_1 D_1}{\omega \Gamma^0} \left\{ \sqrt{\frac{D_1}{\nu_1}} - \sqrt{\frac{D_2}{\nu_2}} \right\} \tag{52d}$$

We will now neglect the surface diffusion (see eq. (44)) since it is of the same order of magnitude as the bulk diffusion coefficient [31].

Both τ and Λ deserve some explanation

a) τ is the sum of two dimensionless $\tau_{(j)}$ each of them linked to one bulk

$$\tau = \tau_{(1)} + \tau_{(2)} \qquad \tau_{(j)} = \frac{1}{K_j}\sqrt{\frac{D_j}{2\omega}} \quad (j=1,2) \tag{53a}$$

Each $\tau_{(j)}$ is specific to mass exchange between bulk j and the surface. It is the ratio of two lengths : the first one is the mass penetration depth defined by

$$\delta_j^D = \sqrt{\frac{D_j}{\omega}} \qquad (j = 1,2) \tag{53b}$$

the other one is understood as being the length to which extent one can consider the equilibrium hypothesis between layer and sublayer to be valid

$$K_j = \frac{d\Gamma}{dc_j} \qquad (j = 1,2) \tag{53c}$$

If the mass penetration depth is small with respect to the slope of the adsorption isotherm, $\tau_{(j)} \ll 1$ and we are very near the equilibrium state. If $\tau_{(j)} \gg 1$ we are very far from an equilibrium state.

b) Λ is linked to the diffusion fluxes $\beta_j D_j$ by the bracket factor in Eq. (52c) related to convection in the bulks. The quantity $(D_1/\nu_1)^{1/2}$ is the classical Schmidt number which is the ratio of the mass to momentum penetration depths. The sign of Λ plays a role in the stability of the system.

5. Application to large wavelengths.

There are three different problems

I $\qquad \varepsilon_c = 0 \quad , \quad \varepsilon_J = 1$

There is no convection in the mass balance equation of the bulk. This problem was analyzed by Van den Tempel [14], [23] and Lucassen [23], [24]. We have then from (52a)

$$\varepsilon = \varepsilon_0 \frac{1 + \tau + i\tau}{1 + 2\tau + 2\tau^2} \tag{54a}$$

Since for a surfactant τ is positive, this case is always stable, so that

$$tg\,\Theta = \frac{\tau}{1+\tau} > 0 \tag{54b}$$

The elasticity modulus of a soluble surfactant has an amplitude which is smaller than the Gibbs elasticity of an unsoluble monolayer since then

$$|\varepsilon| = \frac{\varepsilon_0}{\sqrt{1 + 2\tau + 2\tau^2}} \tag{54c}$$

We find back all the results of Lucassen and Van den Tempel.

II $\varepsilon_G = 1$, $\varepsilon_J = 0$

There is no variation of adsorbed matter at the interface but there is convection in the mass balance equation of the bulk. This is the problem studied previously by Sternling and Scriven [4] .

TABLE 1

Instability conditions after Sternling-Scriven [4] for a transfer $a \rightarrow b$.

Conditions on diffusion and kinematic viscosities	Type of instabilities
$D_a < D_b$	Non Oscillatory Instability
$\frac{\nu_a}{\nu_b} > \frac{D_a}{D_b} > 1$	Non Oscillatory Instability and Oscillatory instability
$\frac{D_a}{D_b} > \frac{\nu_a}{\nu_b} > 1$	Oscillatory Instability
$\frac{D_a}{D_b} > 1 > \frac{\nu_a}{\nu_b}$ $\frac{\nu_a}{\nu_b} > 1 > \frac{D_a}{D_b}$	Possibility of instability with respect to oscillations has not been excluded in a definitive way.

With the approximation (45). If one uses Eq. (52a) and (52b), one has

$$\varepsilon = \varepsilon_o \frac{\Lambda}{\tau} \cdot \frac{1-i}{2} \tag{55a}$$

$$tg\,\Theta = -1 \tag{55b}$$

Provided

$$\Lambda > 0$$

we are always in a marginal oscillatory situation. This condition is discussed below

III $\varepsilon_G = \varepsilon_J = 1$

There are variations of adsorbed matter at the interface and convection in the bulk mass balance. Then we have

$$\frac{\varepsilon}{\varepsilon_o} = \frac{1-i\Lambda}{1+\tau-i\tau} \tag{56a}$$

$$tg\,\Theta = \frac{\tau-\Lambda(1+\tau)}{1+\tau+\tau\Lambda} \tag{56b}$$

The condition (51) of marginal stability is satisfied when

$$\Lambda = 1+2\tau \tag{57a}$$

Then one has

$$\left|\frac{\varepsilon}{\varepsilon_o}\right| = \sqrt{2} \tag{57b}$$

If we represent this oscillatory marginal state in a (τ,Λ) graph we see (fig. 2) , the situation is unstable

$$\Lambda > 1+2\tau \qquad \delta < 0$$

$$\Lambda < 1+2\tau \qquad \delta > 0$$

It is easy to show that when Λ is < 0 the system is stable. Since τ is a positive quantity, Λ has to be at least a positive

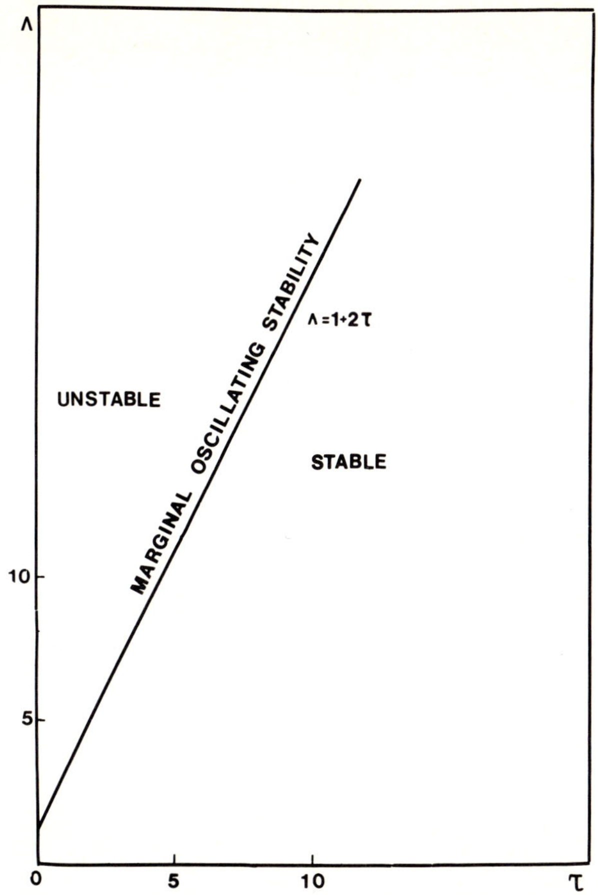

Fig. 2. Condition of oscillatory marginal stability.

quantity in order that unstable situations should be possible. One has to look only at

$$\underset{\sim}{\gamma} = \frac{\beta_1 D_1}{\Gamma_0} \left\{ \sqrt{\frac{D_1}{\nu_1}} - \sqrt{\frac{D_2}{\nu_2}} \right\} > 0 \tag{58}$$

If transfer is from a to b, last inequality is satisfied if one has one of the following inequalities (a being 1 or 2 ; b being then 2 or 1)

$$\frac{D_a}{D_b} > \frac{\nu_a}{\nu_b} > 1 \tag{59}$$

TABLE 2

Instability necessary conditions in our treatment for transfer a→b

Conditions on diffusion and kinematic viscosities	Types of instabilities
$D_a < D_b$	Non oscillating instability
$\frac{D_a \nu_b}{D_b \nu_a} < 1$	Non oscillating instability
$\frac{D_b}{D_a} > 1 > \frac{\nu_b}{\nu_a}$	
$1 < \frac{D_a}{D_b} < \frac{\nu_a}{\nu_b}$	
$\frac{D_b}{D_a} > \frac{\nu_b}{\nu_a} > 1$	
$\frac{D_a \nu_b}{D_b \nu_a} > 1$	Oscillating instability
$\frac{D_a}{D_b} > \frac{\nu_a}{\nu_b} > 1$	
$\frac{D_a}{D_b} > 1 > \frac{\nu_a}{\nu_b}$	
$\frac{\nu_b}{\nu_a} > \frac{D_b}{D_a} > 1$	

$$\frac{\mathcal{D}_a}{\mathcal{D}_b} > 1 > \frac{\nu_a}{\nu_b} \tag{60}$$

$$\frac{\nu_b}{\nu_a} > \frac{\mathcal{D}_b}{\mathcal{D}_a} > 1 \tag{61}$$

The first one (Eq. (59)) corresponds to Scriven and Sternling analysis [4] and to Linde and Keinkel experiments [32]. The second one (Eq. 60) is not excluded from their analysis (see table 1) but they have neglected it. The third possibility (Eq. (61)) was not considered in any previous analysis but seems to be in agreement with experimental facts [33] (see table 2)

Let us now compare our results with the ones previously deduced by Gouda and Joos [18] who also wanted to study pure longitudinal waves with approximation (45). Remember that the Rayleigh phenomenon is at the origin of the capillary ripples and that the Marangoni effect gives rise to longitudinal waves.

In our development we uncouple the two by the condition (48) (or more precisely (19)) with the following consequences :

a) the normal part of the velocity is zero at the interface so that (see Eq.(46))

b) the amplitude of the potential part of the velocity and of the rotational part of it have the same magnitude (see Eq. (17)). On the other hand Gouda and Joos neglected completely the potential part of the velocity with respect to its rotational part. They had thus to ignore all the boundary conditions linked to normal deformations.

Nevertheless both attitudes lead to the same formal secular equation (49) a & b. The difference appears in the explicit definition of the elasticity modulus. In their development,

Gouda and Joos obtained another definition of Λ

$$\Lambda = \frac{\beta_1 D_1}{\Gamma^o \omega}\left\{\sqrt{\frac{\nu_1}{D_1}} - \sqrt{\frac{\nu_2}{D_2}}\right\} \tag{69}$$

Now (34) is the generalization of Scriven and Sternling problem [4] . This is very easy to see if one puts in it $\mathcal{E}_J = 0$ and $\mathcal{E}_C = 1$. Thus we deduced for large wavelengths Eq.(55a) and (55b) by using the definition (52c) and (52d). However the results of Gouda and Joos cannot be obtained from the Sternling and Scriven extended formulation. Nevertheless, at the marginal state, we find both the same numerical value of the elasticity modulus, the marginal values of the frequency and the wavenumber being different.

6. General Discussion.

Let us return to the three cases discussed before in order to study their domain of validity with respect to Λ and to τ . Let us compare the amplitudes and the phase angles of Eq. (54), (55) and (56) (see fig. 3, 4, 5, and 6).

A) We draw $\log_{10}\left|\frac{\mathcal{E}}{\mathcal{E}_o}\right|$ as a function of $\log_{10}\tau$ for various positive values of Λ.

The parallel to the τ axis given by

$$\log_{10}\sqrt{2} = \log_{10}\left|\frac{\mathcal{E}}{\mathcal{E}_o}\right| \tag{63a}$$

is the line of marginal stability for the general case III which separates the stable region from the unstable region. In this representation case II (Sterling-Scriven) is a straight line parallel to the bissectrix since

$$\log_{10}\left|\frac{\mathcal{E}}{\mathcal{E}_o}\right| = \log_{10}(\Lambda\sqrt{2}) - \log_{10}\tau \tag{63b}$$

It intersects thus the marginal stability line. The curve corresponding to case I (Van den Tempel) is independent of Λ and is always in the stable region. For large τ ($\tau > 5$) we are far from equilibrium. The curve describing case I is parallel

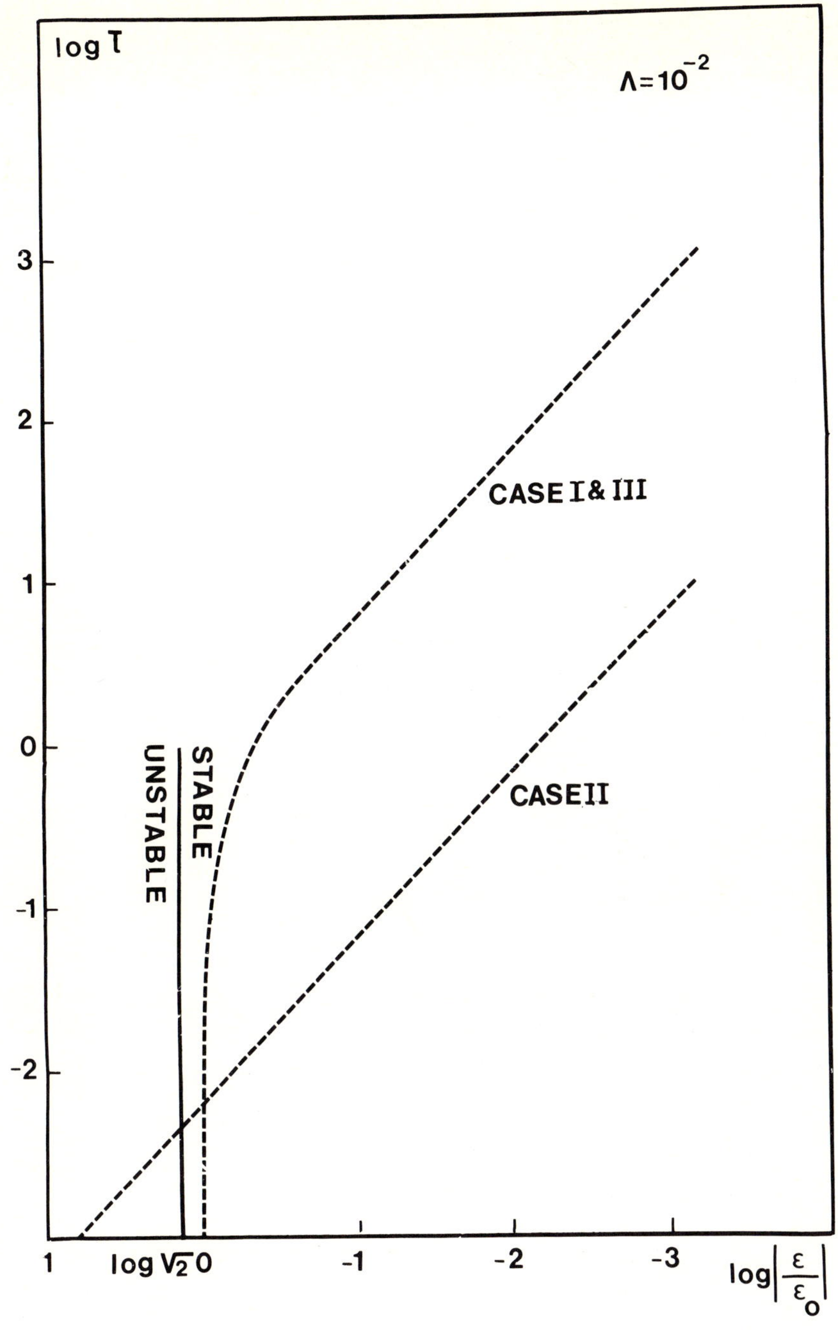

Fig. 3. Parameter τ versus the amplitude of the relative elasticity modulus for small connection effects ($\Lambda = 10^{-2}$).

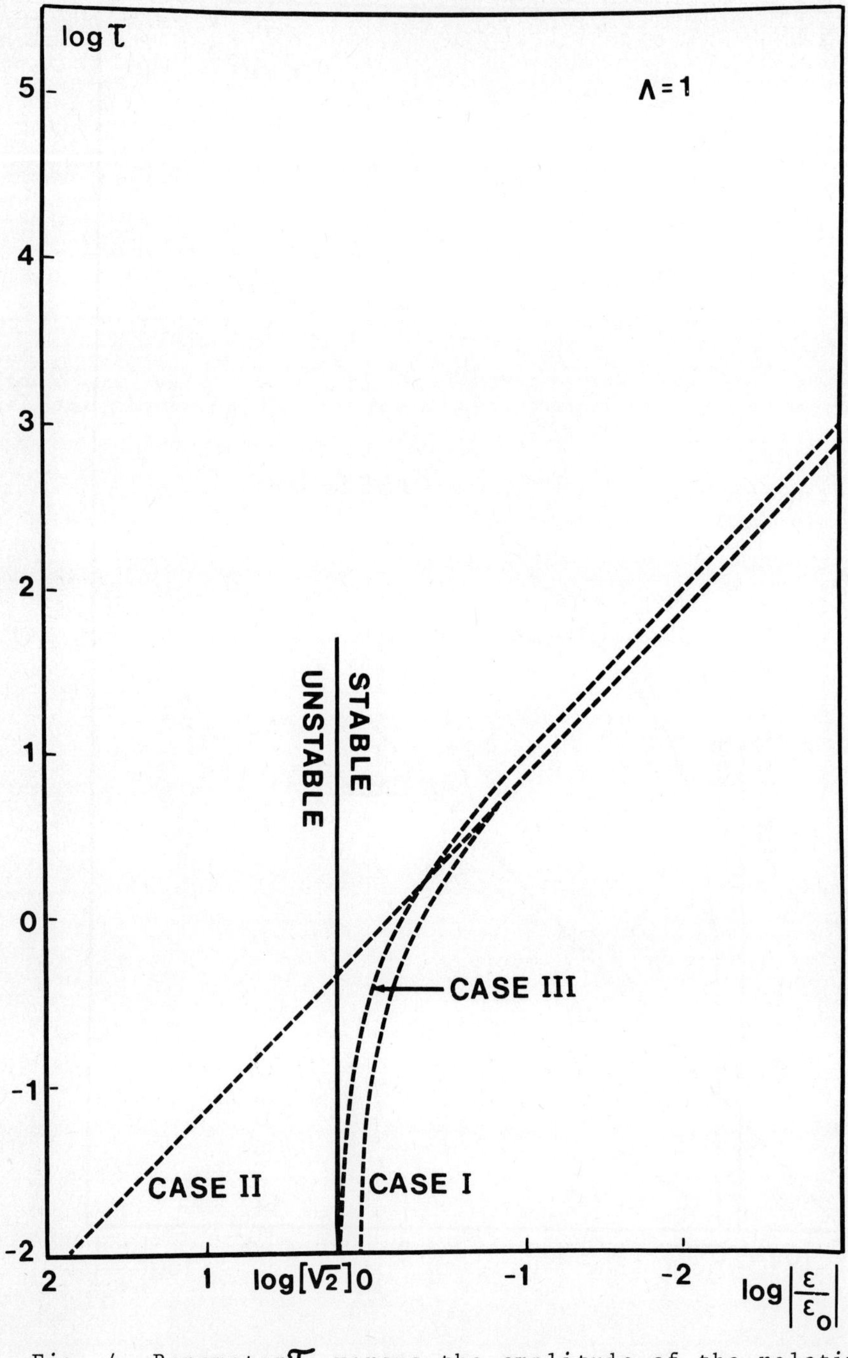

Fig. 4. Parameter τ versus the amplitude of the relative elasticity modulus for mean convection effects ($\Lambda = 1$).

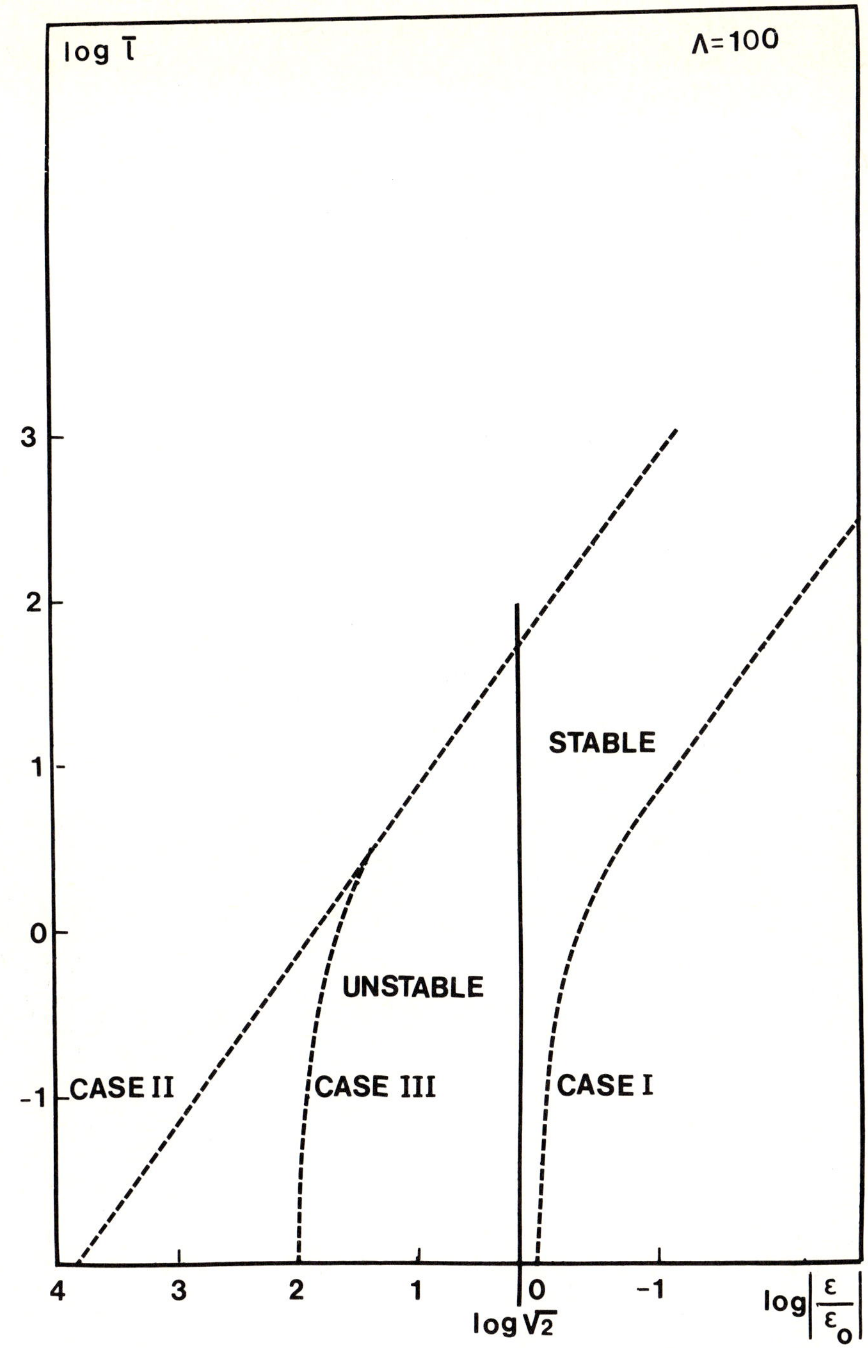

Fig. 5. Parameter τ versus the amplitude of the relative elasticity modulus for large convection effects (Λ= 100)

to the straight line corresponding to case II. For small τ, ($\tau < 5.10^{-2}$) the length of the mass penetration depth is much smaller than the length defining the equilibrium between layer and sublayer. Then the curve describing case I is a parallel to the τ axis. Whatever the values of Λ, the curve of the general case III has always the same profile as the curve given for case I. It is translated depending on the values of Λ. For $0 < \Lambda \leq 1$, the curve describing case III is always located in the stable region between the marginal stability and the curve of case I (fig. 3 and 4).
For $\Lambda \gg 1$, the situation is completely different. Indeed the curve III crosses the marginal line and the system is then unstable.
It is interesting to note here that the case II which characterizes a marginal stability is identical with case III for large values of τ. However curve III has only one marginal point. The curves II and III become mostly identical in the stable domain with respect to case III (see Eq.(63)). This can mislead someone. The figure 6 lifts this ambiguity as discussed below.

B) In the phase angle representation (fig. 6) we put Θ as a function of $\log_{10}\tau$ for various positive Λ. $\Theta = -\frac{\pi}{4}$ represents the marginal oscillatory state of case III but is also the phase angle of case II. $\Theta > -\frac{\pi}{4}$ defines the stable region (upper region) $\Theta < -\frac{\pi}{4}$ defines the unstable region (lower region).

All curves entirely in the upper region correspond to $\Lambda < 1$. The curve of case I (or case III for $\Lambda = 0$) does not intersect the marginal curve $\Theta = -\frac{\pi}{4}$. The curve corresponding to $\Lambda = 1$ is tangent to the marginal phase angle at $\tau = 0$. For $\tau > 0$ it lies entirely in the stable region. Let us note here that the convective effects are then completely compensated by the surface adsorption.

If $\Lambda > 1$ the curves III intersect the marginal line at only one point which gives us by simple lecture the Λ and τ of the marginality. Here case II and case III cannot be confused since curve II is a parallel to the τ axis.

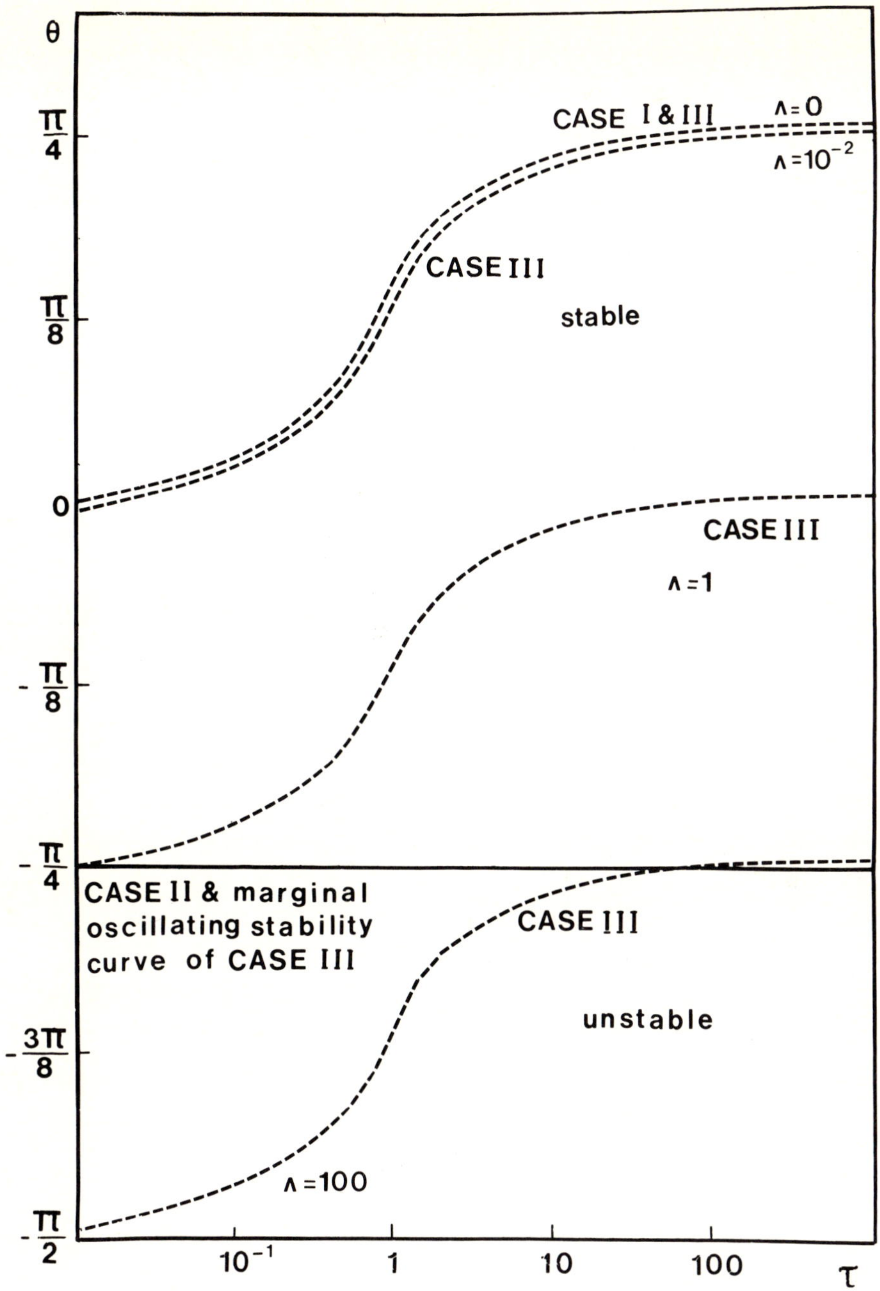

Fig. 6. Phase angle of the surface elasticity versus parameter τ for various convective effects Λ .

In this graph representation even for small Λ, the cases III and I do not give identical curves. Finally the representation of case III is completely valid in the domain analyzed by Gouda and Joos with their explicit definition of Λ (Eq. (60)).

7. Conclusion.

We study the mechanical properties of a surface separating two immiscible Newtonian fluids. Linear diffusion profiles are present in each bulk and the surface is considered as a Newtonian two-dimensional fluid.

Exchanges of mass and momentum between the two bulks in the interfacial region lead to a viscoelastic behaviour of the interface, so that the rheological properties of the layer are changed in a drastic manner. Decoupling the capillary ripples from the longitudinal waves, we restrict our formalism to this last case for large wavelengths. This approach gives us a general formulation where the results of Sternling-Scriven and of Van den Tempel-Lucassen are limiting cases of our development. The analysis of stability is given in terms of two parameters Λ and τ related to convection, diffusion and adsorption. Our treatment shows the role played by the convection in the mass balance equation of the bulks. Diffusion fluxes are stabilizing factors and reduce the surface elasticity. The convection increases the surface elasticity and might destabilize the system.

This set-up helps to choose the various possible experimental conditions in order to predict the onset of instabilities.

Acknowledgements.

We thank Professor J.J. Bernard, Director of the "Laboratoire d'Aerothermique" C.N.R.S. Bellevue for many stimulating discussions. One of us M.H. thanks the C.I.E.S. of the French Foreign Ministry for sponsorship.

Bibliography.

1. Scriven, Chem. Eng. Sci. 12 (1960) p. 98.
2. Boussinesq, J. Ann. Chim. et Phys. 29 (1913) p. 349.
3. Aris, "Vectors Tensor and The Basic Equations of Fluid Mechanics", Prentice-Hall Publishers, Englewood Cliffs, New Jersey 1962.
4. Sternling et Scriven, A.I.Ch.E.J. 5 (1959) p. 514.
5. Slattery, Chem. Eng. Sci. 19 (1964) p. 379.
6. Prigogine, Defay, Bellemans & Everett, "Surface Tension and Adsorption", Longmans and Green Ed. London, 1966.
7. Ghez, Surface Science 4 (1966) p. 125.
8. Moldovan and Georgescu, Surface Science 22 (1970) p. 149.
9. Prudhomme and Barrère, "Equations fondamentales de l'aéro-therm. chimie", Masson & Cie, Paris 1973.
10. Defay, Prigogine & Sanfeld, J. Col. Int. Science 58 (1977) p. 498.
11. Bedeaux, Albano & Mazur, Physica 82A (1975) p.48.
12. Kovac, Physica 86A (1977) p. 1.
13. Levich, Physichochimica U.R.S.S. 14 (1941) p. 391.
14. Van den Tempel, J. of Non Newtonian Fluid Mechanics, 1 (1977) p. 205.
15. Hennenberg, Sørensen, Steinchen and Sanfeld, J. Chim. Phys. (1975) p. 1202.
16. Nakache and Dupeyrat, C.R. Acad. 277C (1973) p. 519.
17. Hennenberg, Sørensen, Sanfeld, Transactions of Faraday Society II (1977) 73 p. 48.
18. Gouda and Joos, Chem. Eng. Sci. 30 (1975) p. 521.
19. Vignes-Adler and Hennenberg, to be published.
20. Dudeck and Prudhomme, J. of Appl. Physics 48 (1977) p.1.
21. Chandrasekhar, Hydrodynamic and Hydromagnetic Stability, Clarendon Press, Oxford 1961.
22. Thiessen, Scheludko, Kolloid Z. 218 (1967) p. 218.
23. Lucassen and van den Tempel, J. of Coll. and Int. Science 41
24. Lucassen-Reynders and Lucassen, Adv. in Colloid and Interface Science 2 (1969) p. 347.
25. Poskanzer and Goodrich, The J. of Phys. Chem. 79 (1975) p. 122.
26. Joly, Encyclopedia of Coll. Science, E. Matijevic Ed. vol. 4.
27. Pipkin, Lectures on viscoelasticity, Springer Verlag, Berlin 1972.
28. Landau and Lifschitz, Mécanique des fluides, Ed. de Moscou, 1972.
29. Schichtling, Boundary Layer Theory, Mac Graw-Hill, 1966.
30. Levich, Physicochemical Hydrodynamics, Prentice and Hall Publishers, Englewood Cliffs, New Jersey, 1962.
31. Hard and Löfgren, J. of Col. and Int. Sci. 60 (1977) p. 529.
32. Linde and Künkel, Wärme und Stoffübertragung 2 (1969) p. 60.
33. Maroudas and Sawistowsi , Chem. Eng. Sci. 19 (1964) p. 919.

DIFFUSION, REACTION AND CONVECTION AT AN INTERFACE : A FEW ASPECTS OF THE STABILITY PROBLEM

by

M. G. VELARDE

Departamento de Fisica, C-3, Universidad Autonoma de Madrid
Cantoblanco (Madrid) Spain

and

Laboratoire de Dynamique et Thermophysique des Fluides,
Université de Provence, Centre de Saint-Jérôme,
13397-Marseille Cedex 4 (France)

1. Introduction

Surface tension-driven flows are among the more interesting not only for their intrinsic academic value and their eventual relevance in small-scale hydrodynamics [1] but also for their potentially technological importance in the containerless proccessing of materials in the reduced-gravity environment of a spacecraft(gravity down to 10^{-6} g on Earth) where container interactions and reactions as well as container contamination can be eliminated. There are other advantages to the containerless handling of liquids such as, for example, the maintanance of longer stable lengths of floating zones. It is also expected that buoyancy-induced flows and sedementation phenomena will be diminished in a reduced-gravity environment. Besides, the traditional problems of evaporative phenomena and many others pertaining to chemical (and oil) engineering demand at present and rather urgently a more complete understanding to help improving the design of many fluid-phase contact devices and operations.

For a horizontal liquid layer open to the ambient air, evaporation causes the surface tension of the liquid right at the top to exceed the value that would correspond to the liquid beneath. On the other hand there is the removal of heat from the interfacial region to the upper environment and this in turn not only reduces the surface temperature but generates temperature gradients along the interface. For virtually all liquids , surface tension rises as temperature falls, and a nonuniform temperature distribution at the interface with the fact that the surface layer has a tension higher than the equilibrium value corresponding to its bulk composition are enough to create a stressed state of the interface and thus a thermomechanically unstable situation , for even under quasiequilibrium conditions the potential energy of the surface layer is not at a minimum. The instability develops as a convective flow, and it can also be generated under isothermal circumstances for the situation just described can be achieved by desorbing at the interface the component of lower surface tension or lower density in a binary solution or by heating nonevaporating liquids from below and allowing them to cool at the top. Convective instability can also be the direct effect of the action of surface active agents or of chemical reactions, or phase transitions at the interface.

Evaporation is doubtless enhanced by the heating from below. On the other hand a vertical thermal gradient in a horizontal liquid mixture helps the development of a concentration gradient through the Soret cross-transport phenomenon which in turn might be enhanced by evaporation and then leading to potentially unstable stratification. Remarkable enough the Soret effect is known to be able to trigger convective instability in nonevaporating mixtures heated from below or from above with even potentially stable stratification [2,3]. The present report aims at summarizing some recent theoretical and experimental work carried out at the Universidad Autonoma de Madrid and at the Université de Provence, Centre Saint-Jérôme on a variety of Bénard-Marangoni problems.

2. Bénard-Marangoni convection in simple liquids:recent experimental findings

A detailed explanation and the most relevant literature dealing with the basic mechanisms capable of inducing convective instability in horizontal fluid layers heated from below can be found in a recent review paper [4].

Here I shall describe some recent work aimed at understanding the coupling of buoyancy to surface tension tractions, and the transition from one to the other dominant mechanism, the role of the thermal properties of the bottom plate and the eventual influence of the geometry in triggering one or other of the possible convective patterns. We have also studied the evolution of an initially imposed roll-pattern to an hexagonal structure. In all cases, and for the conditions of operation used(the Boussinesquian approximations are valid) with an open interface at the top we obtained a polygonally patterned state like those reported by Bénard and Koschmieder [5], if we waited long enough.

Various dimensionless parameters are needed in the thermohydrodynamic description of the problem. One is the Rayleigh number $R = \alpha g d^4 \beta / \varkappa \nu$. Another is the (thermal) Marangoni group $M = -(\partial\sigma/\partial T) d^2 \beta / \eta \varkappa$. The third is the Prandtl number $P = \nu / \varkappa$ which does not seem to play much role for a large neighborhood of point of onset of convection. A fourth parameter is the Biot heat exchange number $L = h\, d / \lambda$. The symbols are respectively: α ,thermal expansion coefficient; g , the acceleration of gravity;d, cell gap; β , temperature gradient(positive when heating from below);$-\partial\sigma/\partial T$,is the derivative of σ , the liquid-air surface tension; η and ν dynamic(shear)and kinematic viscosities; λ and $\varkappa$ heat conductivity and heat diffusivity(thermometric conductivity).

The upper boundary of the oil layer is the ambient air. Thus the temperature is not uniform along it. However, a <u>mean</u> temperature can be defined at the interface by using a thermocouple with an attached light small copper flake. The flake's diameter (half-centimeter) is taken large enough to provide a sensible <u>mean</u> temperature distribution along the horizontal , but smaller indeed than the expected characteristic wavelength of the convective pattern. We checked that such mean temperature did not depend on the location of the thermocouple. For the experiment eight different liquid thicknesses were used : d =0.3-0.4-0.45-0.5-0.55-0.6-0.7-0.8 cm.

We have checked Nield's theory in different ways. For instance Table I provides a comparison between theory and experiment for the wavenumber at onset.

TABLE I : Theoreticallly predicted critical wavenumbers and the experimental results for various liquid depths.

d(cm)	k_c^{th}	k_c^{exp}
0.3	2.046	2.32
0.4	2.067	2.45
0.5	2.102	2.51
0.6	2.121	2.56
0.7	2.116	2.50
0.8	2.157	2.57

For the correlated Marangoni and Rayleigh numbers we have found a satisfactory agreement between theory and experiment. Figs.1 and 2 account for some of our findings and compare with theory.

FIGURE 1 : Correlated Marangoni and Rayleigh numbers for the onset of convection at different liquid depths. O : theory; ● : experiment.

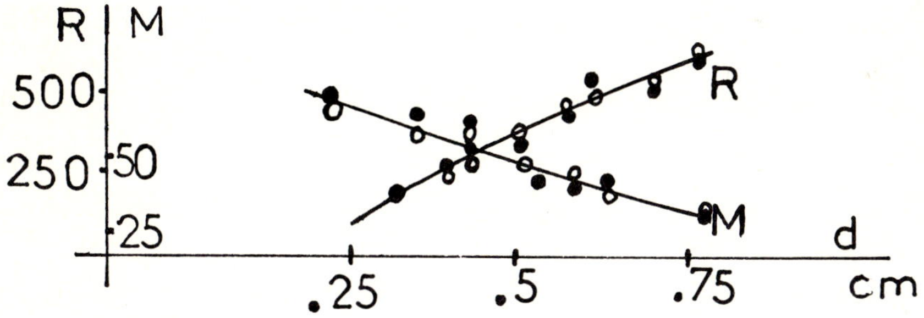

FIGURE 2 : Ratio of Marangoni number to critical Marangoni number versus same type of ratio for Rayleigh number Straight line: theory; ● : experiment.

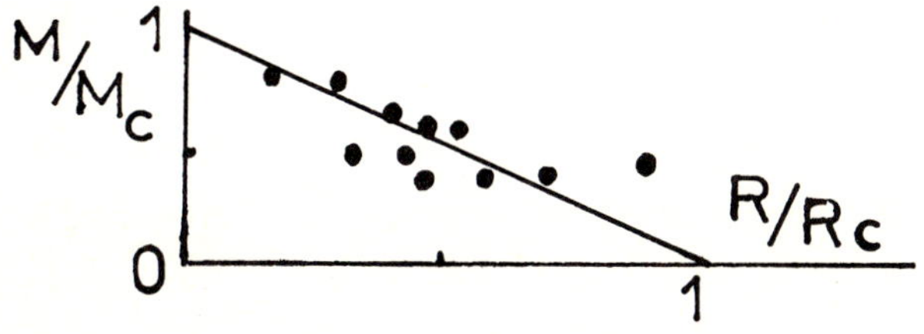

The heat flux across the liquid layer has also been measured and Fig. 3 depicts the results found in various cases.

FIGURE 3 : Normalized heat transport (Nusselt number) across the liquid layer. O–0.3 cm, ▲ –0.4, x–0.45, ●–0.5

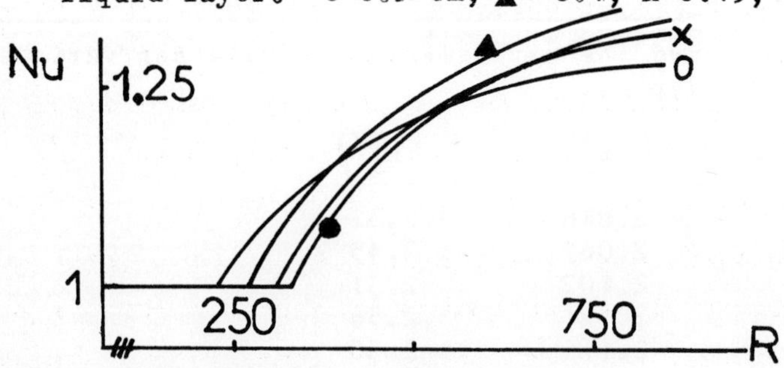

We have also studied the stability of the polygonal convective pattern and the influence of the shape of the container. Polygonal structures seem to be the only pattern in surface tension driven flows[5]. We have performed the following experiment. The thin liquid layer was heated from below and weakly from above by means of an array of wires oriented along the shorter dimension of a rectangularly shaped cell. A convective pattern was then forced which apparently looked very much like the convective structure found in a similar experiment between rigid plates, below and above. Once the roll pattern was well developped the heating was maintained from below but the upper wire array was taken out. Pretty soon one sees a fragmentation of the roll pattern and its evolution to a polygonally patterned structure no matter what the liquid depth was. The only influence of the layer thickness appeared in the time it took a roll pattern to become a polygonal one: the greater the liquid depth the longer the time required.

Finally we have studied the influence of the lower plate. The copper was replaced with a one-cm thick glass. At onset a single roll filling entirely the container appeared. Soon after it broke into a polygonal structure at supercritical Rayleigh numbers. Lowering, then, the thermal gradient to the critical one the single cell appeared again, and it was not an artifact of the lateral boundaries. Verification was thus found of the theoretical prediction that 'insulating' plates yield vanishing wavenumber[4,6].

3. Bénard convection in molten salts: experiments

Non-Boussinesquian effects are known to play a genuine role in the type of convective motions that develop in a Rayleigh-Bénard layer[4,5]. A simple case of non-Boussinesquian convection is that of a layer of molten salt heated from below and left solidifying at the top where the ambient lab temperature (usually less than 30 ^{o}C) is well below its melting point, 140 ^{o}C for the eutectic mixture $Na\ NO_2$-$Na\ NO_3$- $K\ NO_3$ used in our experiments. At the upper interface solidification defines an isothermal plane not level however, and such that in a rather thin layer beneath the (kinematic) viscosity changes from a normal value in the liquid phase (generally 10^{-2}-10^{-3} CGS units) to an extremely large value (ten to twenty orders of magnitude) in the solid phase.

By heating the layer from below, convective instability is expected at the critical Rayleigh number, R_c, corresponding to given boundary conditions. Generally for isotropic Newtonian, and Boussinesquian liquids the structure that develops at and slightly above R_c is a pattern of rolls (concentric donuts for a cylindrically shaped container or straight rolls parallel to the shorter side in rectangular boxes). Non-Boussinesquian fluids however, can develop polygonal structures apparently similar to those of surface tension-driven convection. In reality things are not identical though. In surface tension-driven convection the hot liquid rises where the surface is depleted, as already noted by Bénard[4,5]. With the molten salts the opposite structure was found: where the hot liquid rises there is surelevation of the interface and depletion appears at those places where the liquid, already cooled sinks. Table II gives an account of curvature radii, which is an obvious measure of the elevation of the interface for various layer thicknesses.

TABLE II : Elevation of interface:radii of curvature

d (mm)	radii(mm)
3	13
4.5	14.5
7.2	20

Hexagons, pentagons,etc.were found for both a cylindrically shaped container and a rectangular one. Table III gives a preliminary account of the mean wavelength found at various values of the layer depths. It appears that the ratio of wavelength to liquid depth remains almost constant as it should be. The results were the same whether the liquid was left open to air and thus abruptly solidified or in contact with a transparent glass plate cooled or slightly warmed. The latter arrangement was used in order to be sure that no surface tension phenomena played any role at the upper surface of the liquid layer. A more detailed report can be found in Ref. 7, though Dr. Pantaloni is still carrying a deeper analysis of the problem.

TABLE III : Frozen wavelength at onset of convection in molten salt layers heated from below

d (mm)	wavelength (mm)
4.5	10
6	14
6.5	14.5

Thus polygonally patterned convection is by no means solely due to simple surface tension phenomena. At least it may be triggered by buoyancy forces coupled to strong non-Boussinesquian effects. The theory of convective instability with phase transition at the open surface of a given material is still an open subject of research.

4. Bénard-Marangoni convection in nematic layers : theory

A nematic liquid is still a Newtonian and a Boussinesquian fluid though its material parameters like heat conductivity, viscosities, surface tension,etc. are no longer scalar quantities but have tensorial character. The privilegied alignment of the molecules leads to the definition of a director field[8],$\vec{n}$. Recently Guyon and the present author[9](see also 10)have studied the peculiar effects of surface tension phenomena on the instability of a nematic layer heated from below and open to the ambient air. Figure 4 illustrates the mechanisms of instability in a planar configuration(director, $\vec{n}$, parallel to the bottom plate). Starting from a temperature fluctuation + at the free interface, a surface tension gradient is generated and this induces a convective fluctuation which in turn, already for the isotropic case, enhances the initial temperature change. In the anisotropic liquid the dominant mechanism comes from the distorsion of the director which takes a nonnegligible component along the vertical. Consequently due to the more efficient heat transport along $\vec{n}$ heat is focused (broken lines) in regions of negative curvature of n which reinforces the existing tendency of a fluid parcel to move upwards due to the buoyancy originated by the heating from below. Here are some of the findings for the planar and the homotropic configurations. Furhter details can be found in Refs. 9 & 10. Symbols are taken from Ref. 8.

i) Planar configuration(director parallel to bottom plate)

A suitable Marangoni number is defined $M = -\frac{\partial\sigma}{\partial T} k_a d^2(\lambda_1 + \lambda_2)\beta / k_{//} \rho \nu k_1$ where $\nu = \eta_a/2$, $\eta_a = \alpha_4 + \alpha_3 + \alpha_6$ (viscosities as introduced for instance in Ref. 8). For the surface tension gradient we have approximated by a mean valued quantity.

When heating the layer from below the onset of convection is at $M_c = 17.59$ with a critical wavenumber $a_c = 2.6$(in contrast to the values $M_c = 80$ and $a_c = 2.2$ for an isotropic layer). This figure corresponds for a one-millimeter MBBA layer to a temperature drop of 0.165 °C.

There is also the possibility of destabilizing the layer by heating the planar configuration from above. Then instability starts at $M_c = 511.6$ with $a_c = 0.5$ which for a one-millimeter layer thickness yields a temperature difference of 4.8 °C.

FIGURE 4 : Instability mechanisms in a planar nematic layer heated from below.

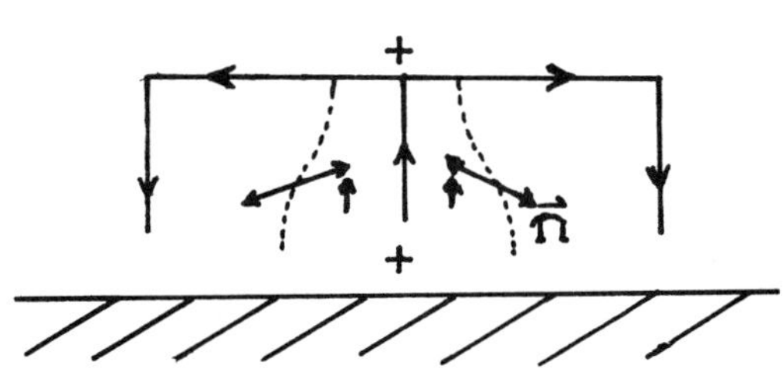

ii) Homotropic configuration (director orthogonal to bottom plate) For this case the Marangoni number is defined as follows

$$M = -2\,\partial\sigma/\partial T\, k_a(\lambda_2 - \lambda_1)\, d^2 \beta \,/\, (k_{//} + k_a)\, k_3 \eta_b \,.$$

When heating the layer from below the onset of convection comes through overstable modes (Hopf bifurcation) at $M_c = 0.32 \times 10^5$ with $a_c = 2.42$, which for a one-millimeter liquid depth corresponds to a temperature difference of +37.7 °C(minus sign indicates the heating from above). Overstability arises with a (linear) period of about one hundred and twenty seconds. Overstability does not mean that oscillatory convection should be seen experimentally. It may very well be a precursor for a finite-amplitude(subcritical or metastable)instability leading to steady convection.

When the heating is from above the onset of steady convection is at $M_c = -40.6$ with $a_c = 2.15$ which for a one-millimeter MBBA layer yields a temperature drop of -0.048 °C.

Both configurations when there is instability leading to steady convection obey Nield's law as in the isotropic case (see Figure 2 above). It may very well be that such small temperature drops as those needed for instability in nematic liquid layers could permit using Bénard-Marangoni convection for I.R. thermography, as suggested by Guyon and Loulergue [9].

5. Bénard-Marangoni convection in mixtures and the Soret effect : theory

In a binary mixture the Soret effect is the mass diffusion cross-transport induced by a temperature gradient. It has a complementary related phenomenon, the heat cross-transport originated by a gradient of chemical potential (Dufour effect). The latter seems of relevance to gases only and we shall comment only on the role of the former when coupled to buoyancy and surface tension phenomena. For the Soret effect to be operating it suffices the pressence of an impurity in a single-component liquid layer(heated from one side)in amounts not even exceeding one part per million. Its role in buoyancy-driven convection has been discussed at length in a recent review by Schechter <u>et al</u>[3] and quite a number of surprising effects can be attributed to its operation even in mechanically stable stratified layers. Authors have also made the Soret effect responsible of the formation of striae,etc. in crystals grown from the liquid phase, even in the reduced-gravity environment of a spacecraft or when dealing with very thin liquid layers for which gravity may be considered irrelevant. We shall comment on the effects of coupling the Soret effect to buoyancy and surface tension-stresses in a Bénard experiment in mixtures.

The following dimensionless groups need to be introduced : a second (concentrational)Marangoni number (also called "elasticity" number) $E = -(\partial\sigma/\partial N)\Delta N\, d / D\eta$, where N accounts for the mass fraction of component one,say; ΔN is its variation; D is the Fickian (direct) mass diffusion constant. The Soret separation number $S = -\gamma\Delta N / \alpha\beta d$, where γ is the volumetric expansion coefficient due to variations in mass fraction. For illustration the numerical results given below correspond to liquids of Lewis number $Le = D/\varkappa = 10^{-2}$ and Prandtl number equals to ten.

For marginally neutral states the following relation generalizes Nield's prediction for single-component liquid layers(see Fig. 2)

$R(S)/R_c(S) + M(S)/M_c(S) + E(S)/E_c(S) = 1$, where the subscript 'c' denotes the corresponding critical value at vanishing values of the other parameters, and arbitrary Soret separation.

Heating the layer from below , the thermal Marangoni number (M is taken always positive) in all cases plays a destabilizing role whatever the positive values taken by R, S, and E. For given positive values of R and M both S and E,when positive are destabilizing too, and tend to induce steady convection. With positive S and negative E convection can be triggered via overstable modes for a critical concentrational Marangoni number that increases in absolute value with increasing Soret separation and decreasing values of M (positive). In the latter case the minimum Rayleigh number corresponding to overstability changes very little with either S or E but strongly depends on the value attained by M.

At negative values of S and positive values of E if , in absolute value S is larger than 0.01, overstability is expected whatever the values taken by E, whereas for absolute values of S smaller than 0.01, overstable modes appear only above a value of E that decreases with decreasing absolute values of S and increasing M. At negative values of both S and E , overstability is also expected whatever the values of M(positive) and then the corresponding Rayleigh number strongly depends on M , but varies very little with either S or E. Figure 5 gives the (linear) stability diagram at vanishing Rayleigh number, at say vanishing gravity. The heavy line depicts the threshold for steady convection. The broken line delineates the onset of convection via overstable modes.

FIGURE 5 : Stability diagram in a two-component liquid layer at vanishing gravitational acceleration ($g = 0$).

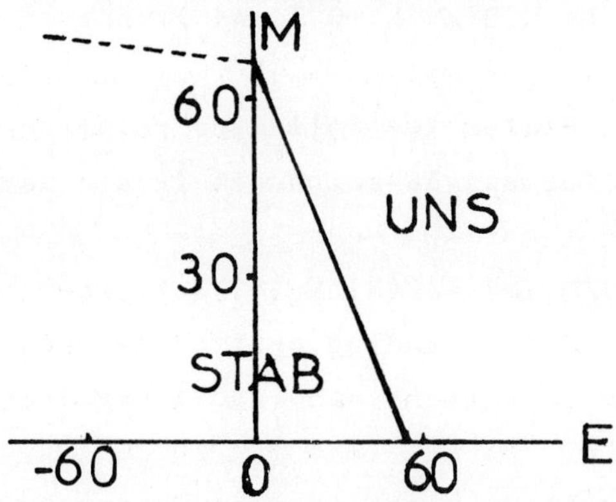

Thus it appears that surface tension-driven convective instability may very well appear at very low thermal or concentrational Marangoni numbers provided the Soret effect couples to surface tension tractions at the open interface of the liquid layer. Further details on this part can be found in a recent paper by Castillo and the present author [11].

6. Rayleigh-Taylor convection and chemical reaction with diffusion at an interface : the role of autocatalysis and the Langmuir adsorption law

Sanfeld and collaborators[12](see the Reports to this Symposium by Sanfeld and Sørensen) have recently studied the coupling of chemical to hydrodynamic phenomena at an interface separating two immiscible liquids. Following their approach and suggestions we have considered the role of various reaction-diffusion schemes operating at the interface.[13,14] Let us recall some of effects in the case of autocatalysis and a saturation law.

We consider two incompressible Newtonian and Boussinesquian fluids separated by an interface $z^I = z^I\ (x,y)$ where $z < z^I$ (respectively $z > z^I$) refers to the lower(respectively the upper) liquid. The coordinates x,y designate the plane orthogonal to z, and supersript '1' will denote quantities belonging to the lower liquid. The interface,I, is taken as a Newtonian two-dimensional layer with mass density Γ_I, over which the following chemical process takes place

$$A \longrightarrow Y\ ,\ X + Y \longrightarrow 2X\ ,\ X \xrightarrow{\quad S \quad} P$$

A and P are reactants whose concentration is kept constant at the interface.'S' stands for the Langmuir-Hinshelwood adsorption which corresponds to the Michaelis-Menten law in enzymatic reactions. Though A and P are soluble in either liquid phase , neither X nor Y are soluble in the bulks. They are present only at the interface where they diffuse according to a two-dimensional Fickian law. When convection takes place both X and Y move also along the interface with the barycentric motion of a surface element.

Under isothermal conditions the intermediate reactants change according to the following reaction-diffusion-and convection equations

$$\partial X/\partial t = k_2XY - k_3X/(1+k_4X) + D_X^I\ \Delta_2X - \partial(v_X^IX)/\partial x - \partial(v_Y^IX)/\partial y$$

$$\partial Y/\partial t = k_1A - k_2XY + D_Y^I\ \Delta_2Y - \partial(v_X^IY)/\partial x - \partial(v_Y^IY)/\partial y$$

where k_i (i=1,...4) are the reaction rate constants, Δ_2 denotes the two-dimensional Laplacian(x,y) along the interface, z^I. D_X^I and D_Y^I are the two Fickian surface diffusion constants. The vector $V^I = (V_X^I, V_Y^I)$ describes the velocity of points belonging to 'I'. Thus variations in the concentration of the intermediate reactants, and consequently possible surface tension gradients may be due either to the chemical reaction or to diffusion or to convective motions along the interface.

The above given evolution equations demand initial and boundary conditions (b.c.). For the b.c. we can follow the suggestions of Ref. 12 :

i) Continuity of velocity and velocity gradient along the normal to z^I, with neglect of cavitation phenomena,

ii) Hydrostatic pressure differences across the interface are balanced by surface tension, according to Laplace condition on normal stresses,

iii) Tangential stresses along the interface are described by a Levich condition on surface tension gradients $\alpha_x = \partial\sigma/\partial x$, $\alpha_y = \partial\sigma/\partial y$ and surface viscosity (both shear, ϵ , and dilational, χ , viscosities are incorporated).

As there is a trivial (primary) solution to the evolution problem ,namely a motionless state with uniform distribution of the inteemediate reactants the question is now to study the non-equilibrium stability of the primary state of the interface under the joint influence of chemical reaction, diffusion, surface tension-tractions and buoyancy(density stratification) should the interface be taken horizontal. The problem is thus a generalisation of Rayleigh-Taylor convection[15].

Following a normal mode approach, instability is decided by the finding of nontrivial solutions to the determinantal equation which arises from the linearized description of disturbances on the primary solution[12]. In the simple, but illustrative case of neglecting the two bulk phase viscosities this corresponds to

$$\left\{\omega\left(\Gamma_I + {}^{(\rho_1 + \rho_2)}/_k\right) + (\chi k^2 - g(\rho_2 - \rho_1))/\omega\right\} \circ$$

$$\begin{vmatrix} -k_3qa(1-qa)+ +\omega+D_X^Ik^2 & -k_3a/(1-qa) & -k_3a/k_2(1-qa) \\ k_3(1-qa) & k_3a/(1-qa)+\omega+ +D_Y^Ik^2 & -k_3(1-qa)/k_2 \\ \alpha_x k^2 & \alpha_y k^2 & k^2(\epsilon+\varkappa)+\omega\Gamma_I \end{vmatrix} = 0$$

where $q = k_3k_4/k_2$, $a = k_1k_2A/k_3^2$, k is a Fourier generic mode along the interface (x,y), and ω is the time constant that determines stability (Re ω positive, unstable; Re$\omega = 0$, neutrally stable, and Reω negative ,(locally) stable).

The first factor in the secular(or dispersion) equation above corresponds to the inviscid Rayleigh-Taylor problem[15]. Instability is predicted when $\rho_2 > \rho_1$, and $\Gamma_I \neq 0$ merely shifts the value of the critical mode at the onset of convection. The second factor(the determinant) incorporates all relevant phenomena. It appears that this factor yields at vanishing wavenumber the stability criterion for the chemical process alone, and thus a neccessary condition for the local(linear) stabiltity of the primary (motionless and uniform)steady state is that the chemistry be stabilizing[16]. The condition ,however, is not sufficient.

For quantitative illustration the following numerical values in CGS units have been considered : $\alpha_x = \alpha_y = 10^8$, $D_X^I = 5.10^{-6}$, $D_Y^I = 10^{-6}$, $\epsilon + \varkappa = 1$, $k_2 = 2.10^3$, $k_3 = 2.10^{-4}$, $\Gamma_I = 10^{-5}$, $\rho_2 = 1.0$, and $\rho_1 = 1.1$; a and q are the state variables. Figure 6 shows the results found. Region I is of (local)asymptotic stability of the primary solution with an interface taken level. The dashed portion corresponds to stable states of the chemical process alone that become unstable with the coupling of the chemistry to the other operating phenomena.

FIGURE 6 : Stability diagram for the reaction, diffusion, and convection problem at an interface separating two immiscible liquids. a(externally controllable product) and q (strength of the saturation law) are the state variables.

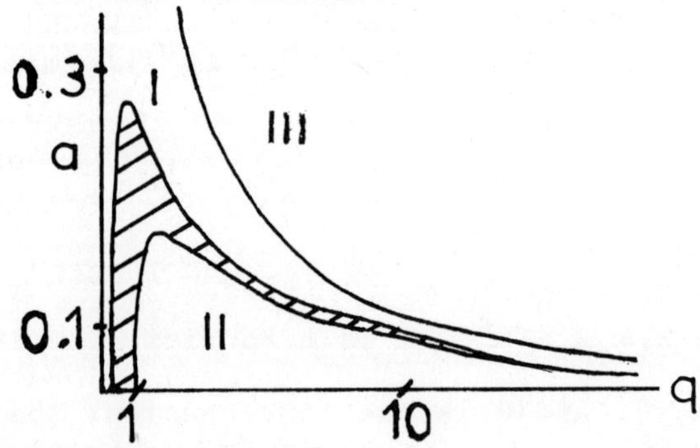

Instability may very well be a genuine consequence of the coupling, as neither mechanical stability nor a chemically stable situation lead the overall operation of the system after the coupling between hydrodynamics and chemistry is switched on. Chemical reaction has been considered here to proceed instantaneously whereas diffusion and convection demand non-negligible time delays to relax the concentration fluctuations due to the reaction.

Acknowledgments

My interest in interfacial convection and reaction was stimulated in discussions with A. Sanfeld, T. S. Sørensen and their collaborators. To Sørensen I wish to express my sincere appreciation and thanks for his invitation to participate at this Symposium. For the experiments I owe much of my understanding to the patience of J. Pantaloni that for the past two years have taught me how to look and manipulate at them. The present report summarizes work done in collaboration with E. Guyon, J. Pantaloni, R. Bailleux, J.L. Castillo, J.L. Ibañez, J. Salan and I. Zuñiga. It was supported by the Instituto de Estudios Nucleares (Spain) and through International Action (A.T.P.) by the C.N.R.S. (France).

REFERENCES

1. B. Levich, Physico-Chemical Hydrodynamics(Prentice-Hall,Englewood Cliffs)1967 (English Translation).
2. M.G. Velarde and R.S. Schechter, Phys. Fluids 12(1972)1707.
3. R.S. Schechter, M.G. Velarde and J.K. Platten, Adv.Chem. Phys. 26(1974)265.
4. Ch. Normand, Y. Pomeau and M.G. Velarde, Rev.Mod.Phys. 49 (1977)581.
5. E.L. Koschmieder, Adv. Chem. Phys. 26(1974)177.
6. J. Pantaloni, R. Bailleux, J. Salan and M.G. Velarde, J. Non-Equilibrium Therm.,submitted for publication.
7. J. Pantaloni, M.G. Velarde, R. Bailleux and E. Guyon, C.R. Acad.Sc. Paris B285(1977)275.
8. P.G. de Gennes,The Physics of Liquid Crystals(Clarendon Press,Oxford)1974.
9. E. Guyon and M.G. Velarde,J. Physique-Lettres 39(1978)205.
10. M.G. Velarde and I. Zuñiga,J. Physique,submitted for publication.
11. J.L. Castillo and M.G. Velarde,Phys.Letters A66(1978)489.See also J.L. Castillo and M.G. Velarde,J. Physique,submitted for publication.
12. M. Hennenberg, T.S. Sørensen, A. Steinchen-Sanfeld and A. Sanfeld,J. Chim.Physique 72(1975)1202.
13. M.G. Velarde, J.L. Ibañez, T.S. Sørensen,A. Sanfeld and M. Hennenberg, in Procs. First Levich Intern. Conf. "Physicochemical Hydrodynamics",edited by D.B. Spalding(Advance Pub., Guernsey)1978.
14. J.L. Ibañez and M.G. Velarde,J. Physique 38(1977)1479.
15. S. Chandrasekhar,Hydrodynamic and Hydromagnetic Stability (Clarendon Press,Oxford)1961.
16. J.L. Ibañez, V. Fairén and M.G. Velarde,Phys.Letters A58 (1976)364.

INSTABILITY OF A SPHERICAL DROP WITH SURFACE CHEMICAL REACTIONS AND TRANSFER OF SURFACTANTS

by

Torben Smith Sørensen* and Marcel Hennenberg†

*Fysisk-Kemisk Institut, Technical University of Denmark, DK 2800 Lyngby.

†Chimie Physique II, Université Libre de Bruxelles, B 1050 Bruxelles.

1. Introduction

Problems of instability of a spherical, fluid drop immersed in another fluid- the fluids being regarded immiscible and incompressible - have received a great deal of attention as well from theoreticians as experimentalists due to the wide range of applications inherent in this problem. The first important observation in the field was made by Lewis and Pratt[1] who noticed "kicking" of a drop under certain circumstances with surfactant present in a non-equilibrium distribution between the two phases. The erratic movements they report were thought to be caused by circulation of the surrounding fluid coupled to Marangoni-effects induced by varying surface concentrations of surfactants.

Davies & Haydon[2,3] attempted to explain the phenomenon by equating the energy release due to Marangoni-effects to the energy dissipated by the viscous damping of the oscillations after a "kick" has taken place. While those authors had some success in explaining certain experimental facts, their model cannot be regarded to be consistent, since the division into a "kick" driven by Marangoni-effects and a subsequent period with damped oscillations and neglect of Marangoni-effects is not clear-cut. The "stages" (I, II and III) giving rise to a kick are poorly defined and even the qualitative explanation on p. 486 in ref.2 (just below eqn.(3)) is incorrect, since an <u>increase</u> in surfactant concentration at one of the hemispheres, and the <u>decrease</u> in interfacial tension resulting from that, will give rise to a <u>decrease</u> of the Laplace overpressure ($2\sigma/R$) just behind that part of the surface. Streaming <u>in</u> the drop will therefore be <u>from</u> the surfactant-poor <u>to</u> the surfactant-rich hemisphere (cf. ref.4, figs.1a-b). Correspondingly the kick will take place towards the higher solute concentration. This is also stated by Haydon in ref.2, but he says that the decrease in surface tension <u>increases</u> the pressure behind the surface giving rise to a kick towards the exterior phase. This explanation is in conflict both with the law of Laplace and the requirement of local equilibrium at the interface even in non-equilibrium situations. Furthermore, surface viscosity - known to be very important for Marangoni-instabilities at plane interfaces[5-8] - is not considered in the surface momentum balance and - finally - the sphere is supposed not to deform from the spherical form which cannot be true.

Miller and Scriven[9] derived a general dispersion equation for the balance of momentum in order to analyse the damping of small oscillations of a fluid droplet making their departure from a Newtonian model of the interface developed earlier by Scriven[10]. They discussed the rate of damping of these oscillations without incorporation of any Marangoni-effect. One of their conclusions on which we fully agree is that determining all the significant features of droplet deformations is scarcely likely to be a trivial matter. But their model showed quite clearly the importance of surface deformation and of tangential momentum balances at the interface, since damping is essentially due to a boundary layer near the interface

Pierson and Whitaker[11,12] have studied the break-off instability of pendent and growing drops in some detail. According to Tate's law[13] the drop weight after the rupture is equal to $2\pi r \sigma$ where r is the capillary radius and σ the interfacial tension. In practice, however, correction factors have to be determined theoretically or experimentally[14]. Whitaker[12] has made evidence for the correctness of deriving the weight of rupture by a <u>static stability criterion</u> instead of a dynamic one, as all previous authors (including Pierson and Whitaker) have tacitly done. However, he gives up to consider a sphere (or a pear shaped pendent drop) because of the computational complexity. Instead, he derives equations for the stability of a <u>cylinder</u> of surfactant solution subjected to surface deformations.

He is then able to show that the critical wavelength of deformation of the liquid cylinder is $\lambda_c = 2\pi r$, r being the cylinder radius. Thus, the wavelength of neutral stability is independent of rheological properties of the interface as well as of the bulk fluids. This fact lends support to the use of static stability criteria also in the drop weight methode for the determination of σ.

The conclusion of Whitaker confirms some of our own studies on deformational instabilities of plane interfaces[6,7,8]. The problem of the rupture of a growing pendent drop may be seen as a "spherical" version of the well-known Rayleigh-Taylor instability of a heavier fluid placed on top of a lighter[15].

In a study of Rayleigh-Taylor instability with mass transfer between the two phases, we found that the critical wave number is only determined by the surface tension and the difference in specific weights of the phases, cf. ref.7, eqn.(10). Although the value of the growth constant should - in principle- be dependent on mass transfer and surface rheological properties, we found numerically for a specific example (ref.7, fig.10) that the whole dispersion curve is independent of mass transfer and surface viscosity in the case of Rayleigh-Taylor instability (in sharp contrast to Marangoni-instabilities where surface viscosity plays a dominant rôle and gravity can be neglected).

Returning to the proper Marangoni-effects in connection with spherical droplets, some introductory words should also be said about the age-old game of comparison between the curious dynamic effects induced in droplets by gradients in surface tension and biological phenomena such as cell division and amoebic motion. Such comparisons were for example proposed by Quincke[16] and Spek[17]. Such analogies influenced the theories of cytokinesis for a long time[18,19]. After the discovery of the highly complex cell membranes those ideas were no longer fashionable among biologists, and surface chemical models are generally thought to be naive. However, studies of the division of animal cells[20-27] have left beyond doubt that the division is initiated by non-uniform stresses near the surface of the cell. Yet, we are not concerned here with the usual "physicochemical" surface tension, since the tension is produced by ATP-consuming microfilaments which form a network in the very viscous layer ("ectoplasm") formed adjacent to the cell membrane during motile events in the cell[28,29]. A discussion of the difficulties connected with a purely surface chemical/ hydrodynamical model of cytokinesis was given in the introduction to our 1975 paper[30] on hydrodynamicochemical instabilities induced by surface chemical reactions. We found in that paper that in the case of only one surface adsorbed specie with variable surface concentration, the surface chemical reaction had to be autocatalytic to trigger instability. However, it was tacitly assumed that an <u>increase</u> in surface concentration would result in a <u>decrease</u> in surface tension.

Greenspan[31-33] has later on - seemingly independently of us - derived a very similar model which can indeed be shown to be equivalent to ours except for the omission of surface viscosity[34]. However, Greenspan introduces the physical idea to incorporate the stress due to the contractile micro-filaments into an "effective surface tension". Then it might very well happen that the "surface tension" will increase with increasing amounts of surface tension elements (e.g. concentration of monomers being able to polymerise into micro-filaments) and we do not have to have autocatalytic surface reactions to get deformational instability of a spherical cell. With that modification our model is more realistic than Greenspan's since the dissipation in the very viscous layer of ectoplam may be simply incorporated into an "effective surface viscosity" in our model.

It is suspected that the initial stages of the cell division process (furrow formation around equator) are linked to chemical activity in the asters (centrosomes)[35]. If chemical species are produced in the asters and diffuse towards the surface where they in some way create mechanical tension, then our previous model (and consequently also Greenspan's) is clearly insufficient, since we have to take into account bulk concentration profiles and the solution of the diffusion-convection eqn. in the bulk fluids[5-8].

The dispersion relation between the exponential growth constant ω and the integer ℓ characteristic of the surface harmonic mode of deformation will be derived in this paper for spheres with mass transfer and surface chemical relations. We shall in the present study neglect gravity, since it is known that gravity has no significant influence on Marangoni-instabilities connected with deformations of plane interfaces. Fluid spheres will of course be more or less deformed by gravity (sessile as well as pendent or buoyant droplets), but stability calculations using the present model will still be approximately correct for not too large drops. The possible practical applications of the model might come out in three different directions: a) A model for kicking drops and the enhanced mass transfer rate linked to such phenomena. Such models are valuable for the design of liquid-liquid extraction apparatus. b) A model for cytokinetic phenomena in the biological cell such as cell division, amoebic motion, pseudopode formation, chemotaxy and phagocytosis. Perhaps also microvilli formation on the surface of cancerous cells[36]. c) A model for division of emulsion droplets due to mass transfer or surface chemical reactions. It is not known at present if there exist such "dynamic instabilities" in emulsions. However, Spek[17] has reported an observation of almost complete division of a single droplet consisting of a mixture of olive oil, chloroform and "Nachtlichteröl" (probably rapeseed oil) initiated by soda crystals placed on opposite sides of the droplet (soap formation). Those experiments have recently been confirmed by Greenspan[32] using a mixture of Gulf gear case lubricant and rapeseed oil and NaOH solution added by means of two opposed syringes.

2. Hydrodynamic and diffusion-convection equations.

The following scheme will be followed in the present paper:
a) The linearised Navier-Stokes eqns. will be solved for an incompressible fluid drop immersed in an infinite amount of surrounding fluid (immiscible with the first fluid). The fluids will be assumed Newtonian and gravity effects neglected.
b) The linearised diffusion-convection eqn. will be solved in

spherical coordinates assuming initial parabolic concentration profiles of the surfactants responsible for fluid motion.
c) No-slip conditions and surface adsorption conditions will be used to eliminate certain of the arbitrary constants appearing in the solutions a) and b).
d) A link between the n+2 remaining arbitrary constants is established through n+2 eqns., viz. the n surface mass balances, the radial momentum balance and the tangential momentum balance of the interfacial layer between the two drops. Marangoni-effects arising from the coupling between surface tension variations and variations in surface adsorption play a vital role in the tangential momentum balance.
e) The condition for having solutions of the n+2, homogeneous and linear eqns. mentioned in d) will be that the coefficient determinant vanishes. This condition constitutes the characteristic equation of the stability analysis, since it gives a relation between the complex growth constant $\omega = \omega_r + i\omega_i$ and all the parameters of the system for each integer value (1,2,3,...) of the "quantum number" ℓ of a specific spherical normal mode.
f) Analytical solutions for the loci of marginal states of non-oscillatory instability will be sought for by putting $\omega_i = 0$ and letting $\omega_r \rightarrow 0$ expanding the characteristic eqn. given above to the lowest power in ω_r only.
g) It will be shown that as well the complete characteristic eqns. as the special result for marginal, non-oscillatory states pass over to eqns. found previously by us for a plane interface, when $\ell \rightarrow \infty$, $R \rightarrow \infty$ and $\ell/R \rightarrow k$, k being the wavenumber of the plane normal mode of fluid motion.

Since a part of the above programme was carried out by us in an earlier paper on the chemical and hydrodynamical analysis of stability of a spherical interface (see Sørensen, Hennenberg, Steinchen and Sanfeld, ref.4), we shall quote the formulae in the paper extensively in the present paper. The solution of the linearised Navier-Stokes eqns. in spherical coordinates for a specific surface harmonic is given by eqns. (1), (16) and (22) in ref. 4 :

$$u_{r,\ell}^{(p)} = \frac{W_\ell^{(p)}}{r} \cdot Y_\ell(\theta,\phi) \cdot e^{\omega t} = \tilde{u}_{\ell,r} \cdot Y_\ell \cdot e^{\omega t} \tag{1}$$

In (1) $u_{r,\ell}^{(p)}$ is the radial velocity, p is the phase number being 1 for the interior phase and 2 for the exterior phase, r is the radial position and Y_ℓ is a surface harmonic of the ℓ'th order of the general type

$$Y_\ell = \sum_{m=-\ell}^{+\ell} a_m \, Y_\ell^m (\theta,\phi) \tag{2}$$

with the a_m's being arbitrary expansion coefficients and θ and ϕ being the usual two angles fixing the position on the surface. For a definition of the surface harmonics Y_ℓ^m see ref.37 section 8.1.1. The "magnetic quantum number" m does not appear explicitly in our stability analysis due to the isotropy in the two fluids and the neglect of gravity. The radial functions in (1) are given by

$$W_\ell^{(1)}(x) = A_\ell^{(1)} x^\ell + B_\ell^{(1)} I^*_{\ell+\frac{1}{2}} (q_1 x) \qquad (x<1) \tag{3a}$$

$$W_\ell^{(2)}(x) = A_\ell^{(2)} x^{-(\ell+1)} + B_\ell^{(2)} K^*_{\ell+\frac{1}{2}} (q_2 x) \quad (x>1) \tag{3b}$$

with (ν_p = kinematical viscosity)

$$q_p = R\,\sqrt{\omega/\nu_p} \quad ; \quad \mathrm{Re}(q_p) \geq 0 \qquad (p = 1,2) \tag{4}$$

and the dimensionless radius

$$x = r/R \tag{5}$$

R being the radius of the drop. Furthermore, in (3a-b)

$$I^*_{\ell+\frac{1}{2}}(q_1 x) = \sqrt{\frac{\pi}{2q_1 x}} \cdot I_{\ell+\frac{1}{2}}(q_1 x) \tag{4a}$$

$$K^*_{\ell+\frac{1}{2}}(q_2 x) = \sqrt{\frac{\pi}{2q_2 x}} \cdot K_{\ell+\frac{1}{2}}(q_2 x) \tag{4b}$$

where $I_{\ell+\frac{1}{2}}$ and $K_{\ell+\frac{1}{2}}$ are the modified, spherical Bessel functions of the first and the third kind discussed in ref.37, section 10.2. By means of the no-slip conditions at the interface in the radial and in the tangential directions the arbitrary A-coefficients can be expressed in terms of the arbitrary B-coefficients in the eqns.(3a-b). The result is

$$A_\ell^{(1)} = -\frac{1}{2\ell+1}\{q_1 I^*_{\ell-\frac{1}{2}}(q_1)B_\ell^{(1)} + q_2 K^*_{\ell-\frac{1}{2}}(q_2)B_\ell^{(2)}\} \quad (5a)$$

$$A_\ell^{(2)} = -\frac{1}{2\ell+1}\{q_1 I^*_{\ell+\frac{3}{2}}(q_1)B_\ell^{(1)} + q_2 K^*_{\ell+\frac{3}{2}}\ B_\ell^{(2)}\} \quad (5b)$$

see ref.4, eqns,(29) and (30).

The linearised diffusion-convection equation in the two bulk solutions for each component are given by ($c_i^{(p)} = c_{oi}^{(p)}(r) + \delta c_i^{(p)}(r,\theta,\phi,t)$)

$$\frac{\partial}{\partial t}\,\delta c_i^{(p)} + u_r\,\frac{dc_{oi}^{(p)}}{dr} = D_{pi}\nabla^2\delta c_i^{(p)} \qquad (p=1,2;\ i=1,n) \quad (6)$$

if there are no chemical reactions in the <u>bulk solutions</u> (there might be chemical reactions on the surface, see later), if the diffusion coefficients can be taken as constants and if there are no cross-diffusion terms. Writing for the i'th concentration in the p'th phase

$$\delta c_{\ell i}^{(p)} = \tilde{c}_{\ell i}^{(p)}\cdot Y_\ell(\theta,\phi)e^{\omega t} \quad (7)$$

the normal mode amplitude $\tilde{c}_i^{(p)}$ should satisfy the following differential eqn.

$$\left[\frac{d^2}{dx^2} + \frac{2}{x}\frac{d}{dx} - \left(s_{pi}^2 + \frac{\ell(\ell+1)}{x^2}\right)\right]\tilde{c}_{\ell i}^{(p)} = \frac{W_\ell^{(p)}}{xD_{pi}}\cdot\frac{dc_{oi}^{(p)}}{dx} \quad (8)$$

with

$$s_{pi} = R\sqrt{\frac{\omega}{D_{pi}}} \qquad (p=1,2\ ;\ i = 1,n) \quad (9)$$

The solutions to the homogeneous eqns. are simply proportional to $I^*_{\ell+\frac{1}{2}}(s_{1i}x)$ and to $K^*_{\ell+\frac{1}{2}}(s_{2i}x)$ in phase 1 and 2, respectively. The inhomogeneous convection terms appearing in (6) and (9) arise due to the possible existence of concentration gradients in the radial direction of the unperturbed concentrations c_{oi}. Those unperturbed concentration distributions might be real stationary distributions. In that case the subsequent calculations are exact. Still, the unperturbed concentration gradients might also be slowly time dependent and the stability analysis will then be almost exact for hydrodynamic motions which are rapid in comparison with the time dependence of the unperturbed state. Thus, we use the technique of "freezing" the concentration profiles which we have used also previously in connection with the stability analysis of a plane, fluid interface with perpendicular exponential concentration profiles of a surfactant (refs.5, 7, 8). For any chosen $dc^{(p)}_{oi}/dx$ the particular solutions to (8) may be found by the method of wronskians. In order not to complicate the analysis unduly, however, we shall here delimit ourselves to the particular simple initial profiles

$$c^{(1)}_{oi} = c^{(1)}_{ooi} + a_i x^2 \quad ; \quad c^{(2)}_{oi} = c^{(2)}_{ooi} + b_i x^2 \qquad (10)$$

since those profiles can be seen to be the true analogues of the linear, profiles in the case of a plane interface. With this profile the complete solutions to (8) can be expressed solely in terms of the Bessel functions already introduced:

$$\tilde{c}^{(1)}_{\ell i} = \mathscr{C}^{(1)}_{\ell i} \cdot I^*_{\ell+\frac{1}{2}}(s_{1i}x) - \frac{2a_i}{D_{1i}} \left\{ \frac{A^{(1)}_{\ell} x^{\ell}}{s^2_{1i}} + \frac{B^{(1)}_{\ell}}{s^2_{1i} - q^2_1} I^*_{\ell+\frac{1}{2}}(q_1 x) \right\} \qquad (11a)$$

$$\tilde{c}^{(2)}_{\ell i} = \mathscr{C}^{(2)}_{\ell i} \cdot K^*_{\ell+\frac{1}{2}}(s_{2i}x) - \frac{2b_i}{D_{2i}} \left\{ \frac{A^{(2)}_{\ell} x^{-\ell-1}}{s^2_{2i}} + \frac{B^{(2)}_{\ell}}{s^2_{2i} - q^2_2} K^*_{\ell+\frac{1}{2}}(q_2 x) \right\} \qquad (11b)$$

The "arbitrary" constants $\mathscr{C}^{(1)}_{\ell i}$ and $\mathscr{C}^{(2)}_{\ell i}$ are fixed by the requirement of local equilibrium between the interface and the adjacent layers of bulk liquid (see especially ref.8, first section after introduction):

$$\delta\Gamma_i = K_{pi}\ \{\delta c_i^{(p)}\Big|_{r=R} + (\frac{dc_{oi}^{(p)}}{dr})_{r=R} \cdot \delta R\}\ (p=1,2) \qquad (12)$$

where $\delta\Gamma_i$ is the perturbation of surface concentration (in <u>mass</u> per area) and δR is the small deflection of the interface from the spherical shape. K_{pi} (i = 1,2) are adsorption "constants" which might depend on the concentrations of the species adjacent to the interface. We have

$$\frac{\partial}{\partial t}\ \delta R = u_r(r=R) \qquad (13)$$

and use the expansion ($\Gamma_i = \Gamma_i^o + \delta\Gamma_i$)

$$\delta\Gamma = \tilde{\Gamma}_{\ell i} \cdot Y_\ell(\theta,\phi)e^{\omega t} \qquad (14)$$

The 2n conditions (12) amount to

$$\tilde{\Gamma}_{\ell i} = K_{1i}\{\tilde{c}_{\ell i}^{(1)} + \frac{2a_i x}{R} \cdot \frac{\tilde{u}_{\ell r}^{(1)}}{\omega}\} = K_{2i}\{\tilde{c}_{\ell i}^{(2)} + \frac{2b_i x}{R} \cdot \frac{\tilde{u}_{\ell r}^{(1)}}{\omega}\} \qquad (15)$$

and hereby we obtain

$$\mathcal{C}_{\ell i}^{(1)} = \frac{1}{I^*_{\ell+\frac{1}{2}}(s_{1i})}\ \{\frac{\tilde{\Gamma}_{\ell i}}{K_{1i}} + \frac{2a_i}{s_{1i}^2}\ B_\ell^{(1)} I^*_{\ell+\frac{1}{2}}(q_1)\ \frac{1}{\nu_1 - D_{1i}}\} \qquad (16a)$$

$$\mathcal{C}_{\ell i}^{(2)} = \frac{1}{K^*_{\ell+\frac{1}{2}}(s_{2i})}\ \{\frac{\tilde{\Gamma}_{\ell i}}{K_{2i}} + \frac{2b_i}{s_{2i}^2}\ B_\ell^{(2)} K^*_{\ell+\frac{1}{2}}(q_2)\ \frac{1}{\nu_2 - D_{2i}}\} \qquad (16b)$$

3. Linearised mass balance on the interface

The linearised mass balance for the i'th chemical specie adsorbed to the interface becomes

$$\frac{\partial}{\partial t}\,\delta\Gamma_i = \sum_{j=1}^{n} c_{ij}\delta\Gamma_j + D_i^s\nabla_s^2\delta\Gamma_i - \Gamma_i^o \mathrm{div}_s \underset{\sim}{u} - \Gamma_i^o\,\frac{\dot{a}}{2a}$$

$$+ \Delta\{D_i \left.\frac{\partial\delta c_i}{\partial r}\right|_{r=R} + D_i \left.\frac{d^2 c_{oi}}{dr^2}\right|_{r=R} \cdot \delta R\} \tag{17}$$

In comparison with ref.4 the only extension is the last term on the r.h.s. where Δ{} stands for "value in phase 2 minus value in phase 1 when crossing the interface". D_i^s is the surface diffusion coefficient of specie no.i. The surface divergence term of the velocity $\underset{\sim}{u}$ can be expressed through the derivative of the radial velocity by means of the incompressibility condition and the term arising from the change in the surface metric determinant (a) due to deformation can be calculated using (ref.4 eqn.(51))

$$\frac{\dot{a}}{2a} = \frac{2}{R}\,(u_r)_{r=R} \tag{18}$$

The "Δ-term" on the r.h.s. of (17) can be visualized as the difference in the influx and the outflux to and from the <u>perturbed</u> interface due to bulk diffusion. It should be noticed that the assumption behind (17) is that cross-diffusion is a passive process, <u>i</u>.<u>e</u>. there is no activation energy barrier for diffusion through the interface. This assumption might not be correct in all cases.

The unperturbed initial state is taken as a state where all surface chemical reactions and all diffusion fluxes to and from the surface just balance. This can be written in the following way:

$$f_k(\underset{\sim}{\Gamma}^o, \underset{\sim}{c}_{fixed}) + J_k^{fixed}(\underset{\sim}{\Gamma}^o, \underset{\sim}{c}_{fixed}) + \Delta\{D_k \frac{dc_{ok}}{dr}\}^{unfixed}_{r=R}$$

$$= 0 \qquad (k = 1,n) \tag{19}$$

Here we have distinguished between "fixed" and "unfixed" components. The first kind are characterised by a permanent, uniform concentration distribution whereas the second kind are characterised by initial concentration profiles. The source terms of the surface chemical reactions are called F and are allowed to depend on $\underset{\sim}{\Gamma}^{o}$ and $\underset{\sim}{c}^{fixed}$ only. For some of the fixed species it might be necessary to add also "buffer" fluxes $\underset{\sim}{J}_k^{fixed}$ from the adjacent bulk phases. In any case we have

$$c_{ij} = \left(\frac{\partial F_i}{\partial \Gamma_j}\right)_o + \left(\frac{\partial J_i^{fixed}}{\partial \Gamma_j}\right)_o \qquad (i=1,n\ ;\quad j=1,n) \tag{20}$$

for the matrix elements c_{ij} appearing in (17), see ref.4 , eqn. (49). Finally, the surface Laplacian in the surface diffusional term of (17) can be expressed as

$$\nabla_s^2 = -\frac{L^2}{r^2} \tag{21}$$

where L^2 is the usual L^2-operator known from quantum mechanics

$$L^2 \equiv -\left\{\frac{1}{\sin\theta}\frac{\partial}{\partial\theta}\sin\theta\frac{\partial}{\partial\theta} + \frac{1}{\sin^2\theta}\frac{\partial^2}{\partial\phi^2}\right\} \tag{22}$$

which has the property

$$L^2 Y_\ell = \ell(\ell+1)\, Y_\ell \tag{23}$$

Introducing the above formulae and the Fourier expansions into the interfacial mass balance, we obtain for component no. i

$$\sum_{j=1}^{n} F_{\ell ij}\tilde{\Gamma}_{\ell i} + E_{\ell i}^{(1)}\cdot B_\ell^{(1)} + E_{\ell i}^{(2)}\cdot B_\ell^{(2)} = 0 \qquad (i=1,n) \tag{24}$$

with

$$F_{\ell ij} = c_{ij} - \delta_{ij}\Big(\omega + \frac{\ell(\ell+1)}{R^2}D_i^s + \frac{D_{2i}}{K_{2i}}\frac{1}{R}\left[\frac{s_{2i}K^*_{\ell-\frac{1}{2}}(s_{2i})}{K^*_{\ell+\frac{1}{2}}(s_{2i})} + \ell+1\right]$$
$$+ \frac{D_{1i}}{K_{1i}}\cdot\frac{1}{R}\cdot\left[\frac{s_{1i}I^*_{\ell+3/2}(s_{1i})}{I^*_{\ell+\frac{1}{2}}(s_{1i})} + \ell\right]\Big) \tag{25a}$$

$$E^{(1)}_{\ell i} = + \frac{\Gamma^{\circ}_{i}}{2\ell+1} \frac{q_1}{R^2} \cdot I^{*}_{\ell+3/2}(q_1)(\ell+2) - \frac{2a_i D_{1i}}{\omega R^3} \cdot \frac{D_{1i}}{\nu_1 - D_{1i}}$$

$$\cdot \{ I^{*}_{\ell+\frac{1}{2}}(q_1) s_{1i} \cdot \frac{I^{*}_{\ell+3/2}(s_{1i})}{I^{*}_{\ell+\frac{1}{2}}(s_{1i})} - q_1 I^{*}_{\ell+3/2}(q_1) \} \tag{25b}$$

$$E^{(2)}_{\ell i} = + \frac{\Gamma^{\circ}_{i}}{2\ell+1} \cdot \frac{q_2}{R^2} \cdot K^{*}_{\ell-\frac{1}{2}}(q_2) \cdot (1-\ell) - \frac{2b_i D_{2i}}{\omega R^3} \cdot \frac{D_{2i}}{\nu_2 - D_{2i}}$$

$$\cdot \{ K^{*}_{\ell+\frac{1}{2}}(q_2) \frac{s_{2i} K^{*}_{\ell-\frac{1}{2}}(s_{2i})}{K^{*}_{\ell+\frac{1}{2}}(s_{2i})} - q_2 K^{*}_{\ell-\frac{1}{2}}(q_2) \} \tag{25c}$$

4. The characteristic equation.

The relations (24) yield n relations for the n+2 arbitrary constants ($\tilde{\Gamma}_{\ell 1}$ ....$\tilde{\Gamma}_{\ell n}$, $B^{(1)}_{\ell}$, $B^{(2)}_{\ell}$). The remaining two relations are supplied by the radial and the tangential momentum balances at the fluid interface. We shall not repeat a discussion of those two equations here, since this was done in ref.4, pp.195-197. The n+2 linear and homogeneous eqns. can only have solutions for vanishing coefficient determinants, however. Therefore, we obtain a characteristic eqn. of the form

$$\begin{vmatrix} \ell(\ell+1)\frac{\alpha_1}{R^2} & \cdots & \ell(\ell+1)\frac{\alpha_n}{R^2} & C^{(1)}_{\ell} & C^{(2)}_{\ell} \\ \frac{2\alpha_1}{R^2} & \cdots & \frac{2\alpha_n}{R^2} & D^{(1)}_{\ell} & D^{(2)}_{\ell} \\ F_{\ell 11} & \cdots & F_{\ell 1n} & E^{(1)}_{\ell 1} & E^{(2)}_{\ell 1} \\ F_{\ell 21} & \cdots & F_{\ell 2n} & E^{(1)}_{\ell 2} & E^{(2)}_{\ell 2} \\ \vdots & & \vdots & \vdots & \vdots \\ F_{\ell n1} & & F_{\ell nn} & E^{(1)}_{\ell n} & E^{(2)}_{\ell n} \end{vmatrix} = 0 \tag{26}$$

The first two rows in the determinant correspond to the tangential and the radial momentum balances, respectively. The coefficients α_k (k=1,n) are expansion coefficients in the linearised equation of state of the surface (temperature assumed constant and pressure effects neglected)

$$\delta\sigma = -\sum_{j=1}^{n} \alpha_j \partial\Gamma_j \tag{27}$$

The α-coefficients may be functions of as well temperature and pressure as of the surface concentrations (or concentrations in one of the adjacent bulk solutions). In extremely diluted solutions of surfactants we would have (two dimensional ideal gas)

$$\alpha_i \rightarrow \frac{RT}{M_i} \tag{28}$$

M_i being the molecular weight of the i'th surfactant. The link between surface tension and surface concentration is of paramount importance for the Marangoni-phenomena studied in this paper. Finally, the C- and D-coefficients in (26) have the precise form given below:

$$C_\ell^{(1)} = \frac{q_1}{2\ell+1} \cdot I^*_{\ell+3/2}(q_1)\cdot\ell\cdot\left[\frac{(\Sigma\Gamma_i^o)\omega}{R^2} + \frac{2(\ell+2)}{R^3}\Delta\eta + \frac{2(\ell+1)}{R^4}(\kappa+\varepsilon) - \frac{2(\ell+2)\varepsilon}{R^4}\right] + \frac{\eta_1 q_1^2}{R^3} I^*_{\ell+\frac{1}{2}}(q_1) \tag{29a}$$

$$C_\ell^{(2)} = \frac{-q_2}{2\ell+1} K^*_{\ell-\frac{1}{2}}(q_2)(\ell+1)\left\{\frac{(\Sigma\Gamma_i^o)\omega}{R^2} + \frac{2(1-\ell)}{R^3}\Delta\eta + \frac{\ell(\ell+1)}{R^4}(\kappa+\varepsilon) - \frac{2(1-\ell)\varepsilon}{R^4}\right\} - \frac{\eta_2 q_2^2}{R^3} K^*_{\ell+\frac{1}{2}}(q_2) \tag{29b}$$

$$D_{\ell}^{(1)} = \frac{q_1}{2\ell+1} I^{*}_{\ell+3/2}(q_1) \left[\left(\frac{\Sigma\Gamma_i^{o}}{R} + \frac{\rho_1}{\ell} + \frac{\rho_2}{\ell+1}\right)\frac{\omega}{R} + \frac{\sigma_o}{R^4\omega}(\ell-1)(\ell+2) + \frac{4}{R^3}\Delta\eta + \frac{2\ell}{R^3}\left(\Delta\eta + \frac{\kappa}{R}\right)\right] + \frac{\eta_1 q_1^2}{R^3\ell} I^{*}_{\ell+\frac{1}{2}}(q_1) \tag{29c}$$

$$D_{\ell}^{(2)} = \frac{q_2}{2\ell+1} K^{*}_{\ell-\frac{1}{2}}(q_2) \left[\left(\frac{\Sigma\Gamma_i^{o}}{R} + \frac{\rho_1}{\ell} + \frac{\rho_2}{\ell+1}\right)\frac{\omega}{R} + \frac{\sigma_o}{R^4\omega}(\ell-1)(\ell+2) + \frac{4}{R^3}\Delta\eta - \frac{2(\ell+1)}{R^3}\left(\Delta\eta + \frac{\kappa}{R}\right)\right] + \frac{\eta_2 q_2^2}{R^3(\ell+1)} K^{*}_{\ell+\frac{1}{2}}(q_2) \tag{29d}$$

In the above formulae η_p is the (shear) viscosity in phase p and κ and ε are the surface dilational and the surface shear viscosity in the extension of the Boussinesq model for a surface fluid made by Scriven[10] and described in the monograph of Aris.[41] The undisturbed surface tension is called σ_o. The density of phase p is denoted by ρ_p. The above expressions are found in ref. 4 , eqns.(43-46), but it should be noticed that a factor R^{-3} has fallen out in ref. 4 , eqns.(45-46) in the last term on the r.h.s. by a misprint. This factor is included in eqns.(29c-29d).

A complete analysis of stationary and oscillatory dispersion curves ω=f (k,parameters) by the methods previously used for the deformation of a plane interface (refs.5-8) will be carried out with computer in the future. In the present paper we shall delimit ourselves to the case of marginal, non-oscillatory instability for which particular simple formulae can be found (cf. refs.5-8).

5. Marginal states of non-oscillatory instability.

We put $\omega_i = 0$ and let $\omega_r \to 0$- Hereby it is possible to make an expansion of the determinantal eqn. in powers of ω_r. By retaining only the lowest power it is possible to obtain simple and <u>exact</u> conditions for having marginal, non-oscillatory states ($\omega_r = \omega_i = 0$). Let us first notice that

$$q_p \to 0 \qquad \text{and} \qquad s_{pi} \to 0 \tag{30}$$

We multiply the (n+1)'th column and the (n+2)'th column in (26) by

$$(q_1 I^{*}_{\ell+3/2}(q_1))^{-1} \qquad \text{and} \qquad (q_2 K^{*}_{\ell-\frac{1}{2}}(q_2))^{-1}$$

We shall also need the eqns.

$$\frac{z I^{*}_{\ell+\frac{1}{2}}(z)}{I^{*}_{\ell+3/2}(z)} \cong (2\ell+3)\ \{1 + \frac{z^2}{(2\ell+3)(2\ell+5)}\} \tag{31}$$

$$\frac{z K^{*}_{\ell+\frac{1}{2}}(z)}{K^{*}_{\ell-\frac{1}{2}}(z)} \cong (2\ell-1)\ \{1 + \frac{z^2}{(2\ell-1)(2\ell-3)}\} \tag{32}$$

where we neglect orders in z (q_p or s_{pi}) higher than two. We then get the following new determinantal elements

$$C^{(1)'}_{\ell} \cong C^{(1)}_{\ell o} + C^{(1)}_{\ell 1} \cdot \omega \tag{33a}$$

$$D^{(1)'}_{\ell} \cong D^{(1)}_{\ell,-1}\omega^{-1} + D^{(1)}_{\ell o} + D^{(1)}_{\ell 1} \cdot \omega \tag{33b}$$

$$E^{(1)'}_{\ell_i} \cong E^{(1)}_{\ell i,-1}\omega^{-1} + E^{(1)}_{\ell io} \tag{33c}$$

$$C^{(2)'}_{\ell} \cong C^{(2)}_{\ell o} + C^{(2)}_{\ell 1} \cdot \omega \tag{33d}$$

$$D^{(2)'}_{\ell} \cong D^{(2)}_{\ell,-1}\omega^{-1} + D^{(2)}_{\ell o} + D^{(2)}_{\ell 1} \cdot \omega \tag{33e}$$

$$E^{(2)'}_{\ell i} \cong E^{(2)}_{\ell i,-1}\omega^{-1} + E^{(2)}_{\ell io} \tag{33f}$$

with the definitions

$$C^{(1)}_{\ell o} = \frac{\ell}{2\ell+1}\left[\frac{2\Delta\eta}{R^3}(\ell+2) + \frac{\ell(\ell+1)}{R^4}(\kappa+\varepsilon) - \frac{2(\ell+2)\varepsilon}{R^4}\right] + \frac{\eta_1}{R^3}(2\ell+3) \qquad (34a)$$

$$C^{(2)}_{\ell o} = -\frac{\ell+1}{2\ell+1}\left[\frac{2\Delta\eta}{R^3}(1-\ell) + \frac{\ell(\ell+1)}{R^4}(\kappa+\varepsilon) - \frac{2(1-\ell)\varepsilon}{R^4}\right] - \frac{\eta_2}{R^3}(2\ell-1) \qquad (34b)$$

$$C^{(1)}_{\ell 1} = \frac{\ell}{2\ell+1}\,\frac{(\Sigma\Gamma^o_i)}{R^2} + \frac{\rho_1}{R}\cdot\frac{1}{2\ell+5} \qquad (34c)$$

$$C^{(2)}_{\ell 1} = -\left[\frac{\ell+1}{2\ell+1}\cdot\frac{(\Sigma\Gamma^o_i)}{R^2} + \frac{\rho_2}{R}\cdot\frac{1}{2\ell-3}\right] \qquad (34d)$$

$$D_{\ell,-1} = D^{(1)}_{\ell,-1} = D^{(2)}_{\ell,-1} = \frac{(\ell-1)(\ell+2)}{2\ell+1}\cdot\frac{\sigma_o}{R^4} \qquad (34e)$$

$$D^{(1)}_{\ell o} = \frac{4}{(2\ell+1)R^3}\Delta\eta + \frac{2\ell}{(2\ell+1)R^3}\left(\Delta\eta + \frac{\kappa}{R}\right) + \frac{\eta_1}{R^3\ell}\cdot(2\ell+3) \qquad (34f)$$

$$D^{(2)}_{\ell o} = \frac{4}{(2\ell+1)R^3}\Delta\eta - \frac{2(\ell+1)}{(2\ell+1)R^3}\left(\Delta\eta + \frac{\kappa}{R}\right) + \frac{\eta_2}{R^3(\ell+1)}\cdot(2\ell-1) \qquad (34g)$$

$$D^{(1)}_{\ell 1} = \frac{1}{(2\ell+1)R}\left(\frac{\Sigma\Gamma^o_i}{R} + \frac{\rho_1}{\ell} + \frac{\rho_2}{\ell+1}\right) + \frac{\rho_1}{\ell(2\ell+5)R} \qquad (34h)$$

$$D^{(2)}_{\ell 1} = \frac{1}{(2\ell+1)R}\left(\frac{\Sigma\Gamma^o_i}{R} + \frac{\rho_1}{\ell} + \frac{\rho_2}{\ell+1}\right) + \frac{\rho_2}{(\ell+1)(2\ell-3)R} \qquad (34i)$$

$$E^{(1)}_{\ell i,-1} = -\frac{2a_i D_{1i}}{R^3} \quad ; \quad E^{(2)}_{\ell i,-1} = -\frac{2b_i D_{2i}}{R^3} \qquad (34j)$$

$$E^{(1)}_{\ell io} = \frac{\Gamma^o_i}{R^2}\cdot\frac{\ell+2}{2\ell+1} + \frac{2a_i}{R}\cdot\frac{1}{(2\ell+3)(2\ell+5)} \qquad (34k)$$

$$E^{(2)}_{\ell io} = \frac{\Gamma^o_i}{R^2}\cdot\frac{(1-\ell)}{2\ell+1} + \frac{2b_i}{R}\cdot\frac{1}{(2\ell-1)(2\ell-3)} \qquad (34\ell)$$

Finally, the $F_{\ell ij}$ can be approximated by the expression

$$F_{\ell ij} \cong F_{\ell ijo} + F_{\ell ij1}\omega \tag{35}$$

with

$$F_{\ell ijo} = c_{ij} - \delta_{ij} \{\frac{\ell(\ell+1)}{R^2} D_i^s + (\ell+1)\frac{D_{2i}}{K_{2i}R} + \ell \frac{D_{1i}}{K_{1i}R}\} \tag{36a}$$

$$F_{\ell ij1} = - \delta_{ij} \{1 + \frac{R}{(2\ell+3)K_{1i}} + \frac{R}{(2\ell-1)K_{2i}}\} \tag{36b}$$

6. One fluctuating component joining a surface chemical reaction. No bulk profiles (Case I).

We shall consider here the special case where there are only variations in surface concentration (and in adjacent bulk concentrations) of one of the chemical species present. The component considered may join one or more surface chemical reactions together with the other species present, but the other species are assumed to have their concentrations "buffered". The component considered has no _initial_ concentration profile. The simplified characteristic eqn. is immediately obtained from (26), (33), (34), (35) and (36) (for brevity we omit indices ℓ and i)

$$\begin{vmatrix} \ell(\ell+1)\frac{\alpha}{R^2} & C_o^{(1)} + C_1^{(1)}\omega & C_o^{(2)} + C_1^{(2)}\omega \\ \frac{2\alpha}{R^2} & D_{-1}\omega^{-1} + D_o^{(1)} + D_1^{(1)}\omega & D_{-1}\omega^{-1} + D_o^{(2)} + D_1^{(2)}\omega \\ F_o + F_1\omega & \frac{\ell+2}{2\ell+1}\cdot\frac{\Gamma^o}{R^2} & \frac{1-\ell}{2\ell+1}\cdot\frac{\Gamma^o}{R^2} \end{vmatrix} = 0 \tag{37}$$

In (37) we replace column 3 with column 3 minus column 2 whereby the D_{-1} term in element 2,3 disappears. The second column in (37) is multiplied by ω and terms in ω^2 are discarded. Calculating the terms proportional to ω^o in the determinant, we get as an _exact_ equation for the marginal states of stationary instability:

$$D_{-1} \cdot \{F_o \cdot \Delta C_o + \ell(\ell+1) \frac{\alpha\Gamma^o}{R^4}\} = 0 \tag{38}$$

Using the definitions of $C_o^{(1)}$ and $C_o^{(2)}$ given in eqns,(34a-b) we have

$$\Delta C_o = - \frac{\ell(\ell+1)}{R^4}(\kappa+\varepsilon) + \frac{2\varepsilon}{R^4} - \frac{2\ell+1}{R^3} \Sigma\eta_p < 0 \quad (\ell=1,2,3...) \tag{39}$$

and the critical value for instability of the chemical matrix element c_{11} can be written (for $D_{-1} \neq 0$)

$$c_{11}^{cr} = \frac{\ell(\ell+1)}{R^2} D^s + (\ell+1)\frac{D_2}{K_2 R} + \ell \frac{D_1}{K_1 R} - \frac{\ell(\ell+1)}{R^4} \frac{\alpha\Gamma^o}{\Delta C_o} \quad (\ell=2,3,4..) \tag{40}$$

The formulae (39) and (40) are exactly the same as eqns.(59c) and (63) given in ref.4 (apart from the addition of the bulk diffusion terms), but here the derivaiton is made more consistently. In ref.4, eqn.(63) was described as an approximate result for small drops, but that restriction is unnecessary. Since the determinant (37) can be shown to be proportional to D_{-1} (_vide_ eqn.(38)) which is zero for ℓ=1, we have to reconsider the expansion for that special case. The mode ℓ=1 corresponds to a (non-uniform) infinitesimal translation of the drop leaving the spherical form of the surface invariant (see ref. 4 , fig.1). The mode ℓ=0 is always excluded from our analysis, since it corresponds to a uniform dilatation of the drop which is not possible for an incompressible fluid. The determinant for the case ℓ=1 is

$$\begin{vmatrix} \frac{2\alpha}{R^2} & C_o^{(1)} + C_1^{(1)}\omega & \Delta C_o + \Delta C_1\omega \\ \frac{2\alpha}{R^2} & D_o^{(1)} + D_1^{(1)}\omega & \Delta D_o + \Delta D_1\omega \\ F_o + F_1\omega & \frac{\Gamma o}{R^2} & -\frac{\Gamma o}{R^2} \end{vmatrix} = 0 \tag{41}$$

Here it is not necessary to multiply the second column by ω since the $D_{-1}\omega^{-1}$ term in element 2,2 has disappeared for $\ell=1$. Taking only terns of the order ω^{o} in (41) we obtain

$$F_o\{C_o^{(1)}\Delta D_o - D_o^{(1)}\Delta C_o\} = \frac{2\alpha\Gamma^o}{R^4}\{D_o^{(1)} - C_o^{(1)} + \Delta D_o - \Delta C_o\} \qquad (42)$$

Introducing into (42) the definitions (34) and (36) we obtain for the critical coefficient of autocatalysis

$$c_{11}^{cr} = \frac{2}{R^2} D^s + \frac{2D_2}{RK_2} + \frac{D_1}{RK_1} + \frac{2\alpha\Gamma^o}{R} \cdot \frac{\frac{4}{3}\frac{\varepsilon}{R} + \frac{3\eta_2}{2}}{\frac{2\kappa}{R}\left(\frac{4\varepsilon}{3R} + \frac{\eta_2}{2}\right) + 3\eta_1\left(\frac{4\varepsilon}{3R} + \frac{3\eta_2}{2}\right) + \eta_2\left(\frac{2\varepsilon}{R} + 3\eta_2\right)} \qquad (\ell=1) \qquad (43)$$

Eqn.(43) has principally the same form as eqn.(40) for the other modes, but the last term on r.h.s. is slightly more complicated. In ref.4 we stated as eqn.(65) a special formula for $\ell=1$. Since there is a somewhat confusing misprint in that formula, we reproduce it correctly here:

$$c_{11}^{cr} = \frac{2}{R^2}D^s + \frac{2\alpha\Gamma^o(8\varepsilon + 9\eta_2 R)}{2\left[2\kappa + 3\left(\eta_1 + \frac{\eta_2}{2}\right)R\right]\left(4\varepsilon + \frac{3}{2}\eta_2 R\right) + 9R^2\eta_2\left(2\eta_1 + \frac{3}{2}\eta_2\right)}$$

Apart from the bulk diffusion terms omitted in the analysis of ref. 4 the above formula is seen to be exactly equivalent to eqn.(43).

The formulae (40) and (43) can be physically interpreted in the following way:

<u>The coefficient of autocatalysis c_{11} of the fluctuating substance has to exceed a certain critical threshold composed by a part due to the damping effect of diffusion and a part due to the combined effects of surfactivity and viscosities</u>.

For normal surfactants $\alpha\Gamma^o$ will be positive and since ΔC_o is always negative (*vide* eqn.(39)) both the mentioned components of the threshold will be positive for all ℓ. Thus, the reaction has to exceed a certain autocatalyticity for the combined hydrodynamic and chemical problem to be unstable. There exist, however, certain "inverse surfactants" (*e.g.* many anorganic salts) which increase surface tension and where $\alpha\Gamma^{o} < 0$. For such inverse surfactants it is seen from (40) and (41) that hydrodynamic instabilities may be produced even for *stable* chemical reactions (*i.e.* negative value of the coefficient of autocatalysis). It should be remarked here, that for inverse surfactants Γ^{o} will become negative (from the Gibbs adsorption law) and K_1 and K_2 will be negative whereas α stays positive. The bulk diffusion terms $(\ell+1)D_2/RK_2$ and $\ell D_1/RK_1$ will therefore *destabilise* the system in the case with an inverse surfactant in contrast to the effect of those terms for normal surfactants.

Another situation is the one considered by Greenspan in his model of the division of biological cells. In that model the very viscous layer of cytoplasm near the cell membranes containing contractile actomysin microfilaments is treated as a "surface" with an "effective surface tension" encompassing the traction due to the network of microfilaments. Since the effective surface tension increases with increasing concentration of chemical "fuel", we have in this case a situation reminiscent of the inverse surfactant case, but the Gibbs law of adsorption does of course not apply to the variations in effective surface tension.

To conclude this section it should be pointed out that with more than one fluctuating species there should be possibilities of having unstable hydrodynamicochemical states even in situations of chemical stability and for normal surfactants, see ref.4, pp.203-204 and ref.38, fig.1, dashed area).

7. The case of n components without surface accumulation and without chemical reactions (case II).

We shall here treat the case where there are no surface chemical reactions between the n species present. In the stationary state each of the n species taken separately has therefore to satisfy a no-accumulation boundary condition on the interface:

$$a_i D_{1i} = b_i D_{2i} \qquad (i = 1,n) \tag{44}$$

In that case we have to take care when we evaluate the determinant, since

$$E_{i,-1}^{(1)} = E_{i,-1}^{(2)} = E_{i,-1} \tag{45}$$

We replace the (n+2)'th column in the determinant with the difference between the (n+2)'th and the (n+1)'th column and multiply the (n+1)'th column by ω. The determinant thus obtained has the form (terms of order ω^2 are discarded)

$$\begin{vmatrix} \frac{\ell(\ell+1)\alpha_1}{R^2} & \cdots\cdots & \frac{\ell(\ell+1)\alpha_n}{R^2} & C_o^{(1)}\omega & \Delta C_o + \Delta C_1\omega \\ \frac{2\alpha_1}{R^2} & \cdots\cdots\cdots & \frac{2\alpha_n}{R^2} & D_{-1} + D_o^{(1)}\omega & \Delta D_o + \Delta D_1\omega \\ F_{11,1}\omega + F_{11,o} & \cdots & 0 & E_{1,-1} + E_{1,o}^{(1)}\omega & \Delta E_{1,o} \\ 0 & & \vdots & \vdots & \vdots \\ \vdots & & \vdots & \vdots & \vdots \\ 0 & & F_{nnm1}\omega + F_{nn,o} & E_{m,-1} + E_{n,o}^{(1)}\omega & \Delta E_{n,o} \end{vmatrix} = 0 \tag{46}$$

To derive an equation for the marginal states of stationary instability we could just drop all the terms proportional to ω in (46) and calculate the determinant. We shall limit ourselves to the case of one component in the rest of this section, however. The sought eqn. is then given by

$$\frac{\ell(\ell+1)\alpha}{R^2} D_{-1}\Delta E_o + \Delta C_o \frac{2\alpha}{R^2} E_{-1} - F_o D_{-1}\Delta C_o - E_{-1}\Delta D_o \frac{\ell(\ell+1)\alpha}{R^2} = 0 \tag{47}$$

From the definitions (34) we calculate

$$\Delta E_o = -\frac{\Gamma^o}{R^2} + \frac{2a}{R}\left\{\frac{D_1}{D_2}\cdot\frac{1}{(2\ell-1)(2\ell-3)} - \frac{1}{(2\ell+3)(2\ell+5)}\right\} \tag{48}$$

$$\Delta D_o = -\frac{2\kappa}{R^4} - \frac{3}{R^3}\left(\frac{\eta_2}{\ell+1} + \frac{\eta_1}{\ell}\right) \tag{49}$$

and ΔC_o is given by (39). Inserting also the definitions of D_{-1}, E_{-1} and F_o in (47) together with the definition

$$\beta_1 = -\frac{2a}{R} \tag{50}$$

we obtain after some manipulation the following eqn. for the marginal states

$$\frac{\beta_1}{\Gamma o}\left\{\frac{R^2}{(2\ell+3)(2\ell+5)} - \frac{D_1}{D_2}\cdot\frac{R^2}{(2\ell-1)(2\ell-3)}\right\} = \tag{51}$$

$$1 + \frac{1}{\alpha\Gamma^o}\left\{\frac{\ell+1}{R}D^s + \frac{\ell+1}{\ell}\frac{D_2}{K_2} + \frac{D_1}{K_1}\right\}\cdot\left\{\frac{\ell}{R}(\kappa+\varepsilon) - \frac{2\varepsilon}{R(\ell+1)} + \frac{2\ell+1}{\ell+1}\Sigma\eta_p\right\}$$

$$+ \frac{\beta_1 D_1(2\ell+1)}{(\ell-1)(\ell+2)\sigma_o\Gamma_o\ell(\ell+1)}\left[2\varepsilon\{\ell(\ell+1)-2\} + R\eta_1\{\ell-1\} + R\eta_2\{\ell+2\}\right]$$

$$(\ell=2,3,4\ldots.)$$

Eqn.(51) is a highly interesting extension of the eqn. for marginal Marangoni-instability of plane interfaces derived previously by Sternling and Scriven and in a somewhat broader context by the present authors (refs.39 and 5-8). The transition to the plane case may be obtained taking the limits

$$\ell \to \infty \quad ; \quad R \to \infty \quad ; \quad \ell/r \to k \tag{52}$$

with k being the <u>wavenumber</u> of the two-dimensional waves on the plane surface. In the limit (52) the eqn.(51) reduces to

$$\frac{\beta_1}{4k^2\Gamma^o}\{1-\frac{D_1}{D_2}\} = 1+\frac{1}{\alpha\Gamma^o}\{kD^s+\frac{D_2}{K_2}+\frac{D_1}{K_1}\}\cdot\{k(\kappa+\varepsilon)+2\Sigma\eta_P\} \qquad (53)$$

which is exactly the same as the eqn. derived previously for the plane case (see ref.6, eqns.(91) and (94)). The precise relation between our treatment of Marangoni-instability and the treatment of Sternling and Scriven is discussed in length in another chapter in this book.

The normal mode $\ell = 1$ again requires special treatment, since $D_{-1} = 0$. The determinant is still given by (46), however, because we have to multiply the (n+1)'th column by ω even with $D_{-1} = 0$ since $E_{i,-1}$ (i=1,n) are not zero (in contrast to case I). Limiting ourselves to n=1, the equation sought for can now be found directly from (47) putting $D_{-1} = 0$. We obtain

$$\frac{2\alpha}{R^2} E_{-1}\{\Delta C_o - \Delta D_o\} = 0 \qquad (54)$$

If there is mass transfer ($E_{-1} \neq 0$) and the diffusion component is a surfactant ($\alpha \neq 0$), the condition for non-oscillatory marginal instability in the ℓ=1 mode will be

$$\Delta C_o - \Delta D_o = 0 \qquad (55)$$

or

$$\frac{3}{2}\cdot\frac{\eta_2}{R^3} = 0 \qquad (56)$$

So either the shear viscosity of the exterior phase vanishes and then the present method of expansion gives no information about marginal states for the ℓ=1 mode. Or the exterior viscosity does not vanish and (56) is a contradiction. We have therefore the result: Marginal states of Marangoni-instability do not exist for the ℓ=1 normal mode when the viscosity of the exterior liquid is different from zero. Therefore, the phenomena observed in connection with "kicking drops" cannot be treated by the method given in this section. The kicking drop phenomena might still be described by the general determinant (26), but those motions linked to diffusion of surfactants must then be connected with oscillatory solutions which have to be found by computer analysis (cf. refs.7 and 8).

8. The case of one component with bulk concentration profiles and chemical surface reactions with total accumulation zero. (Case III).

We consider the case with one diffusing surfactant joining a surface chemical reaction. The other concentrations and surface concentrations are buffered. The stationary state with no accumulation at the interface is described by the equation

$$-\Delta E_{-1} = -\frac{2a}{R}D_1 + \frac{2b}{R}D_2 = \text{rate of disapp. by surface chemical reactions.} \tag{57}$$

The determinant becomes

$$\begin{vmatrix} \ell(\ell+1)\frac{\alpha}{R^2} & C_o^{(1)} + C_1^{(1)}\omega & C_o^{(2)} + C_1^{(2)}\omega \\ \frac{2\alpha}{R^2} & D_{-1}\omega^{-1} + D_o^{(1)} + D_1^{(1)}\omega & D_{-1}\omega^{-1} + D_o^{(2)} + D_1^{(2)}\omega \\ F_1\omega + F_o & E_{-1}^{(1)}\omega^{-1} + E_o^{(1)} & E_{-1}^{(2)}\omega^{-1} + E_o^{(2)} \end{vmatrix} = 0 \tag{58}$$

and the big difference between this case and the previous cases is that $E_{-1}^{(p)}$ $(p = 1,2)$ are neither zero (case I) nor identical (case II). Developping (58) to the lowest order in ω (*i.e.* ω^{-2}) we obtain

$$\frac{\alpha}{R^2}\ell(\ell+1)\,D_{-1}\,\Delta E_{-1} = 0 \tag{59}$$

Since $\Delta E_{-1} \neq 0$ by assumption (57) and $D_{-1} \neq 0$ for $\ell \neq 1$, (59) will be a contradiction for real surfactants ($\alpha \neq 0$) and we have the following result:

<u>There can be no marginal states of non-oscillatory instability in the case of one diffusing surfactant participating in surface chemical reactions for the normal modes $\ell = 2,3,4,$ ..</u>

In case of $\ell = 1$ we have $D_{-1} = 0$ and (59) gives us no information. We have then to go to the next power (ω^{-1}). The result is:

$$\frac{2\alpha}{R^2}\,[E^{(2)}_{-1}\,\{D^{(1)}_{o} - C^{(1)}_{o}\} - E^{(1)}_{-1}\,\{D^{(2)}_{o} - C^{(2)}_{o}\}] = 0 \tag{60}$$

For a surfactant $\alpha \neq 0$ and the condition (60) amounts to the following generalisation of eqn.(56):

$$\frac{2aD_1}{R^3}\,\{\frac{4\varepsilon}{3R^4} + \frac{3}{2}\,\frac{\mu_2}{R^3}\} - \frac{2bD_2}{R^3}\,\{\frac{4\varepsilon}{3R^4}\} = 0 \tag{61}$$

Thus, there appear to be instability for the $\ell = 1$ mode for a very special combination of the fluxes of mass transfer, but the the case seems quite pathological.

Ruckenstein and Berbente[40] have investigated Marangoni-instabilities in the case of diffusion and chemical reactions. Their conclusions can not be transferred to the present case, since they a) have used the simplified Sternling and Scriven model for a plane interface without surface deformation and b) have considered the case of <u>bulk</u> chemical reactions. Their conclusion that "the criteria of instability have a totally different structure than in the case of diffusion unaccompanied by a chemical reaction, so that even small values of the reaction rate constant change essentially the conditions in which instability occur" is, however, similar to the conclusion reached here: Just the slightest reaction rate on the surface destroys the identity of the coefficients $E^{(1)}_{i,-1}$ and $E^{(2)}_{i,-1}$ and thus changes the structure of the determinantal eqn.

9. Complete transition from the spherical to the plane case.

We have seen a special example, <u>i.e.</u> eqn.(51), where the limit (52) turns the "spherical" formula over to a previously derived formula for the plane case. It should be demonstrated clearly, however, that the <u>complete</u> <u>determinant</u> passes over to the determinant derived earlier for deformational instabilities of a plane interface. This can easily be done by application of the uniform asymptotic expansions for large orders of the Bessel functions mentioned in Abramowitz and Stegun[37] §§ 9.7.7 and 9.7.8. We have

$$I^{*}_{\ell}(\ell z) \cong \frac{1}{2\ell\sqrt{z}} \; \frac{e^{\ell\eta}}{\sqrt[4]{1+z^2}} \qquad \ell \to \infty \tag{62}$$

with

$$\eta \equiv \sqrt{1+z^2} \; + \; \ln \frac{z}{1+\sqrt{1+z^2}} \tag{63}$$

and

$$K^{*}_{\ell}(\ell z) \cong \frac{\pi}{2\ell\sqrt{z}} \; \frac{e^{-\ell\eta}}{\sqrt[4]{1+z^2}} \tag{64}$$

The relevant arguments in the present case are

$$\ell z = R\sqrt{\frac{\omega}{\phi}} \quad ; \qquad \phi = \left\{ \begin{array}{l} \text{kinematical viscosity } \nu \\ \text{diffusion coefficient } D \end{array} \right. \tag{65}$$

From the above formulae it is easy to derive

$$\lim \left[\frac{\ell z}{R} \cdot \frac{K^{*}_{\ell-\frac{1}{2}}(\ell z)}{K^{*}_{\ell+\frac{1}{2}}(\ell z)}\right] = \lim \left[\frac{\ell z}{R} \cdot \frac{I^{*}_{\ell+3/2}(\ell z)}{I^{*}_{\ell+\frac{1}{2}}(\ell z)}\right] = \sqrt{k^2 + \frac{\omega}{\phi}} - k$$

$$\ell/R = k \; ; \; \ell \to \infty ; \; R \to \infty \tag{66}$$

where we have used that

$$\exp\;\{(n+1)\sqrt{1+z^2}\} \to \exp\;\{n\sqrt{1+z^2}\} \; ; \qquad n \to \infty \tag{67}$$

Taking our departure in the determinant (26) where we divide the (n+1)'th column by $q_1 I^*_{\ell+3/2}(q_1)$ and the (n+2)'th column by $q_2 K^*_{\ell-\frac{1}{2}}(q_2)$ and using (66) we obtain one by one the determinantal elements found previously by completely independent studies of deformational instabilities of plane interface with transfer of matter and surface chemical reactions (refs.5-7 with some corrections in ref.8).

10. The first and second mode of surface deformation.

The velocity at the surface of the drop is given by:

$$u^{(1,2)}_{R,\ell} = \frac{W^{(1,2)}_{\ell}(R)}{R} \cdot Y^m_{\ell}(\theta,\phi) \cdot e^{\omega t} \tag{68}$$

When ω is real, the shape of the deformed surface after a time dt will therefore be given by the shape of the real part of the surface harmonic $Y^m_{\ell}(\theta,\phi)$. We have

$$Y^m_{\ell}(\theta,\phi) = P^m_{\ell}(\cos\theta)\ e^{im\phi} \qquad (m=-\ \ell/+\ell) \tag{69}$$

where $P^m_{\ell}(\cos\theta)$ are the associated Legendre functions of the first kind. They are given by (ref.37, §8.6.6)(m=1,2,..)

$$P^m_{\ell}(x) = (-\ 1)^m(1\ -\ x^2)^{m/2}\ \frac{d^m P_{\ell}(x)}{dx^m} \tag{70}$$

where $x=\cos\theta$ and $P_{\ell}(x)$ are the Legendre polyomials:

$$\begin{aligned} P_o(x) &= 1 \\ P_1(x) &= x \\ P_2(x) &= \tfrac{1}{2}(3x^2 - 1) \\ &\vdots \end{aligned} \tag{71}$$

We have then

$$Y^o_o\ (\theta,\phi) = 1 \tag{72a}$$

$$Y^o_1 = P_1(\cos\theta) = \cos\theta \tag{72b}$$

$$\mathrm{Re}(Y_1^1) = -\sqrt{1-\cos^2\theta}\cdot\cos\phi = -\sin\theta\cos\phi \qquad (72c)$$

$$Y_2^o = P_2(\cos\theta) = \tfrac{1}{2}(3\cos^2\theta - 1) \qquad (72d)$$

$$\mathrm{Re}(Y_2^1) = -3\cos\theta\sqrt{1-\cos^2\Theta}\,\cos\Phi \qquad (72e)$$

$$= -\frac{3}{2}\sin 2\Theta\,\cos\Phi \qquad (72f)$$

$$\mathrm{Re}(Y_2^2) = 3(1-\cos^2\Theta)\cos 2\Phi = 3\sin^2\Theta\cos 2\Phi \qquad (72g)$$

The P_ℓ^{-m} functions are given by

$$P_\ell^{-m}(x) = (-1)^m \frac{\Gamma(\ell - m + 1)}{\Gamma(\ell + m + 1)} P_\ell^m(x) \qquad (73)$$

(see ref.42, § 8.752,2). We obtain

$$Y_1^{-1} = \sin\theta\cdot\cos\phi \qquad (74a)$$

$$Y_2^{-1} = \frac{1}{4}\sin 2\theta\cdot\cos\phi \qquad (74b)$$

$$Y_2^{-2} = \frac{1}{8}\sin^2\theta\cdot\cos 2\phi \qquad (74c)$$

In the following figures some cross sections are drawn through the deformed drops after a time dt for the ℓ=1 and ℓ=2 normal modes. The amplitudes are arbitrary, but their size has been exaggerated - of course - on the figures, since $\frac{\delta R}{R}$ has to be << 1 for the linearisation to be valid.

Figure 1 shows that the mode ℓ=1, m=0 corresponds to a translation of the drop in the z-direction. It is only the <u>surface</u> of the drop which is uniformly translated. Streaming inside and outside the drop will be non-uniform, so the case is <u>not</u> a trivial Galilei-transformation. The modes for m = - 1 and m = 1 correspond to translation in the other two directions.

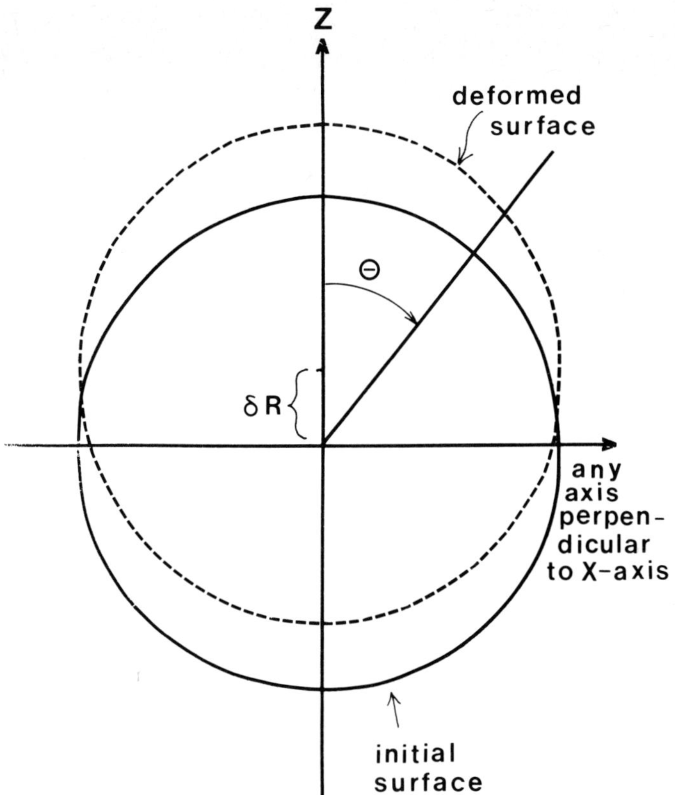

Fig.1 Visualisation of the mode $R+aY_1^0$ with a/R=0.3. $\ell=1$, m=0 corresponds to translation in the z-direction

The axi-symmetric mode $\ell=2$, m=0 is shown on figure 2. This is the mode which is interesting from the point of view of cell division and non-equilibrium stability of emulsions. The almost complete correspondence between the flow due to this normal mode and observed motion of the protoplasm during cleavage of a sea-urchin egg[26] (as pointed out by Greenspan[31]) is indeed fascinating.

Figure 3 and 4 show different cross-sections through the $\ell=1$ mode with "magnetic quantum number" m=1 and 2, respectively. The mode Y_2^1 has the appearance of a "parcel of 4 closed-packed spheres" (corresponding to the initial stages of the formation of the blastula in biological morphogenesis?), whereas the Y_2^2-mode has the appearance of a flat "post-parcel tied with two perpendicular strings". Biologically, such modes have significance for phagocytosis and amoebic movements. Here we have delimited ourselves, however, to show graphically how the most fundamental normal modes are like, since it is an advantage to know precisely what we are talking about in the prededing theory, when we take ℓ = 1,2, ... etc. The higher modes might have significance for the formation of small vesicles (pinocytosis) or microvilli formation in connection with some forms of cancer.

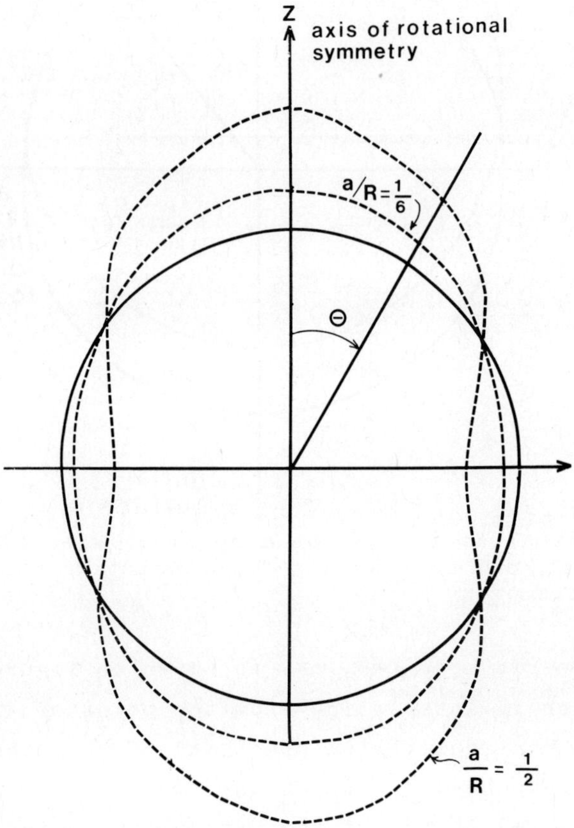

Fig.2. The mode $R + aY_2^o$ with $\frac{a}{R} = \frac{1}{6}$ and $\frac{1}{2}$, respectively.

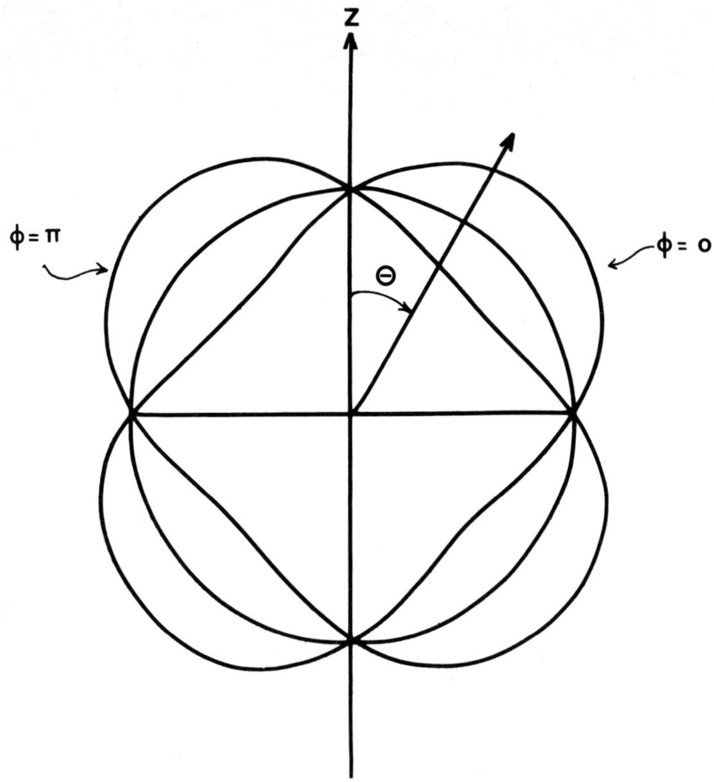

Fig.3. Different cross-sections through the mode $R + aY_2^1$ along the z-axis with $a/R = 0.2$

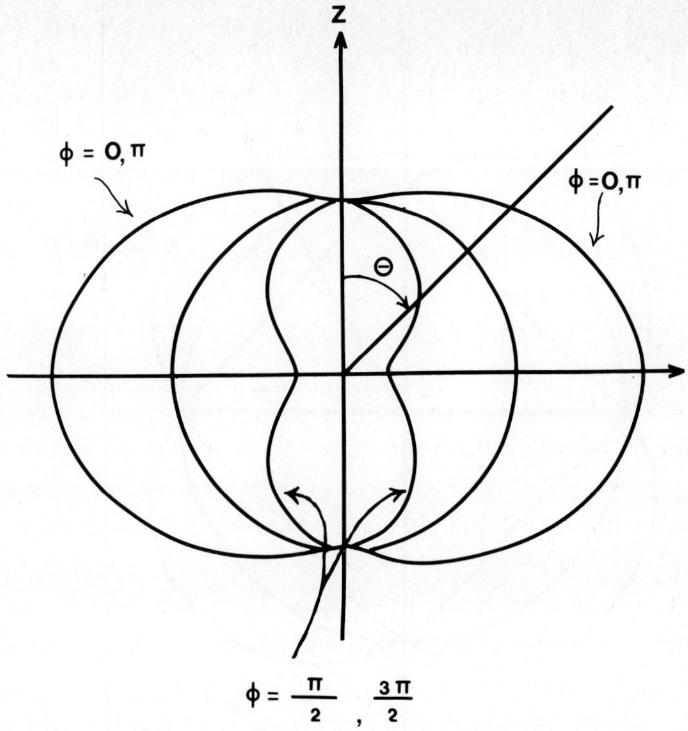

Fig.4. Different cross-sections through the mode $R + aY_2^2$ along the z-axis with a/R = 0.25

11. Discussion.

The present model may be used in different directions: a) "Kicking drops" may be explained by looking at the stability of the ℓ=1 surface harmonic in situations with mass transfer of a surfactant. b) Mechanisms of spontaneous emulsification may be proposed by looking at the stability of the ℓ=2 mode and the higher modes, especially at low surface tensions. c) Cell division and other cytokinetic phenomena may be described by the model using the concepts of "effective surface tension" and "effective surface viscosity". The mode $\ell=1$ will be relevant for chemotaxy (motion of protozoas in a gradient in concentration), but it cannot be the mechanism of chemotaxy for bacteria with rigid cell walls. Cell cleavage is approximately described by the $\ell=2$ mode and formation of small vesicles in connection with phagocytosis etc. corresponds to higher modes.

With regard to kicking drops we are not able at the present stage to give analytic results for the marginal stability, since eqn.(51) is not valid for the mode $\ell=1$, and we have shown in § 7 that there do not exist any non-oscillatory marginal states for the mode $\ell=1$ when we have only transfer of mass. However, oscillatory unstable solutions are not excluded, and they have to be found from the general determinant using the general computer method described in ref.7, Appendix. Probably the frequencies of oscillation will be close to the frequencies which can be found by solving the simplified equations of Davies & Haydon[3] for the (translational) oscillations of a viscous droplet in a viscous medium of infinite extension. (in spite of the main title of the papers 2 and 3 - "An investigation of droplet oscillation during mass transfer" - mass transfer and Marangoni-effects are <u>not</u> included in these formulae). The formulae of Davies & Haydon may then serve as a guideline for finding the roots (ω_r, ω_i) to our more complex determinantal equation.

It should be stressed here that whereas it is invariably found for <u>plane interfaces</u> that diffusion has to be <u>out</u> of the phase with the <u>least</u> diffusion coefficient in order to have a <u>possibility</u> of non-oscillatory instabilities, this is <u>not</u> the case for spherical interfaces. If we for a moment neglect the term proportional to $\frac{1}{\sigma_o}$ in eqn.(51), we see that the r.h.s. is positive and continuously growing with ℓ for normal surfactants ($\alpha\Gamma^o > 0$, $K_1 > 0$, $K_2 > 0$). L.h.s. must then also be positive, and if $\beta_1 > 0$ (transfer from the drop to the exterior phase) we must have

$$\frac{D_1}{D_2} < \left(\frac{D_1}{D_2}\right)_{cr} = \frac{(2\ell-1)(2\ell-3)}{(2\ell+3)(2\ell+5)} \qquad (75)$$

for having any possibility of instability of the non-oscillating kind, whereas $\frac{D_1}{D_2}$ must be greater than $(\frac{D_1}{D_2})_{cr}$ for mass transfer in the reverse direction. For plane - but deformable - interfaces the corresponding value of $(\frac{D_1}{D_2})_{cr}$ is equal to 1, but in the spherical case $(\frac{D_1}{D_2})_{cr}$ converges surprisingly slowly towards 1 for $\ell \to \infty$ see fig.5. Even for $\ell = 100$ the deviation from unity is almost 10%.

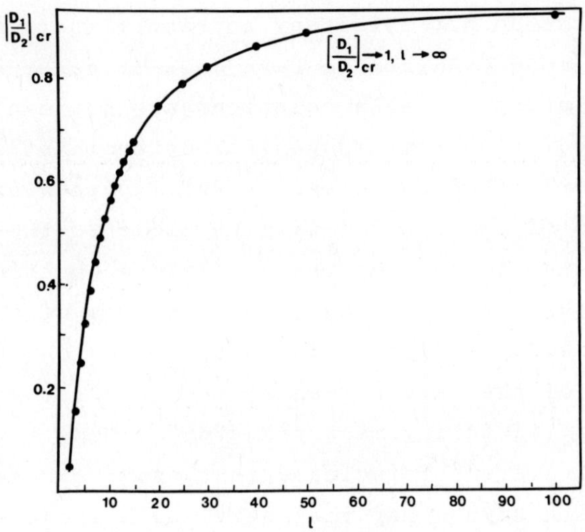

Fig.5 $(D_1/D_2)_{cr}$ as a function of the surface harmonic number ℓ.

For ℓ=2 we have $(\frac{D_1}{D_2})_{cr} = \frac{1}{21}$ so for $\beta_1 > 0$ the restriction $\frac{D_1}{D_2} < \frac{1}{21}$ is quite severe, whereas for $\beta_1 < 0$ the restriction $\frac{D_1}{D_2} > \frac{1}{21}$ will be fulfilled by most systems. For ℓ=1 we get $(\frac{D_1}{D_2})_{cr} = -\frac{1}{35}$, but the formula (51) is not valid in that case. It would not be surprising, however, if it is possible to have oscillatory instabilities with ℓ=1 where it is quite immaterial for the appearance of instabilities whether the ratio $\frac{D_1}{D_2}$ is greater than or less than unity. This is also shown in TABLE I where approximate values of $\frac{D_1}{D_2}$ and $\frac{\eta_2}{\eta_1}$ are given for the "kicking drop" systems observed by Haydon[2] with data found in Landolt-Börnstein[43].

TABLE I. Kicking drops observed by Haydon[2]. Direction of diffusion in all cases 2 → 1

Interior phase (1)	Exterior phase (2)	Diffusing substance	D_1/D_2	$\frac{\eta_2}{\eta_1}$
H_2O	Toluene	Acetic acid	0.57	0.62
H_2O	Benzene	Acetone	0.48	0.68
CCl_4	H_2O	Acetone	1.37	1.01
H_2O	Ethyl acetate	Acetone	probably	0.48
H_2O	Ethyl acetate	Acetic acid	less than 1	0.48
H_2O	Butyl acetate	Acetone		0.77

Since $\beta_1 < 0$ for all the cases in TABLE I, $\frac{D_1}{D_2}$ should be greater than unity to have instabilities in the plane case, but this is only the case for the system CCl_4 (1), H_2O(2) with acetone diffusing from 2 to 1.

Until now we have neglected the term in eqn.(51) inversely proportional to the surface tension σ. We want now to discuss also the influence of that term, since it is known that spontaneous emulsification is very often promoted by very low surface tension. It will turn out appropriate, however, to transform eqn.(51) to the following dimensionless form. We have marginal states of non-oscillatory instabilities if:

$$\frac{\hat{D}\hat{V}_2}{\hat{\alpha}} = \left(\frac{D_1}{D_2}\right)_{cr} - \frac{D_1}{D_2} - \frac{\hat{V}_1}{\hat{\sigma}} \tag{76}$$

where we have neglected the first term equal to unity at the r.h.s. of eqn.(51). (This amounts to neglecting the inertia of the mass adsorbed on the surface and this term is usually much less than the other terms on the r.h.s. of (51)).

The dimensionless <u>Marangoni-parameter</u> $\hat{\alpha}$ is given by

$$\hat{\alpha} = \frac{\alpha\beta_1 R^3}{D_2\eta_1(2\ell-1)(2\ell-3)} \tag{77a}$$

the dimensionless diffusion number by

$$\hat{D} = (\ell+1)\frac{D_s}{D_2} + \frac{\ell+1}{\ell}\frac{R}{K_2} + \frac{R}{K_1}\frac{D_1}{D_2} \tag{77b}$$

the first dimensionless viscosity number by

$$\hat{V}_1 = \frac{(2\ell+1)(2\ell-1)(2\ell-3)}{(\ell+2)(\ell+1)\ell(\ell-1)}\left[\frac{2\varepsilon}{R\eta_1}\{\ell(\ell+1)-2\} + (\ell-1)+(\ell+2)\frac{\eta_2}{\eta_1}\right] \tag{77c}$$

the second dimensionless viscosity number by

$$\hat{V}_2 = \frac{\ell(\kappa+\varepsilon)}{R\eta_1} - \frac{2\varepsilon}{R(\ell+1)\eta_1} + \frac{2\ell+1}{\ell+1}\left(1+\frac{\eta_2}{\eta_1}\right) \tag{77d}$$

and the dimensionless surface tension by

$$\hat{\sigma} = \frac{\sigma_o R}{D_1\eta_1} \tag{77e}$$

The quantities $\hat{D}_1\hat{V}_1$, $\hat{V}_2$ and $\hat{\sigma}$ in (76) are positive for $\ell\geq 2$, positive surface tension and normal surfactants ($K_1 > 0$, $K_2 > 0$) whereas $\hat{\alpha}$ is positive for mass transfer from the interior of the drop to the exterior fluid ($1 \rightarrow 2$) and negative for transfer in the reverse direction. If the $\hat{V}_1\hat{\sigma}^{-1}$ term in (76) is neglected, it is seen that D_1/D_2 has to be less than the critical value for $1 \rightarrow 2$ transfer and greater for $2 \rightarrow 1$ transfer to have possibility of instabilities. When $\ell/R \rightarrow \infty$ we pass to the plane case with the wavenumber $k = \ell/R$. In the Sternling & Scriven case with a linear concentration profile we know that we have instability for $k < k_{cr}$ and stability for $k > k_{cr}$. We shall therefore also expect instability for $\ell < \ell_{cr}$, where ℓ_{cr} satisfies (76). The quantity on l.h.s. of (76) increases roughly as ℓ^4 (ℓ^3 if $D_s = 0$, ℓ^2 if $D_s = \kappa + \varepsilon = 0$). Greater surfactivity or steepness of the concentration profile enhances $\hat{\alpha}$ and ℓ_{cr} has to be higher so that more modes are destabilized. Viscosity and diffusion act naturally in the opposite direction through $\hat{V}_2$ and $\hat{D}$. The term $\hat{V}_2\hat{\sigma}^{-1}$ increases roughly as ℓ^1 (ℓ^0 if $\varepsilon = 0$) and this term will become dominant at low surface tensions. For $1 \rightarrow 2$ transfer, decreasing surface tension will decrease the positive threshold for the parameter $\hat{D}\hat{V}_2\hat{\alpha}^{-1}$ and less modes will be unstable for a given value of $\hat{D}\hat{V}_2\hat{\alpha}^{-1}{}^2$.

Finally, r.h.s. becomes negative and no modes are instable. With mass transfer in the $2 \rightarrow 1$ direction, $\hat{\alpha}$ is negative. When the surface tension goes to zero some modes become destabilised when r.h.s. becomes as negative as l.h.s. in eqn.(76). The lower the surface tension σ is, the higher is $\hat{V}_1/\hat{\sigma}$, and the more negative becomes the r.h.s. Then ℓ_{cr} is displaced towards higher values and more modes are destabilised. Further discussion must await a detailed computer study, but the present model seems promising for the purpose of explaining spontaneous emulsification phenomena at low surface tensions. Instability of plane, deformable interfaces seems only to be promoted by low σ-values in very special cases (oscillatory instabilities with curved, initial concentration profiles[8]). But for spheres the effect seems very pronounced, and one might suggest that in reality low surface tensions also play a great role in the destabilisation of deformable, plane interfaces, since the conditions around the maximum and the minimum of the surface deformation is closer to spherical than to plane geometry. Emulsification should then arise as a non-linear effect in those cases.

Finally, our model may as mentioned be applied to describe cytokinetic effects in biological cells. The "surfactants" - whether it is ATP molecules or protein monomers for the formation of teh contractile microfilaments - will then <u>increase</u> the "effective surface tension" instead of decreasing it, and the above arguments should be reversed since α becomes negative. However, Γ and therefore K_1 and K_2 will stay positive in contrast to the case of "inverse surfactants" (<u>e.g</u>. many inorganic salts at oil-water or air-water interfaces) where the surface adsorption is demanded negative by the Gibbs' adsorption law. Realistic models of cell division and other cytokinetic phenomena require more experimental information about the dynamical and chemical mechanisms taking place in the cytoplasm than present available, however.

REFERENCES

1. J. B. Lewis and H. R. C. Pratt, Nature, London, 1953, 171, 1155
2. D. A. Haydon, Proc.Roy.Soc., 1958, A 243, 483
3. T. V. Davies and D. A. Haydon, Proc.Roy.Soc., 1958, A243,492
4. T. S. Sørensen, M. Hennenberg, A. Steinchen & A. Sanfeld, J.Coll.Int.Sci., 1976, 56, 191
5. T. S. Sørensen, M. Hennenberg & A. Sanfeld, J.Coll.Int.Sci., 1977, 61, 62
6. M.Hennenberg, T. S. Sørensen & A. Sanfeld, J.Chem.Soc., Faraday Trans., 1977, 73, 48
7. T. S. Sørensen, F. Y. Hansen, J. Nielsen & M. Hennenberg, ibid., 1977, 73, 1589
8. T. S. Sørensen, M.Hennenberg & F. Y. Hansen, ibid., 1978, 74, 1005
9. C. A. Miller & L. E. Scriven, J.Fluid Mech., 1968, 32, 417
10. L. E. Scriven, Chem.Eng.Sci., 1960, 12, 98
11. F. W. Pierson & S. Whitaker, J.Coll.Int.Sci., 1976, 54, 203; 219
12. S. Whitaker, J.Coll.Int.Sci., 1976, 54, 231
13. T. Tate, Phil.Mag., 1864, 27, 176
14. M. C. Wilkinson, J.Coll.Int.Sci., 1972, 40, 14
15. S. Chandrasekhar, "Hydrodynamic and Hydromagnetic Stability" Oxford University Press,(1961) chap.X
16. G. H. Quincke, Wied.Ann.Phys.N.F., 1888, 35, 580
17. J. Spek, Arch.f.Entw.Mech., 1918, 44, 5
18. E. B. Wilson, "The Cell in Development and Heredity", McMillan Co., New York, 1925
19. E. N. Harvey, Protoplasmatologia II, 1
20. N. M. Swann & J. M. Mitchison, J.exp.Biol., 1953, 30, 506
21. Y. Hiramoto, Expl.Cell Res., 1956, 11, 630
22. M. Mota, Expl.Cell Res., 1959, 17, 76
23. L. Wolpert, Int.Rev.Cytol., 1960, 10, 163
24. R. Rappaport, J.exp.Zool., 1961, 148, 81
25. Y. Hiramoto, Expl.Cell Res., 1963, 30, 59
26. Y. Hiramoto, Symp.Soc.exp.Bio., 1968, 22, 311
27. R. Rappaport, J.Exp.Zool., 1969, 171, 59
28. L. Wolpert & D. Gingell, Symp.Soc.exp.Bio., 1968, 22, 169
29. H. Komnick, W. Stockem & K. E. Wohlfarth-Bottermann, Int. Rev.Cytol., 1974, 34, 169
30. T. S. Sørensen, M. Hennenberg, A. Steinchen & A. Sanfeld, J.Coll.Int.Sci., 1976, 56, 191
31. H. P. Greenspan, J.Theoret.Biol., 1977, 65, 79
32. H. P. Greenspan, J.Theoret.Biol., 1978, 70, 125

33. H.P.Greenspan, Stud.Appl.Math., 1977, 57, 45
34. T. S. Sørensen & M. Hennenberg, "Some comments on Greenspan's model of cytokinesis", Manuscript received by J.Theoret.Biol. August 8, 1978
35. R. Rappaport, Int.Rev.Cytol., 1971, 31, 169
36. R. K. Jain, Lecture given at the meeting "Dynamics and Instability of Fluid Interfaces" May 2, 1978. Technical University Copenhagen.
37. M. Abramowitz and I. A. Stegun, "Handbook of Mathematical Functions" Dover, New York, 1965
38. M. G. Velarde, J. L. Ibañez, T. S. Sørensen, A. Sanfeld and M. Hennenberg, in Proceedings First Levich Int.Conf. "Physiochemical Hydrodynamics", (D. B. Spalding, ed.) Advance Pub., Guernsey, 1978
39. C. V. Sternling and L. E. Scriven, A.I.Ch.E. Journal, 1959, 5, 514
40. E. Ruckenstein and C. Berbente, Chem.Eng.Sci., 1964, 19, 329
41. R. Aris, "Vectors, Tensors and the Basic Equations of Fluid Mechanics", (Prentice-Hall, London, 1962)
42. I. S. Gradshteyn and I. W. Ryzhik, "Table of Integrals, Series and Products", Academic Press, New York, San Francisco, London, 1965
43. Landolt-Börnstein, "Zahlenwerte und Funktionen aus Physik, Chemie, Astronomie, Geophysik und Technik", 6.Auflage, 5.Teil, Bandteil a. Transportphänomene I. Springer-Verlag, Berlin, Heidelberg, New York, 1969 §§ 25121 & 25222.

Selected Issues from

Lecture Notes in Mathematics